Water Resources Development and Management

Indexed by Scopus

Each book of this multidisciplinary series covers a critical or emerging water issue. Authors and contributors are leading experts of international repute. The readers of the series will be professionals from different disciplines and development sectors from different parts of the world. They will include civil engineers, economists, geographers, geoscientists, sociologists, lawyers, environmental scientists and biologists. The books will be of direct interest to universities, research institutions, private and public sector institutions, international organisations and NGOs. In addition, all the books will be standard reference books for the water and the associated resource sectors.

More information about this series at http://www.springer.com/series/7009

Asit K. Biswas · Cecilia Tortajada
Editors

Water Security Under Climate Change

 Springer

Editors
Asit K. Biswas🆔
University of Glasgow
Glasgow, UK

Water Management International Pte Ltd.
Singapore, Singapore

Third World Centre for Water Management
Ciudad López Mateos, Atizapán, Estado de
México, Mexico

Cecilia Tortajada🆔
School of Interdisciplinary Studies
College of Social Sciences
University of Glasgow
Glasgow, UK

ISSN 1614-810X   ISSN 2198-316X (electronic)
Water Resources Development and Management
ISBN 978-981-16-5495-4   ISBN 978-981-16-5493-0 (eBook)
https://doi.org/10.1007/978-981-16-5493-0

Foreword by The Right Honourable Nicola Sturgeon MSP, First Minister of Scotland

Water is the crucial underpinning of all life on earth, with huge social, environmental and economic significance. And water is undoubtedly where many of the most challenging impacts of climate change are being seen and felt. Like the rest of the world, Scotland too is increasingly experiencing the impacts of climate change on our water environment.

In this year when COP26 comes to our great city of Glasgow, no-one can be in any doubt about the importance of action on climate change. Not only has the world had to contend with the impacts of the global COVID-19 pandemic in 2021, but yet again we have seen record high temperatures in many parts of the globe, including Scotland, as well as devastating wildfires and lives tragically lost through flooding, all commonly agreed to be exacerbated, if not directly caused, by climate change.

As a Hydro Nation, Scotland is a country that recognises the sustainable, responsible management of our water resources is crucial to developing a flourishing low-carbon, climate-resilient economy. Water is a critical resource in most key sectors of the Scottish economy, particularly manufacturing, agriculture, food and drink, tourism and energy. But it is also a high-performing sector in its own right with a diverse supply chain, an established innovation support ecosystem, world-leading research base and a highly regarded governance and regulatory framework.

While we are fortunate in Scotland to have an abundance of water, we recognise the pressing importance of action to improve water security across the globe. As a small, but responsible nation, Scotland is helping to lead the way on the sustainable management of water resources and sharing our knowledge with and learning from our partners around the world.

Urgent, collective action on climate change is required not just across Scotland but right across the world. Governments, organisations, individuals and businesses all need to work together to meet the challenge of climate change. No one can doubt the importance of COP26 — we need to harness the best ideas to improve our understanding, galvanise ever more effective action and provide inspiration that we can build on.

I am delighted that this publication has been brought forward to coincide with COP26, and that work in Scotland is reflected in the book. I am sure you will agree it provides an important contribution to the discourse and understanding of water security issues in the light of climate change.

Nicola Sturgeon
First Minister of Scotland
Edinburgh, Scotland

Foreword

Climate change is undoubtedly one of the greatest challenges of our time. It is transgenerational, transcends borders, and if we do not act urgently, we will continue to see the devastating impact of global warming on our planet. From more intense and frequent droughts and dangerous weather, to rising sea levels and warming oceans, this is having an impact on us all.

Now, more than ever, we must prioritize our oceans, seas, rivers, lochs and lakes. Water is precious and of vital importance to ecosystems and human societies, from sustaining life, cooking and agriculture to energy, sanitation and transport. We often have the luxury in the Western world of forgetting how much we rely on safe, clean and sustainable sources of water. The effects of human activity on land and water are now extensive, and if we do not protect and prioritize clean and green water management, then some of these changes caused by humanity may be irreversible.

The impact of climate change on water is also compounded by a number of other factors. As our global population grows and we see increased urbanization, and as we look to transition from finite energy resources, we are placing unprecedented pressure on water systems. The only way to achieve sustainable development and preserve safe and clean water resources for our planet is to achieve water security. In *Water Security under Climate Change*, the editors Profs. Asit K. Biswas and Cecilia Tortajada use their extensive knowledge in water management and policy to bring together world-leading authorities and policy-makers to guide us through the challenges and risks our water systems face and discuss how societies can work towards the wider goal of sustainability through effective water security.

The United Nations declared 2018–2028 the Water Action Decade, with the overarching goal to avoid a global water crisis. This will require humanity to be innovative and research-driven and learn from those who are leading the charge for environmental change. My own home country of Scotland is one of these leaders in sustainability. Scotland has some of the most ambitious climate targets in the world, and as a first step in response to the climate emergency, the Scottish Government introduced Scotland's Climate Change Bill with a net zero target for all greenhouse gases by 2045. In 2020, Scotland generated the equivalent of 97.4% of its electricity demand from renewables (Scottish Government 2021) and industry body Scottish

Renewables said output had tripled in the last decade, with enough power for the equivalent of seven million homes (Scottish Renewables 2021). In regard to water security and sustainable water management, Scotland has developed an innovative Hydro Nation strategy to make Scotland a nation where water resources are developed sustainably to bring maximum benefit to the Scottish economy.

Indeed, it is the city of Glasgow, home to my institution the University of Glasgow, which has the greatest potential to support Scotland as a Hydro Nation by becoming a global knowledge hub for research, teaching and technological innovation for water-related issues. This year, 2021, the city of Glasgow is due to play host to the 26th United Nations Climate Change Conference (COP26). This is an exciting opportunity to demonstrate that Glasgow—and Scotland—are committed to play their part in tackling climate change and protecting the planet for future generations. The venue for COP26 will be the Scottish Event Campus, aptly located on the banks of the River Clyde, the most important waterway in Scottish history and identity, and the focal point of the Clyde Mission. The Scottish Government's Clyde Mission has ambitions to use the Clyde to drive sustainable and inclusive growth for the city, the region and Scotland. The River Clyde was once synonymous with shipbuilding, engineering and industrial growth, and communities have settled along the riverbanks since the Palaeolithic era, with prehistoric canoes unearthed in the river. The Clyde sustained communities in the Kingdom of Strathclyde during the Roman occupation of Britain. In the thirteenth century, the first bridge was built over the river: an important step in Glasgow's development into a city and in the centuries that followed transforming the Clyde into a vital trade route for exporting and importing resources from the rest of the world.

Since 1451, the University of Glasgow's history has been intertwined with that of the River Clyde, and even today, the University's archives are home to a collection of hundreds of business documents detailing vessels built on the Clyde when Glasgow was at the height of its shipbuilding legacy. It is incredibly important for the University of Glasgow to maintain this relationship with the river flowing through the heart of our city and to ensure we are playing our part in unlocking the potential of the river to be transformational for local communities. Key to the University's contribution in this endeavour is the development of the Glasgow Riverside Innovation District (GRID). Encompassing both banks of the River Clyde, GRID offers the city the chance to reimagine our proud industrial heritage for the twenty-first century and to establish Glasgow's leadership in the hi-tech industries of the future. The GRID represents the changing civic nature and role of Universities beyond their traditional academic one, as a key anchor organization supporting the delivery of national government ambitions for inclusive and sustainable growth, by strengthening links between place, academia and research. The areas surrounding the River Clyde were hit hard by deindustrialization and still feel the effects of the loss of shipbuilding and heavy industry today. By reinvigorating the River Clyde and its surrounding areas, the river can once again become a key driver in Glasgow and Scotland's economic development, recognizing the riverbank as a vital organ of regeneration and learning from experiences in comparable cities across the world.

The Clyde Mission and regeneration of the Clyde Waterfront is a major pillar of the University of Glasgow's work. However, our institution is also involved in a multitude of initiatives aimed at preserving and protecting water as a vital natural resource and in supporting those communities who live alongside watercourses and coastlines across the world.

One of the most concerning factors of climate change is the impact it has on shifting ecosystems, and our rivers and fluvial ecosystems are particularly vulnerable. According to the UK Rivers Network, climate change impacts on rivers will include changes in water quality, impacts on wetland plants and animals, biodiversity loss and an increase in periods of intense, heavy rainfall leading to an increased risk of flooding (LSE 2021). Indeed, the 2017 Committee on Climate Change's UK Climate Change Risk Assessment 2017 Summary for Scotland notes that annual rainfall over Scotland has increased since the 1970s to a level about 13% above the average for the early decades of the twentieth century (Climate Change Committee 2017a). At the University of Glasgow, experts are working on a variety of research strands to establish how societies can adapt to climate risks, especially flooding. For example, experts from the University's Water and Climate Research group have been probing a novel artificial intelligence (AI) approach for flood hazard mapping, and colleagues at the University are leads partner in the EU-funded OPERANDUM project (OPEn-air laboRAtories for Nature baseD solUtions to Manage environmental risks) which focuses on disaster risk reduction to hydro-climatic hazards. Colleagues are also involved in research into the potential to use Nature-based Solutions (NbS) to mitigate flooding and have also been working across borders with counterparts in countries such as Sweden and the Philippines to understand and monitor the effects of climate change on rivers and flooding. The Living Deltas Hub, led at the University of Glasgow, also researches coastal tipping points in deltas across the world to understand when a social–ecological system can tip from one state, to another. If a piece of coastline erodes rapidly, the land disappears and so do any social activities on this land, including the loss of homes and livelihoods. The Hub is working to understand the biophysical, social and economic mechanisms that explain erosion, with a specific focus on mangrove systems, with an aim to propose solutions that improve the conservation or restoration of mangrove systems in the deltas covered by the project. There are several lessons to be learnt from this research, including the approach used by the Hub to understand the impact of plant life on waterways and developing Nature-based Solutions to strengthen resilience of waterways.

The University of Glasgow is also home to the National Centre for Resilience (NCR) at our Dumfries campus, which works with Scotland's Universities, researchers, policy-makers, emergency responders, volunteers and communities to build Scotland's resilience to natural hazards. Furthermore, the University's James Watt School of Engineering Water and Environment research team works closely with partners across the world to investigate sustainable management of the environment, extracting energy and resources from waste and making our rivers, water infrastructure, water and wastewater services resilient to climate change. The University has invested heavily in some of the best equipped and environmental laboratories and hydraulic facilities in the UK to support this research, and we believe we have an important role to play in sharing our knowledge and expertise and contribute to Scotland's place as a Hydro Nation.

The University of Glasgow was the first university in Scotland to declare a climate emergency in May 2019, and to further reflect our commitment to the sustainable agenda, we also launched a Centre for Sustainable Solutions in 2020 to support interdisciplinary, cross-campus and cross-sectoral solutions to climate change. We are also a key contributor to the COP26 Universities Network, a growing group of more than 40 universities working together to raise ambition for tangible outcomes from COP26. There is an appetite in Glasgow, in Scotland and across the international higher education sector to do more together to reverse climate change. For us, in Glasgow, we have ambitions to secure our place on the world map as leaders in water and climate research, alongside our world-changing teaching, training and innovation.

Universities have shown our capacity to innovate rapidly during the pandemic (through vaccine development, diagnostics and genomic sequencing of the virus), and given the right support and backing, the sector will also be capable of developing rapid and innovative solutions to the climate emergency. The post-pandemic environment presents the opportunity for accelerated collaboration and the establishment of alliances between academia, industry, government and other bodies to ensure efficient use of resources and the sharing of skills, strengths and best practices. Climate change is a complex issue that cannot be solved by one institution or actor alone and instead requires a diverse set of knowledge, expertise, skills and experience, particularly when it comes to water security and preserving our most important global resource.

Scotland may only be small geographically, but we have around 19,000 km of coastline (approximately 8% of Europe's coast) (Marine Scotland Information 2021). We are home to 90% of the UK's surface freshwater (Climate Change Committee 2017b) and have some of the world's cleanest drinking water, with 99.1% of public drinking water of a high quality (Health Protection Scotland 2019). Our water is not just of great importance to our natural heritage and identity, but is also a precious economic resource and a major portion of our overall natural capital worth £291 billion (Scottish Government 2019). All of this, coupled with a global reputation for research excellence and some of the brightest minds working to produce creative solutions to our most pressing global challenges, means that we are well-placed to pave the way for a sustainable blue future and protect our water for future generations to come.

Professor Sir Anton Muscatelli
Principal and Vice-Chancellor
University of Glasgow
Glasgow, UK

References

Climate Change Committee (2017a) National summaries. https://www.theccc.org.uk/uk-climate-change-risk-assessment-2017/national-summaries/scotland/. Accessed 19 June 2021

Climate Change Committee (2017b) UK climate change risk assessment 2017 evidence report. Summary for Scotland. https://www.theccc.org.uk/wp-content/uploads/2016/07/UK-CCRA-2017-Scotland-National-Summary.pdf. Accessed 21 June 2021

Health Protection Scotland (2019) Drinking water quality in Scotland, 2019. https://www.hps.scot.nhs.uk/publications/hps-weekly-report/volume-54/issue-33/drinking-water-quality-in-scotland-2019/#: ~ :text=The%20Drinking%20Water%20Quality%20Regulator,remains%20high%2C%20at%2099.2%25. Accessed 21 June 2021

LSE (2021) How is climate change affecting river and surface water flooding in the UK? https://www.lse.ac.uk/granthaminstitute/explainers/how-is-climate-change-affecting-river-and-surface-water-flooding-in-the-uk/. Accessed 18 June 2021

Marine Scotland Information (2021) Facts and figures about Scotland's sea area (coastline length, sea area in sq kms). Scottish Government. https://marine.gov.scot/data/facts-and-figures-about-scotlands-sea-area-coastline-length-sea-area-sq-kms. Accessed 20 June 2021

Scottish Government (2019) Scotland's natural capital worth £291 billion. https://www.gov.scot/news/scotlands-natural-capital-worth-gbp-273-billion. Accessed 23 June 2021

Scottish Government (2021) Energy statistics for Scotland Q4 2020 figures. https://www.gov.scot/binaries/content/documents/govscot/publications/statistics/2018/10/quarterly-energy-statistics-bulletins/documents/energy-statistics-summary—march-2021/energy-statistics-summary—march-2021/govscot%3Adocument/Scotland%2BEnergy%2BStatistics%2BQ4%2B2020.pdf. Accessed 16 June 2021

Scottish Renewables (2021) 97.4% of Scotland's electricity consumption met by renewables in 2020. https://www.scottishrenewables.com/news/829-974-of-scotland-s-electricity-consumption-met-by-renewables-in-2020. Accessed 17 June 2021

Preface

Over the long term, water security and climate change could become serious existential issues for many parts of the world unless water governance practices are improved substantially and effective climate change adoption measures are implemented in timely manner. At the time of writing this preface, in July 2021, the world is witnessing unprecedented extreme climatic events, including an extensive heat dome and very high temperatures and extensive forest fires in British Columbia and the Western United States; heavy flooding in Germany, Belgium and the Netherlands; and repeat serious droughts in California and other Western states of the USA, parts of Brazil and many other parts of the world. Between 1998 and 2017, droughts have affected at least 1.5 billion people and have resulted in global economic losses of at least $124 billion. It is highly likely that during the next two decades lives lost due to floods, droughts, forest fires and landslides and economic losses suffered due to extreme climatic events would increase to new heights that were not considered to be likely even five years ago.

In terms of water security, much of the world has been on an unsustainable development path for the last several decades. In earlier times, levels of population and urbanization were lower than what they are at present, total water demands could be met most of the times, and provision of adequate quantity of water of reasonable quality was not a serious and unmanageable issue in most parts of the world. However, consistently poor governance practices, including inefficient functioning of water-related institutions, have contributed to the steady worsening of water security conditions all over the globe.

Water security of individual nations has been further exacerbated since water has never been high up in the political agenda on a sustainable basis in any country of the world, except Singapore. As a general rule, water becomes a politically priority issue only when there is a serious drought or flood. As soon as these extreme hydrological events are over and the media loses its interest in covering them, water invariably disappears from the political agenda.

Water security over the long term can only be achieved when there is sustainable interest of senior policy-makers to resolve related problems. It takes time to prepare a realistic and implementable plan to ensure water security, build necessary water-related infrastructure, make certain it is managed and operated properly and on a regular basis, establish necessary legal, economic and behavioural instruments and functioning institutions to promote good governance. To achieve water security, it is also necessary to establish enabling conditions that will allow adoption of technological advances.

The world has enough knowledge, experience, technology and investment funds to ensure it becomes water secure within the next 10–20 years. A main concern is that these are not necessarily utilized. Regrettably, there are no signs that the current poor governance practices are likely to change any time soon in most countries of the world. Incremental improvements have been the order of the past decades, which have prevented formulation and implementation of business unusual practices.

In contrast to water security, climate change is full of uncertainties that are unlikely to be resolved with any degree of certainty over the next several decades. While much of the current discussions have been based primarily on greenhouse gas emissions, average global temperature rises and average sea level rise, there are many other factors which could and would affect the climate in specific regions. This, in turn, will affect their water security. Among these factors are number and structure of population, extent of urbanization, rates and types of economic growth, types and extent of energy use, agricultural practices and production, policies of individual countries, technological developments and rate and extent of their adoption, changing social norms, perceptions and value systems of different societies, formation of different feedback loops and a host of other associated factors.

Even if perfect climate models can be developed, which is unlikely for decades to come, uncertainties will always exist because of social, economic, political, institutional and human factors. Uncertainties, however, do not mean that it is not possible to formulate and successfully implement effective adoption measures. What this means is that countries need to make long-term adoption plans for water security due to climate change, which should be rigorously and regularly reassessed, in the light of new knowledge, availability of more and reliable data, technological developments, scientific breakthroughs and changes in societal perceptions and values and other similar factors.

A very good example is the case of Singapore. It already has a water security plan for 2060 which explicitly considers climate change impacts. This plan will be regularly and rigorously reassessed every five years by the cabinet, and necessary changes will be made in all relevant policies.

The present book contains a series of papers on water security under climate change which were especially prepared by invited authors from different parts of the world, for publication before the COP26 event, which will be held in Glasgow, UK, in October-November 2021. We are most grateful to Sir Anton Muscatelli, Principal and Vice Chancellor, and Ms. Bonnie Dean, Vice Principal, University of Glasgow, and Barry Greig, Hydro Nation Manager, at the Scottish Government

Water Industry Team, for their encouragement and support which made this book possible. We also would like to express our appreciation to Ms. Thania Gomez, Third World Centre for Water Management, Mexico, for her help with the editing and formatting of the book.

Asit K. Biswas
Distinguished Visiting Professor, University of Glasgow
Glasgow, UK

Director, Water Management International Pte Ltd.
Singapore

Chief Executive, Third World Centre for Water Management
Ciudad López Mateos, Atizapán, Estado de México, Mexico

Cecilia Tortajada
Professor, School of Interdisciplinary Studies, College of Social Sciences
University of Glasgow
Glasgow, UK

Contents

Contributors

Ioanna Akoumianaki Centre of Expertise for Waters (CREW), Hydro Nation International Centre, James Hutton Institute, Aberdeen, UK

Mohamed Abdel Aty Government of Egypt, Cairo, Egypt

M. Sophie Beier Centre of Expertise for Waters (CREW), Hydro Nation International Centre, James Hutton Institute, Aberdeen, UK

Asit K. Biswas University of Glasgow, Glasgow, UK;
Water Management International Pte Ltd., Singapore, Singapore;
Third World Centre for Water Management, Ciudad López Mateos, Atizapán, Estado de México, Mexico

Marius Claassen Centre for Environmental Studies, Department of Geography Geoinformatics and Meteorology, University of Pretoria, Pretoria, South Africa

Bruce Currie-Alder International Development Research Centre, Ottawa, Canada

Diane D'Arras International Water Association, London, UK

Ken De Souza Foreign, Commonwealth & Development Office, London, UK

Cristina Díez Santos International Hydropower Association (IHA), London, UK

Nikki H. Dodd Centre of Expertise for Waters (CREW), Hydro Nation International Centre, James Hutton Institute, Aberdeen, UK

Zeynep Kisoglu Erdal Black & Veatch, Irvine, CA, USA

David Faichney Scottish Environment Protection Agency (Seconded To Scottish Government To Develop Water Resilient Places Policy 2019–2021), Stirling, Scotland, UK

Robert C. Ferrier Centre of Expertise for Waters (CREW), Hydro Nation International Centre, James Hutton Institute, Aberdeen, UK

R. Quentin Grafton Crawford School of Public Policy, The Australian National University, Acton, ACT, Australia

Barry Greig Scottish Government, Water Industry Division, Edinburgh, Scotland, UK

Rachel C. Helliwell Centre of Expertise for Waters (CREW), Hydro Nation International Centre, James Hutton Institute, Aberdeen, UK

Walter W. Immerzeel Utrecht University, Utrecht, The Netherlands

Shaofeng Jia Water Resources Research Department, Institute of Geographic Sciences and Natural Resources Research, Chinese Academy of Sciences, Beijing, China

Helen M. Jones Scottish Government Rural and Environment Science and Analytical Services Division, Edinburgh, UK

Aifeng Lv Water Resources Research Department, Institute of Geographic Sciences and Natural Resources Research, Chinese Academy of Sciences, Beijing, China

Amina Maharjan International Centre for Integrated Mountain Development (ICIMOD), Kathmandu, Nepal

Sara Mercier-Blais UNESCO Chair in Global Environmental Change, University of Quebec at Montreal (UQAM), Montreal, Canada

David J. Molden International Centre for Integrated Mountain Development (ICIMOD), Kathmandu, Nepal

Megumi Muto Japan International Cooperation Agency (JICA), Tokyo, Japan

Santosh Nepal International Centre for Integrated Mountain Development (ICIMOD), Kathmandu, Nepal

Peter Joo Hee Ng PUB, Singapore's National Water Agency, Singapore, Singapore

Ashwin B. Pandya International Commission on Irrigation and Drainage (ICID), New Delhi, India

Simon A. Parsons Scottish Water, Dunfermline, UK

Saurav Pradhananga International Centre for Integrated Mountain Development (ICIMOD), Kathmandu, Nepal

Golam Rasul International Centre for Integrated Mountain Development (ICIMOD), Kathmandu, Nepal

Gordon Reid Scottish Water, Dunfermline, UK

Prachi Sharma International Commission on Irrigation and Drainage (ICID), New Delhi, India

Arun B. Shrestha International Centre for Integrated Mountain Development (ICIMOD), Kathmandu, Nepal

Sahdev Singh International Commission on Irrigation and Drainage (ICID), New Delhi, India

Debra Tan China Water Risk (CWR), Hong Kong, China

Cecilia Tortajada School of Interdisciplinary Studies, College of Social Sciences, University of Glasgow, Glasgow, UK

María Ubierna International Hydropower Association (IHA), London, UK

Nisha Wagle International Centre for Integrated Mountain Development (ICIMOD), Kathmandu, Nepal

Cindy Wallis-Lage Black & Veatch, Kansas City, MO, USA

Philippus Wester International Centre for Integrated Mountain Development (ICIMOD), Kathmandu, Nepal

Mark E. Williams Scottish Water, Dunfermline, UK

Glyn Wittwer Victoria University (Melbourne), Footscray, Australia

Sharon Zheng PUB, Singapore's National Water Agency, Singapore, Singapore

Part I
Perspectives

Chapter 1
Ensuring Water Security Under Climate Change

Asit K. Biswas and Cecilia Tortajada

Abstract Water security and climate change are only two of the major long-term problems the world is facing at present. Increasing population, urbanisation and demands for a better quality of life all over the world mean more food, energy and other resources will be necessary in the future. Increasing food and energy supplies will require more efficient water management all over the production and supply chains. All these requirements have to be met in a way so that significantly less greenhouse gasses are emitted into the atmosphere which are contributing to climate change at an increasing scale. Historically, the total global water demands have steadily increased. Currently, about 70% of global water is used by agriculture, 20% by industry and 10% by domestic. In all these three use areas, there is enough knowledge available to reduce water requirements very significantly. Agricultural production can be very substantially increased with much-lower water requirements. Domestic and industrial wastewaters can be collected, treated and reused. With proper management, this virtuous cycle can be a reality. While conceptually global water security can be assured by using current knowledge, climate change considerations have made ensuring global water security a very complex task. This is because major uncertainties are associated with any forecast of future extreme rainfalls and then translating them into runoffs in river basins and sub-basins which often are units of planning. This chapter reviews and assesses what can be done to ensure water security for individual countries as well as the world as a whole. Thereafter it analyses the risks and uncertainties that policymakers and water professionals are likely to face in dealing with climate change through the lens of water security.

A. K. Biswas (✉)
University of Glasgow, Glasgow, UK
e-mail: prof.asit.k.biswas@gmail.com

Water Management International Pte Ltd., Singapore, Singapore

Third World Centre for Water Management, Ciudad López Mateos, Atizapán, Estado de México, Mexico

C. Tortajada
School of Interdisciplinary Studies, College of Social Sciences, University of Glasgow, Glasgow, UK
e-mail: cecilia.tortajada@glasgow.ac.uk

Keywords Water security · Climate change · Water use efficiencies ·
Hydro-climatic models

1.1 Introduction

Water security and climate change are only two of the major problems humankind
is facing at present. These two issues will continue to be serious global problems for
decades to come because most of the issues that are associated with them will
continue to change over time and space. Important though these two long-term
problems are, there are also many other serious problems that the world is facing at
present. Probably an important first-order problem is the steady increase in the
global population. The current global population of around 7.85 billion is likely to
increase to 9.7 billion by 2050, and 11 billion by 2100 (UN Population Division
2019a). In 2019, 55.7% of the population lived in urban areas, and this percentage
figure is estimated to increase to 68% by 2050, and then further to 85% by 2100
(UN Population Division 2019b). This means an increasingly larger percentage of
global population is likely to be concentrated in and around urban centres. This will
put increasing and serious stress in and around these areas in terms of reliable
provision of every type of major human needs like food, energy, water, other
natural resources, environment, public health, medical and all other forms of social
services, housing, land use, transportation and numerous other associated issues.

In addition, in the coming years, there will be considerable emphasis on poverty
alleviation focusing on increasing standards of living of all people, industrialisation,
all forms of social requirements, including employment generation, better and more
efficient connectivities and a better environment (Biswas 2021).

All these major issues facing the world are interrelated. The dynamics of the
future of humankind will be determined not by any one or two issues, but by the
interactions and impacts of a multitude of them. For example, increasing population
and demand for steadily improving standard and quality of life will mean that more
food, energy and other materials will be required, unless there are major changes in
efficiencies in terms of how they are produced and used. Augmenting and ensuring
appropriate food and energy supplies will necessitate sustainable and increasingly
more efficient water management in their production, supply and use. Equally,
many of these developments may mean, unless special measures are taken, more
greenhouse gases may be emitted into the atmosphere which could contribute to
global warming and result in climate change through numerous pathways, some
known and identifiable at present, but others not known or fully appreciated. This
may further precipitate a host of additional second-order problems which could
seriously affect existing food production and supply arrangements, energy
requirements and use patterns as well as water management practices and processes.

In the coming years it will be important to ensure that the solutions that are
deemed to be effective to solve one major problem do not create problems in other
areas and/or regions. This has often been a recurring problem in the past: solutions

to solve one problem have mostly created serious problems in other areas. Accordingly, it is essential that solutions be sought after considering and assessing the overall problematique rather than focusing exclusively on solution of any one specific individual problem. Taking such a holistic analytical and assessing framework is becoming an increasingly complex and difficult task, technically, institutionally and nationally.

Climate change is already having serious impacts on the world. Figure 1.1 shows estimated economic losses due to climate-related disasters as a percentage of GDP for countries at different income levels, during the period 1998–2007. It categorically shows that economic losses as a percentage of GDPs were much higher for low-income countries compared to high-income countries, by nearly a factor of 4.5. This means low-income countries not only face much higher and widespread losses but also, they have less funds, management and administrative expertise, lower access to adaptation and use of technology and inadequate institutional capacities. These constraints are unlikely to change markedly in the future. Accordingly, economic losses from climate-related costs are highly likely to increase significantly in the low-income countries. This may further exacerbate inequalities between rich and poor countries, as well as between rich and the poor in the low income and lower middle-income countries. This would aggravate already serious situations even more even further.

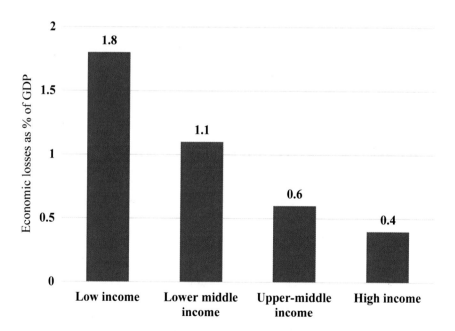

Fig. 1.1 Economic losses due climate-related disasters as % of GDP, 1998–2007. *Source* Adapted from Wallemacq and House (2018)

1.2 Water Security

The Ministerial Declaration during the Second World Water Forum held in The Hague, in March 2000, defined water security as "ensuring freshwater, coastal and related ecosystems are protected and improved, that sustainable development and political stability are improved, that every person has access to enough safe water at an affordable cost to lead a healthy and productive life, and that vulnerable population are protected from the risks of water-related hazards."

While there are numerous other definitions of water security, the overall context of all of them is somewhat similar. They all, directly or indirectly, refer to the fact that water security means everyone has reliable and ready access to adequate quantity and proper quality of water, and enough water is available on a reliable and timely basis for all social and economic needs, proper water quality is maintained, and people are protected from water-related disasters.

While conceptually it may be relatively easy to agree on a definition of water security at national or subnational levels, the problems associated with assuring water security for a specific region or nation are inordinately complex. In addition, it is difficult to formulate policies addressing all components of water security, let alone implement them properly on a timely basis. Difficulties are further compounded because different parts of the water security landscape invariably change over space and time. Additionally, societal attitudes and perceptions of the various factors that contribute to water security are continually evolving. Thus, on an operational level, it has been exceedingly difficult to formulate policies to assure water security, let alone implement them.

There are many factors which, in the final analysis, contribute to define water security. They cover many issues, including extent and structure of population growth; rate and extent of urbanisation; climate, soil and land use characteristics; institutional and governance capacities of institutions involved; level of sustained political interest in water; economic and behavioural aspects of how water is managed and used and people's attitudes to and perceptions of different water-related issues; technological advances and their adoption rates, as well as a host of other factors. Water security is ultimately the net result of all these interacting and often interrelated or even conflicting issues. These complexities and the fact that historically senior policymakers in almost all countries seldom have water as a priority item over the long-term in their political agendas, have ensured that water security has been hard to achieve in nearly all countries (Biswas and Tortajada 2019). The only major exception has been Singapore.

Historically there have been three main types of water uses all over the world. These are for domestic, industrial and agricultural purposes. Water allocation for environmental purposes has been a relative newcomer during the past 3–4 decades. It is becoming an increasingly important issue, but, as yet, very few countries have allocated water for environmental uses. If water security is to be assured for any country, the demands for all these four types of water requirements have to be met over the long-term in all countries.

In this chapter, the complexities of assessing demands to ensure water securities for the main three historical uses will be briefly considered, especially as environment has not been allocated in any specific quantum of water in most countries.

1.2.1 Water Security for Domestic Sector

Since human survival depends on adequate availability of water for domestic uses, it is undoubtedly the most important socio-political consideration in all countries. The Holy Quran explicitly stipulates that the humans should have first priority among water uses. It should also be noted that on 28 July 2010, United Nations General Assembly explicitly recognised the human right to water and sanitation and opined that clean drinking water is essential for the realisation of all other human rights. The resolution, however, does not address the industrial and agricultural uses of water, except in passing (Brooks 2008).

Complexities of ensuring water security can be realised by considering only a few essential issues embedded in its definition noted earlier. For example, what is exactly meant by "adequate quantity and quality" of water needed for an individual to lead a healthy and productive life? At a first glance, this may appear to be a rather simple and straightforward question to which most people have not given much serious thought or attention. This issue, in reality, is rather complicated and difficult to answer meaningfully.

Empirical studies available at present unambiguously indicate that the quantity of water used has important bearing on human health and well-being. However, there are no simple answers to the simple question like what is the daily water requirement of a human being to lead a healthy and productive life? Even the water needs for basic human survival is not easy to define. It depends on numerous factors, including body size, climate, type of work done by individuals, as well as their socio-cultural backgrounds and lifestyles.

Normally, the basic survival water requirement per person is around four litres per day. However, survival needs are very different from the water needed for leading a healthy and productive life. Unfortunately, very limited actual studies have been conducted on what are the daily water needs for human beings to lead a healthy life.

To our knowledge, only one global study is available at present on this important and fundamental question. Even this particular study was carried out over half a century ago, in Singapore, from 1960 to 1970. It attempted to correlate the quantity of water used in relation to incidences of waterborne diseases reported in all the Singaporean hospitals. Not surprisingly, it concluded that as domestic water use went up, disease incidents went down. However, there did not seem to be any noticeable improvement in health conditions beyond 75 L of water use per capita per day (lpcd). Hence, it may be concluded that this quantity represented a minimum level, at least for the Singaporean conditions at that time (Biswas 1981). Any additional water uses beyond 75 lpcd were found not to produce any perceptible health benefits: they were primarily of aesthetic nature and the result of personal preferences or convenience.

Unfortunately, similar studies have not been conducted in other parts of the world, especially in recent years, so that appropriate conclusions can be drawn. Without such definitive knowledge, it is very difficult to estimate what should be the per capita daily use that should be used in terms of estimating water security for the domestic sector for a city or a nation.

Some current data and trends indicate that the Singaporean results of around 75 L of water use per capita per day may be valid in other parts of the world even now. Assessments of the latest information available on per capita daily water use from various European cities indicate that perhaps 70–80 L may be adequate for a person to lead a healthy and productive life. Several Belgian urban centres have managed to bring their current daily per capita water consumption within the 70–80 L range. These levels for water use do not seem to have any adverse health impacts on their inhabitants. Spanish cities like Barcelona, Zaragoza, Valencia, Seville and Murcia have witnessed a steady decline in per capita daily use from around the year 2000 (Sauri 2019). For these cities, it is now less than 100 L at present. Current information from these cities is still show declining trends in per capita daily water use. Tallinn's water use is now below 90 lpcd.

The regulator of water for England and Wales, Ofwat, has already indicated that all the water utilities should try to reduce the per capita water consumption, by 2050, to half of what it was in 2020, which was 141 L. This means the average per capita daily water consumption should be around 70–71 lpcd by 2050, a figure that is similar to the level that was found in Singapore some five decades ago.

Not surprisingly, global trends in per capita water use are not uniform. For most countries, the general trend in per capita water use in recent years has been downward. This includes countries like the United States, Australia, Japan, all European countries and Singapore. The extent of the decline often varies from one country to another, and also from one city to another even in the same country. However, this decline is not a universal trend. Per capita water use in some countries and cities of the world has been increasing, like in Qatar and Phnom Penh, Cambodia (Biswas et al. 2021). Per capita water consumption during the past five years in Qatar has increased steadily. It is at present 590 lpcd for an average Qatari national. This is probably one of the highest domestic daily water consumption rates in any urban centre of the world.

Per capita water consumption is only one of the considerations for the domestic sector for assessing domestic water security. Equally important is the amount of water that is lost from the water utilities due to leakages, burst pipes, unauthorised connections and for other reasons which cannot be accounted for. Such losses are often quite high. In many countries, ranging from India, Mexico, Nigeria and Sri Lanka, losses of over 50% in many of their cities are not exactly uncommon. In a significant number of cities of the developing world, and even desert countries of the Middle East where water is in short supply, unaccounted for water losses of 35% or more are fairly common.

Even in highly developed countries like the United States and Canada a comprehensive study of water main break rates, indicated that they increased by 26% between 2012 and 2018 (Folkman 2018). For cast iron and asbestos cement pipes, which represent around 41% of installed water mains in these two countries, breakage rates increased by more than 40% over this 6-year period (Kolman 2018). In the world's most economically and technologically advanced country, the United States, well over eight billion litres of water are being lost each year. This represents about 14–18% of all water treated.

Thames Water, has been one of the very few water utilities that is completely in private hands for well over four decades and the largest water utility of England and Wales, was fined £120 million in 2018 by Ofwat, the water regulator, to compensate customers for consistent poor management of leaks. Privatised water utilities of England and Wales are losing through leakages some 3170 million litres per day, accounting to nearly 21% of their total production (PwC 2019). Progress in leakage reduction in England and Wales, for a variety of reasons, has basically stalled during the past two decades (Fig. 1.2).

Thus, a fundamental question that neither the policymakers nor the water profession has generally not asked, let alone answered, is should future water security assessments automatically accept these types of very significant losses, or should future estimates consider much lower levels of losses that can be achieved using present knowledge, technology and management practices? For example, a city like Tokyo now loses only 3.9% of its water, one of the very best performances in the world. In Singapore, the losses are around 5%. Whether future water security assessments consider 50, 20 or 5% losses for domestic water use sector, the resource requirements would be radically different.

The issues that water professionals and policymakers need to answer are should water security assessments consider that an average citizen should have access to 70–80 lpcd, as is now the case in many urban centres of Europe, or consider only incremental improvements in the coming years of their current per capita water use rates for estimating water security?

Depending upon what is decided, water requirements to ensure societal security will be very different. The efficient estimates may easily be only about 30–35% of

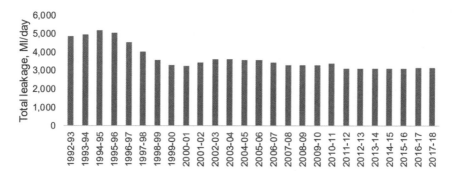

Fig. 1.2 Historical leakage in England and Wales. *Source* PwC 2019

using business-as-usual scenarios. It is not only in the domestic water use sector where there are major uncertainties in terms of assessing water security but also in the two other major water use sectors as well.

1.2.2 Water Security for Industrial Sector

Uncertainties associated with estimating water requirements for the domestic sector are also similar for the industrial sector. There are many factors which ultimately will ensure water security for the industrial sector. Water requirements for the industrial sector can be very significantly reduced from their current situations by better management practices, adopting new and cost-effective technologies that are already available to improve manufacturing processes, and realisation of their CEOs that if their businesses are to survive and thrive, their individual perceptions of importance and relevance of water have to be radically revised. While there are indications that this has happened, or is happening for some companies, regrettably an overwhelming number of CEOs still do not appreciate the extent of water and climate risks that are currently embedded in their existing business models. While more and more corporations are becoming aware of the importance of water security for their businesses, unfortunately a very significant percentage of industries are still neither aware nor considering such risks seriously and regularly.

While, globally, the percentage shares of industrial water requirements have steadily increased in recent decades, water requirements per unit of industrial outputs for several companies, especially multinational ones, have been steadily declining in recent decades. All the current trends indicate that these trends are not only likely to continue to decrease significantly in the coming years but also more and more multinational as well as national companies will follow these trends in the future out of both necessity and reputational reasons. If this happens, and it is likely to happen, water security estimates for industry would radically decline steadily in the coming years.

Water and energy are two absolutely essential requirements for any industry, anywhere in the world, so that they can operate. They are also closely interlinked. Energy cannot be generated without water, and, equally, water cannot be produced and used without energy. Improving the efficiency of production and use of one will positively impact on the other.

While it is essential for industry to have water to function, water requirements to produce same industrial goods often vary very significantly, depending upon the companies that manufacture them, their geographical locations, processes uses and management practices. Those companies that are aware of the importance and value of water, and the increasing climate and water risks that are likely to affect their existing business models, mostly require much smaller quantities of water to manufacture the same products compared to other more profligate companies that may require 2–10 times more water for similar operations.

During the post-2000 period, CEOs of numerous multinational companies became aware of the risks to their future expansion plans, and even eventual survival, unless they became increasingly water-efficient in their manufacturing processes, raw materials procurement practices and explicitly consider water and climate-related risks. Those multinational and national companies that have become aware of the importance of water security and climate change for smooth functioning of their businesses have started to realise that for their own long-term survival and growth, they must consider a holistic and coordinated approach which should adequately consider the risks posed by these new emerging factors. Accordingly, they have started to steadily improve their efficiencies of use of all resources needed in their manufacturing processes, including water.

Processes associated with energy generation and all types of manufacturing invariably contribute to greenhouse gas emissions which ultimately affect the climate. Corporations use all types of natural resources and chemicals to manufacture different types of products. Improving the efficiencies of resource use is an important consideration, but this alone is unlikely to be enough. In the final analysis, it is essential for each industry to have a clear understanding and appreciation of the interlinkages and inter-relationships between energy, water, other resources, greenhouse gas emissions and climate change. Each industry needs to manage these intricate interrelationships in a continually improving holistic and strategic manner. Focusing on improving the efficiency of one specific resource, which may be the most important resource for that company, may not contribute to an optimal and economically and environmentally efficient solution for the long-term.

From early 2000, many major multinational companies started to consider their water use patterns and management practices within their manufacturing plants. They also have shown an increasing interest in water use practices of their supply chains. These sustained interests have resulted in significant water savings during their manufacturing processes as well as the water requirements for the raw materials they need to produce various products. The net result of all these types of improvements has meant that water security for the industrial sector is constantly evolving in positive ways. Thus, what may have been the water requirements of any specific industry to assure security in 2000, is likely to be significantly higher than what is required now, some two decades later. The current indications are these estimates are likely to be even less in 2030, compared to what they are now.

A company like Nestlé, one of the world's largest 100 companies, is a good example to illustrate the above point. It significantly reduced its direct water withdrawals in every product category between 2005 and 2013. It has successfully managed to reduce its overall water requirements per tonne of products manufactured during this period by 33.3% (Brabeck-Letmathe 2016). It has further reduced wastewater discharges per tonne of product manufactured by 60.1% between 2003 and 2013. It also reduced total discharges by 37.2%. In addition, Nestlé recycled or reused 6.7 million m^3 of water in 2013, further reducing its water footprints. These numbers, since then, have progressively become even better.

In 2010, Nestlé started a 3-year study to measure consumptive use of water at farm levels. The company formulated a series of good practices for different crops

needed for its factories in different parts of the world. It then proactively dissem-
inated these findings to the farmers to improve their water use practices. Further,
thousands of Nestlé agronomists worked with the farmers to improve their water
use practices so that less water is used for farming. In addition, qualities of agricultural runoff were improved by ensuring efficient and timely use of fertilisers and
pesticides (Biswas-Tortajada and Biswas 2015).

By investing continuously in technology and process development and
improving management practices, Nestlé has continued to reduce its water
requirements regularly and progressively. For example, between 2010 and 2020, it
reduced direct water withdrawals per tonne of product in its manufacturing operations by 32%. In some countries like Spain, Nestlé has managed to reduce its water
requirements per tonne of manufactured products by nearly 50% during the same
period (Nestlé 2020).

There are many major industrial companies, both international and national, that
have similarly reduced their water footprints since 2005, both in terms of quantity
and quality, and are also practising circular economy in terms of water use. Nestlé is
only one example. Other multinational companies with similar good water management records during the past 15 years include Unilever, Coca-Cola, Pepsi-Cola,
Procter and Gamble, Nike, General Mills and Givaudan.

Generally, major industrial corporations are now increasingly adopting a 3-stage
process for their water management.

During the first stage, companies are critically assessing all aspects of their
current production processes to see where there are scopes for reducing water
consumption. Second stage includes properly treating the wastewater produced in
their factories so that it can be reused for different purposes within the factory walls.
In the third stage, wherever and whenever possible, they are considering
out-of-the-box unconventional techniques to extract water from the raw materials
they use in their factories to manufacture products.

A good example of the third stage is Nestlé's effort to extract water from milk
from their processing factories. Milk contains about 87% water. When it is heated
to produce powdered or condensed milk, normally water in the milk is simply
evaporated and then escapes into the atmosphere. However, when an appropriate
shadow price for water is used, it makes very good economic sense to develop a
process to catch the water vapour from milk processing and then condense it to
produce water. This water is then properly treated to take out impurities like
minerals and bacteria. Thereafter, this reclaimed water is used for water requirements of the plants.

Nestlé first tried this process in its milk factory in Lagos de Moreno, which is
located in the arid region of north eastern of Jalisco, Mexico. This "zero-water"
factory no longer needs to obtain water from local water sources. In 2018, which
was a drought year, this factory even managed to sell extra reclaimed water to a
nearby factory. After this success, Nestlé has similarly transformed some of its
other milk factories in arid water scarce regions like Modesto, California; Mossel
Bay, South Africa; and Qingdao, China. The plan is to transform its other milk
processing factories located in water-scarce arid areas into zero-water factories.

Such innovative and unusual practices very significantly improve water security of the company over the long-term.

More and more similar out-of-the-box transformational practices are likely to occur in the future which would significantly reduce industry's water requirements, and contribute to improve their water security.

Major corporations have at least two important reasons to progressively improve their water management practices and processes, which could contribute to their water security requirements. First, they are becoming increasingly aware of the risks posed by not having reliable access to adequate quantities of water of acceptable qualities so that their suppliers can source the raw materials which their plants need to manufacture the necessary products, and then have enough water of right quality so that the products can be manufactured.

Second, good water management has become a reputational issue for major corporations. During the past decade, the way companies handle their economic, social and governance (ESG) issues has become an increasingly important factor for the investors. Shareholder activism to ensure that all major corporations follow good ESG practices have become an increasingly important consideration for them. Both the national and international media are increasingly assessing the ESG ratings of individual companies. Their stock prices may depend on their ESG performances, which was not an important consideration even a decade ago.

Both the reasons are likely to become progressively more important for corporations in the coming years. Thus, the senior-most company officials are likely to be forced to continually improve their resource use efficiencies, including for water, and reduce their greenhouse gas emissions significantly, in order that they could be considered to be good corporate entities. This will mean that their water and energy use practices will continue to become more and more efficient, quantities of wastewaters produced will become progressively lower, and reuse and recycling would progressively become normal practices. In addition, wastewaters are likely to be treated to higher degrees, and reused within the factory. All these developments would help industry to become progressively more water secure because their water requirements would become less and less in the future.

All these developments mean that water requirements for industrial purposes are likely to be significantly less for increasingly more major corporations during the second quarter of the twenty-first century, compared to the first quarter.

1.2.3 Water Use for Agricultural Sector

Agricultural water use is another area where the importance and relevance of current discussions on water security need to be reconsidered. At present, agriculture is the largest user of water, at around 70% of the total global use. In recent years, on a percentage basis, agriculture's share of total water use has been steadily declining. However, in terms of total quantity of water used, agriculture's water requirements are still increasing.

The fact remains that in nearly all countries similar, or even significantly more, quantities of agricultural products can be produced with substantially less quantities of water by better management practices and adoption of more efficient technological developments. Agricultural water use practices in most countries, including those in most developed ones, are inefficient, often significantly. Thus, considerable scope exists to decrease agricultural water requirements very substantially without affecting productivities or total production.

China is a good example. From about 2010, China has made significant improvements in agricultural water use management practices which have enabled it to increase its total agricultural production, but with steadily decreasing water use per hectare of irrigated land. Between 1990 and 2012, China's production of cereals increased by 35% and over 80% for cotton. During this 22-year period, the country's total agricultural water use increased only marginally, from 374 to 388 billion m^3 per year, an increase of around 4%. However, during this period water used per hectare of irrigated land declined by 22% (Doczi et al. 2014).

China's plans are to significantly increase agricultural production even further by 2030. However, this will be achieved with steadily decreasing water use per hectare of irrigated land. All practices and processes associated with water use and agricultural production would become increasingly more efficient through increasing investments, institutional modifications and strengthening and steadily adopting latest technological developments, including information technology, artificial intelligence and data analytics. Other countries can, and should, follow China's example and reduce their agricultural water use in the future very substantially. Such steps will substantially reduce water requirements for agricultural production, thus enhancing water security.

In terms of water security assessments for agricultural practices, the current thinking of the water profession, at least implicitly, is that the future will be an extension of the present but only with marginal improvements. However, this assumption is likely to prove to be fundamentally flawed. Agricultural water requirements, the biggest water user of the world, are highly likely to decline very significantly in most major countries during the coming decades. This will make the current water security assessments increasingly irrelevant. In reality, the world has simply no other choice if water security for agricultural production is to be assured in the future.

1.2.4 Circularity and Water Security

An important issue that has not received much attention in recent years is the implications for water security considerations when the circularity of water use is considered. Non-renewable energy-producing materials like coal, oil and gas break down into different components after they are used. This means energy-producing materials cannot be reused following their first and only use.

Water, however, is a renewable resource. It can be used, wastewater can be collected and treated properly, and then reused. This cycle can continue ad infinitum with good planning and management and steadily advancing technology. Depending upon the extent and quality of wastewater treatment, and reliabilities of the treatment processes, treated wastewater can be used for all purposes, including direct use for potable purposes.

Using properly treated wastewater as a direct source for domestic water is not a revolutionary idea, even though it has not been used in more than one city. The capital of Namibia, Windhoek, has been using reclaimed water as a direct source for potable use for over 50 years. This landlocked and very arid country and its capital do not have access to adequate sources of water. Thus, for over a half-century, the city has collected its domestic wastewater, treating it adequately and then using it directly for potable use (Tortajada and van Rensburg 2019). The city has not faced any health problem thus far for this long-term reclaimed water use.

Windhoek, in Africa, is an unlikely global pioneer for using reclaimed water for direct potable use. This confirms the old adage that necessity is the mother of invention. It was several decades later that the city-state of Singapore followed Windhoek's footsteps and started to use its treated wastewater. This is termed as NEWater. The quality of treated wastewater is better than the tap water currently supplied to its citizens. Thus, a major part of this high-quality treated wastewater is supplied for wafer industry which requires very high-quality water for manufacturing. The balance of reclaimed water is added to the reservoirs which supply Singapore's domestic water (Tortajada et al. 2013).

Water scarcity and uncertainty of water availability have resulted in several urban centres of the United States considering using reclaimed water, including for potable purposes (Smith and EPA 2017). Windhoek has shown for over 50 years, and Singapore and Orange County in California, United States, for nearly two decades, that reclaimed water can be safely used for drinking purposes without any health concerns. Technical considerations have been resolved decades ago. The main problem now is psychological. Many people are still not comfortable with the idea of drinking reclaimed water. However, as the total water demands increase, and cost-effective new sources of water are not available, many urban centres in different parts of the world will have no other alternative but to seriously consider this one to ensure their water security over the long-term. It is likely to be a challenge to convince people that reclaimed wastewater is safe to drink.

Climate change is making droughts in many parts of the world more common than ever before. In addition, droughts are often lasting for longer durations and often are becoming more intense. The millennium drought in Australia and the California drought of 2012–2016 are good examples. This means many urban centres, whether they prefer it or not, will be forced to consider the alternative of potable reuse since they are unlikely to have much choice if they wish to ensure their water security. Such developments will mean water use for domestic and industrial purposes are likely to become increasingly more circular in the future. Water will be used, wastewater will be collected and treated and then will be reused in homes and industries. This virtuous cycle is likely to become increasingly the norm in many cities in the coming 1–2 decades. Such developments, and their gradual acceptance by the general public as a source of domestic water supply, will significantly add to the future water security of urban areas in a positive manner.

Singapore has further strengthened its water security by capturing rainfall that falls in nearly 2/3rd of the city-state that is considered to be its urban catchment. The rain that falls in this large area is collected through an extensive network of drains, canals, rivers, stormwater collection ponds and reservoirs. Altogether, such drains, canals and rivers cover some 8000 km in length. All the rainwater collected is treated and then used as an important source for domestic and industrial water for the city-state. This has significantly contributed to its water security.

As climatic patterns become more extreme and unpredictable, stormwater drainage systems cannot only harvest rainwater but can also play an important role in controlling floods. For example, in Singapore, since January 2014, all new developments and redevelopments of 0.2 ha or more must implement "source" solutions to slow down stormwater runoff that could enter the public drainage system. These onsite storage measures include alternatives like detention tanks and raingardens.

Such special arrangements are helping Singapore to not only collect much of the rainwater that falls in its designated catchment area but also slow release of stormwater after heavy rainfalls to public drainage system is enabling it to manage urban flooding more effectively and efficiently.

All the above discussion indicates that the concept of water security is somewhat amorphous. Countries, and very specifically cities, have many policy options which can help them to strengthen their water security by steadily decreasing demands for domestic and industrial water uses, as well as expanding their supply by extensive treatment and reuse of wastewater, and pursuing non-conventional policies like collecting rainwater over much of any city and then using it for productive purposes. Significant reductions in agricultural water requirements in the future mean water security in 2030 can be assured with much less water than was necessary in 2010.

1.3   Climate Change

An important issue that the policymakers in general and the water professionals in particular, has not given much attention thus far is how likely future climate change characteristics can be seamlessly integrated with climatic fluctuations factors that have been regularly witnessed in the past and will continue to occur in the future. Historically, climatic fluctuations have been accepted as given facts, and the science of hydrology, over the past several decades, has advanced sufficiently to incorporate climatic fluctuations considerations adequately in terms of water resources planning, operations and management. We now have enough knowledge, experience, technology and management expertise to handle climatic fluctuations effectively. However, at present we do not have enough knowledge, experience and background to handle the risks and uncertainties that are inherent in the forecasts of climate change factors to plan, operate and successfully manage specific water resources projects adequately in a cost-effective manner.

One of the important reasons for this inability to incorporate climate change considerations in water projects is the continuing and prevailing disconnect between hydrologists and climatologists. One of the main mediums through which climate change impacts are often felt is through the lens of water. Climatologists dealing with climate change issues have been primarily concerned with averages, such as how average temperature may increase or how mean sea level may rise in the future. In contrast, water professionals are not much interested in average values: they are only of limited interest to them. Planning, design and management of water projects are based on extreme hydrological events. For example, designs for urban flood management are not based on average rainfall, but on extreme rainfalls, like maximum rainfall expected in 50 or 100 years, depending on cities and damages they may cause. Estimating the maximum likely rainfall over a city, even by 2070, to design an efficient and cost-effective stormwater disposal system under changing climate as well as other factors like land-use changes and extent of urbanisation, is now more of an art rather than exact science. Cities that currently design stormwater drainage systems so that they can withstand one in a hundred-year rainfall face even more uncertainties than ever before in history because of climate change.

The situation is even worse for major hydraulic infrastructures like large dams for which the current practice is to estimate the maximum probable flood which may occur once in 500 years, or even once in 1000 years for a few large and strategic dams. Estimating such floods over a very long return period with any degree of certainty has always been difficult in the past. Changing climate and other major factors that could influence their design, like vegetation covers, rainfall patterns, levels of urbanisation, land use patterns, afforestation/deforestation, evolving societal attitudes to and perceptions of acceptable risks, and host of other hydro-climatic, social, economic and political factors, have made it an exceedingly complex and difficult process. Consequently, the risks of over-design or under-design of major hydraulic structures have increased manifold because of the high levels of uncertainties associated with each individual factor that needs to be considered. At our present state of knowledge, climate change has made a difficult task almost impossible.

Probable maximum floods of large dams are generally estimated by considering the worst hydrological and meteorological conditions that may occur simultaneously over the catchment area. Such estimates have always been conservative. The final estimates always depended on the judgment, experience and risk-tolerance levels of the specialists concerned, even before the climate change era.

If to all these earlier complexities, uncertainties of climate change are superimposed, the reliability of the final estimates are difficult to predict. As the risks of over-design or under-design have increased significantly, the final estimates of costs of water projects may vary significantly depending on the levels of the design floods. This may result in a decision not to construct the project because of high-estimated costs. Alternatively, if maximum probable floods are under-estimated, and the structure is built, there is a real danger that it may fail at some time creating catastrophic consequences. At the current stage of knowledge, it is not possible to say whether a structure is properly designed, over-designed or under-designed because of estimating unprecedented runoffs of rivers due to climate and other change factors. The risks and uncertainties associated with such estimates are significant. This is one of many issues which has not received adequate attention, either from climatologists or water specialists.

Currently, the primary means of forecasting future climate is through the use of global circulation models (GCMs). There are now some 40 different such models. Even at the levels of large countries, the various GCMs do not give similar and consistent results (Strzepek and Smith 1996). This is not surprising since the climate is a coupled and nonlinear chaotic system. Even though our understanding of hydro-climatological factors has increased significantly in recent years, there are many climate processes that are not yet fully understood. Accordingly, these complex processes have to be simplified for constructing the GCMs, thus introducing uncertainties in the predicted results.

In addition, water planning is generally done at river basin or sub-basin levels. Unfortunately, at present state of knowledge, downcasting GCMs to forecast river flows at river basin or sub-basin levels, whichever may be the unit of planning, give results whose reliability is basically unknown. Urban centres consider stormwater management at city levels, which are even at a much smaller scale than river basins. The GCMs are ill-suited to forecast possible future climate scenarios at river basin, sub-basin or city levels (IPCC 2014).

1.4   Concluding Remarks

There is no question that changes in climate in the coming decades will affect how water resources activities are planned, designed, constructed and operated. Social, economic, political, environmental and institutional implications of climate change will be very significant in the future, both over space and time. These are, and will continue to be, complex, nonlinear and interrelated issues, with numerous known and unknown feedback loops which are not fully understood, appreciated, or even known at present. Lack of understanding of how various physical, social, economic, environmental and political forces may interact with each other, over space and time, as well as lack of reliable data and information over a reasonable period of time, are only two factors which currently prevent most national and international organisations to make reasonably reliable and actionable predictions of future climate scenarios for water planning and management.

At our present state of knowledge, there are numerous uncertainties which will continue to handicap the water profession to ensure water security of individual nations and cities due to the risks posed by climate change. These include, but are not limited to, future global emission scenarios, predicting how these are likely to affect future global, regional and local rainfall and temperature patterns over time and space, interpreting their overall impacts on the hydrologic cycle at different scales, assessing types and extent of scientific and technological breakthroughs that may be expected in the coming years which would enable us to better understand, predict, ameliorate and adapt to the various climate change scenarios, and their potential impacts. These are likely to be some of the major challenges that will continue to confront the water and the climate professions for many years to come.

While climatic uncertainties are many, and it will take time to understand and appreciate them properly, future water security can be reasonably assured by making water management and use practices and processes increasingly more efficient and equitable. As discussed earlier in this chapter, the world's water management practices have been on an unsustainable path for decades. These can be substantially improved so that significantly more can be done with less quantities of water. There are numerous examples from different parts of the world where good practices are delivering excellent results with less amounts of water. All such good practices need to be identified and properly documented especially in terms of the enabling conditions which made such results possible. These attempts would very significantly improve the water securities of individual cities, states and nations.

For successful and intelligent planning of the water sector within an overall framework of sustainable development, knowledge needs to advance much more than what is available at present. Meeting these challenges successfully and within a reasonable period of time will depend on going well beyond climatological-hydrologic modelling. Policies for adaptation and viable strategies for cost-effective mitigation measures have to be formulated on the basis of their overall effectiveness. The policies formulated have to be properly implemented. Equally, in the areas of water security and climate change, there are many factors which affect them that are likely to change over time and space. Thus, it will be essential that after a long-term plan is formulated, to ensure water security, it be reviewed rigorously to incorporate any changes necessary, every 3–5 years.

Above all, formulating long-term plans for ensuring water security and implementing them properly will need political leadership in all countries to put water higher-up in the political agenda over the long-term. Water security cannot be assured by only short-term and ad-hoc interest on the issues only when a serious flood, drought or natural disaster occur, as has been the case in nearly all parts of the world. It will require long-term sustained commitment from the national and state political leaders who should consider water security to be an important issue. This, sadly, is missing at present in nearly all countries of the world.

References

Biswas AK (1981) Water for the third world. Foreign Aff 60(1):148–166. https://doi.org/10.2307/20040994

Biswas AK (2021) Water as an engine for regional development. Int J Water Resour Dev 37 (3):359–361. https://doi.org/10.1080/07900627.2021.1890409

Biswas AK, Tortajada C (2019) Water crisis and water wars: myths and realities. Int J Water Resour Dev 35(5):727–731. https://doi.org/10.1080/07900627.2019.1636502

Biswas AK, Sachdeva PK, Tortajada C (2021) Phnom Penh water story: remarkable transformation of an urban water utility. Springer, Singapore

Biswas-Tortajada A, Biswas AK (2015) Sustainability in coffee production: creating shared value chains in Colombia. Routledge, Abingdon

Brabeck-Letmathe P (2016) Climate change, resource efficiency and sustainability. In: Biswas AK, Tortajada C (eds) Water security, climate change and sustainable development. Springer, Singapore, pp 7–26

Brooks DB (2008) Human rights to water in North Africa and the Middle East: what is new and what is not; what is important and what is not. In: Biswas AK, Rached E, Tortajada C (eds) Water as a human right for the Middle East and North Africa. Routledge, Abingdon, pp 19–34

Doczi J, Calow R, d'Alançon V (2014) Growing more with less: China's progress in agricultural water management and reallocation. Overseas Development Institute, London. https://cdn.odi.org/media/documents/9151.pdf

Folkman S (2018) Water main break rates in the USA and Canada: a comprehensive study. Utah State University, Logan

IPCC (2014) AR5 climate change 2014: impacts, adaptation and vulnerability. https://www.ipcc.ch/report/ar5/wg2/

Nestlé (2020) Creating shared value and sustainability report 2020. Nestlé, Vevey, pp 44–48

PwC (2019) Funding approaches for leakage reduction. Report for Ofwat (December). https://www.ofwat.gov.uk/wp-content/uploads/2019/12/PwC-%E2%80%93-Funding-approaches-for-leakage-reduction.pdf

Sauri D (2019) The decline of water consumption in Spanish cities: structural and contingent factors. Int J Water Resour Dev 36(6):909–925. https://doi.org/10.1080/07900627.2019.1634999

Smith CDM and EPA (2017) Potable reuse compendium. https://www.epa.gov/sites/production/files/2018-01/documents/potablereusecompendium_3.pdf

Strzepek KM, Smith JB (1996) As climate changes: international impacts and implications. Cambridge University Press, Cambridge

Tortajada C, van Rensburg P (2019) Drink more recycled wastewater. Nature 577:26–28. https://doi.org/10.1038/d41586-019-03913-6

Tortajada C, Joshi Y, Biswas AK (2013) The Singapore water story: sustainable development in an urban city state. Routledge, Abingdon ·

UN Population Division (2019a) World urbanization prospects: the 2018 revision. Department of Economic and Social Affairs, United Nations, New York. https://digitallibrary.un.org/record/3833745?ln=en

UN Population Division (2019b) 2019 Revision of world population prospects. Department of Economic and Social Affairs, United Nations, New York. https://population.un.org/wpp/

Wallemacq P, House R (2018) Economic losses, poverty & disasters 1998-2017. Centre for Research on the Epidemiology of Disasters and United Nations Office for Disaster Risk Reduction. https://www.preventionweb.net/files/61119_credeconomiclosses.pdf

Chapter 2
Water Security in the Face of Climate Change: Singapore's Way

Peter Joo Hee Ng and Sharon Zheng

Abstract The effects of climate change are ominous. If ignored, devastation is guaranteed. To avoid tragedy, action is required and must be taken well before disaster arrives. Singapore's way is to find the best means of reducing the impacts of rising seas, coping with flooding that accompanies extreme and unpredictable rainfall, and further reinforcing an already weather resilient water system. Singapore seeks to guarantee its own water security in the face of climate change by remaining highly pragmatic, taking the long view, planning well in advance and executing relentlessly.

Keywords Singapore · Water scarcity · Coastal protection · Drainage · Desalination · Water reuse

2.1 Climate Change

The current climate crisis is entirely manmade. It is not brought on by a change in the earth's tilt and orbit, which triggered the last ice age. Neither is it the consequence of a meteorite collision, like the catastrophic Chicxulub impact, that so altered the earth's weather it wiped out the dinosaurs and brought an end to the Cretaceous. It is not even because of a super volcanic explosion, like Krakatoa in 1883, that darkened the sky, cooled the seas, and lowered global temperatures for years afterwards.

The climate is changing, and this time around, it is us, of the extant species *Homo sapiens*, who are causing it. It is human activity that has driven atmospheric CO_2 levels from about 280 parts per million (ppm) in 1850 to the 410 ppm today, thanks to industrial emissions and the unrestrained burning of fossil fuels (Bereiter et al. 2015;

P. J. H. Ng (✉) · S. Zheng
PUB, Singapore's National Water Agency, Singapore, Singapore
e-mail: ng_joo_hee@pub.gov.sg

S. Zheng
e-mail: sharon_zheng@pub.gov.sg

© The Author(s), under exclusive license to Springer Nature Singapore Pte Ltd. 2022   21
A. K. Biswas and C. Tortajada (eds.), *Water Security Under Climate Change*,
Water Resources Development and Management,
https://doi.org/10.1007/978-981-16-5493-0_2

Tans and Keeling n.d.). Because CO_2 traps heat, its rapid accumulation in the atmosphere can only lead to an inexorable rise in air and sea temperatures, as the world is now witnessing, and then to still not-fully-understood effects on the earth's climate.

CO_2 is not the only greenhouse gas getting released as detritus of modern civilisation. There is also methane, and NOx, and CFCs, all trapped in the air and slowly cooking the planet. As a result, the polar ice caps melt, glaciers shrink and eventually disappear; the tundra thaws, Greenland turns greener, and the oceans rise. It gets worse: because of more moisture in the air, storms become more violent, hurricanes and typhoons increase in power, and flooding is more common. The opposite, droughts and heatwaves, also become more ordinary. And year-on-year record temperatures shall no longer be unusual.

2.2 Suffer Little Singapore?

The foregoing can sound alarmist. And there are some, scientists among them, who continue to deny climate change. Even so, for those who govern countries, make public policy, administer cities or operate water systems, keeping fingers firmly crossed cannot be a recommended method.

When the full impact of unmitigated climate change arrives, in a matter of decades if not sooner, the unprepared will see its water infrastructure devastated, its water supply in jeopardy, and its water scarcity gravely imperilled. Inundation from frequent and extreme weather events can only lead to colossal flood damage. The challenge of draining a dense metropolis, like Singapore, then becomes wickedly complex. And rising seas may just make it altogether impossible.

Should nothing be done, Singapore, an island-state, pancake-flat, will surely become modern-day Atlantis. Unsurprisingly then for its prime minister Lee Hsien Loong to declare in his 2019 state-of-the-nation address that: '*We should treat climate change defences…with utmost seriousness. These are life and death matters. Everything else must bend at the knee to safeguard the existence of our island nation*' (Prime Minister's Office Singapore 2019). And he put money where his mouth is, pledging upwards of US$75 billion, some 20% of national GDP, towards Singapore's coastal protection. National water agency PUB was promptly charged, in April 2020, with leading this effort.

2.3 No Retreat

When the Centre for Climate Research Singapore (CCRS) applied the findings of the International Panel for Climate Change's Fifth Assessment Report to Singapore, it found that sea levels would likely rise one metre by 2100, and annual average rainfall could increase by up to 27% for the 2070–2099 period (CCRS and Met Office Hadley Centre 2015).

To fully appreciate the existential challenge that sea level rise poses to Singapore, one just needs to know that much of pint-sized Singapore—in area, half of greater London—stands barely five metres above mean sea level (National Climate Change Secretariat, n.d.). Almost six million people call the island home, making it one of the most densely populated corners in the world. Of course, unlike in other places where there is an option of just giving up and moving to higher ground, Singapore does not enjoy this luxury.

Jakarta, capital of neighbouring Indonesia, is already slowly getting consumed by the ocean. The Indonesian response is to build a spanking new capital city altogether, in an entirely different location in Borneo, and much farther inland this time. In bleak contrast, Singapore does not have this choice to make. For the island-state, there is just no retreat when the ocean swells. For Singapore to endure, it has to hold back the rising seas. The country and its leaders are crystal clear about this, and PUB has to get this done before it is too late.

2.4 How to Build a Seawall

Only five metres above water and facing 3-m tides, parts of Singapore are already vulnerable to coastal flooding sans climate change. To keep the country dry when the surrounding sea is permanently higher will require a barrier that encloses the most susceptible portions of its coastline. Even for petite Singapore, constructing this will be no mean feat. Sizable expenditure aside, the form and format of such a barrier demand careful and deliberate consideration. In this respect, PUB's planners are cognisant that pouring concrete would be the easy bit. Seawalls can come in many guises, and may not even require brick and mortar. Where terrain and topography permit, a biophilic approach could prove effective and be infinitely more sustainable. Restoring mangrove forests, for example, can attenuate waves and mitigate erosion, and be effective coastal defence for Singapore (Spalding et al. 2014). Hence, considerable thought and research, together with extensive community consultations prior to engineering design, will prove crucial to success.

Empoldering is likely to become a key tactic in Singapore's strategy for coastal protection. Attractive because it creates additional land for the crowded city-state, it is also a less resource-intensive way to reclaim land while simultaneously resisting higher seas. The opportunity to fashion extensive polders off Singapore's southern coast is currently under intensive study. At the same time, work on a starter polder is well underway off Pulau Tekong, Singapore's military island. When completed in 2022, the Tekong polder, Singapore's first, will feature a 10 km dyke enclosing 810 ha of new land for military use (The Straits Times 2016). PUB's engineers expect to maintain this dyke, and to operate the water body, pumping stations and some 45 km of drainage within. In the process, they gain valuable experience in polder management and ready themselves for bigger ones to come.

2.5 Taming Stormwater

Keeping the sea at bay is, literally, only one side of the equation. Water inevitably accumulates on the landward side of any barrier, natural or artificial. And so, polders, dykes and seawalls always require barrages and pumps in order to prevent flooding inland. As such, successful coastal protection will certainly require mastery of the intersection between seawater and stormwater. Holding back the seas must also allow for the taming of stormwater. If not, Singapore will still flood. Hence, PUB's planners and engineers need to know and understand how tides, storm surges and heavy rainfall will interact so that they may take the right action to keep deluge at bay.

Taming stormwater is not novel to PUB, as it is one of three legs in its water mission. The other two being to supply good quality water and to reclaim used water.

It can rain a lot in Singapore. Smack on the equator, in the tropics, it is the regular recipient of 2200 mm in annual precipitation. Increasingly, the rain threatens to come down all at once. Climate change may yet parch Singapore but studies suggest a trend towards higher rainfall intensities, and a greater frequency of such occurring in the future.

If space and money are plentiful, ordering bigger drains to cope with bigger storms might work. For Singapore though, upsizing every drain and canal would be prohibitively disruptive and expensive. To cope with the heaviest rain, stormwater channels would have to be the size of freeways, and necessarily be empty nearly all of the time. By themselves, larger storm drains will not keep Singapore's streets dry. But upgrading drainage infrastructure in tandem with source and receptor enhancements just might.

To tame stormwater effectively, Singapore's preferred method is a far more holistic one, and embraces much more than 'plain vanilla' drains and canals. The "source-pathway-receptor" approach (Fig. 2.1) dictates that flood protection does not just encompass drains and canals (the pathways), but also upstream areas generating stormwater run-off (the source), and parts downstream which might be inundated (the receptors). The whole idea is to slow the accumulation of stormwater at every turn, even during the heaviest bouts of rain, by detaining, retarding, delaying and diverting it, so that it does not overwhelm and then spill.

Since 2014, PUB has required real estate developers to implement "source" measures that reduce run-off from their plots by up to 35%. The less imaginative may just put in a detention tank, but rain gardens and bioswales (landscape elements used to slow, collect and filter stormwater) can also help, and are prettier to boot. One example is Tengah Pond (Fig. 2.2), built as part of a new 700 ha, 42,000-household residential precinct in northern Singapore. The pond, integrated into surrounding parkland, serves to attenuate peak run-off discharging from the Tengah Town catchment, while doubling up as recreational space for residents.

Singapore's latest "pathway" measure has to be the Stamford Detention Tank. The thousands who visit Singapore's sole UNESCO World Heritage site are

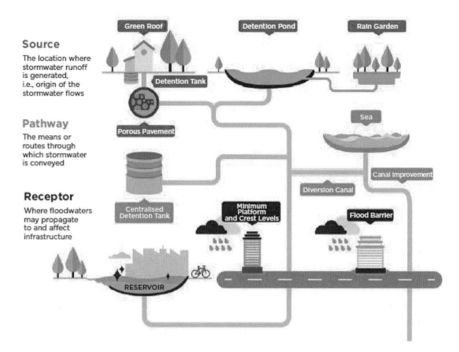

Fig. 2.1   The source-pathway-receptor approach. *Source* PUB Singapore

Fig. 2.2   Tengah Pond. *Source* Housing and Development Board, Singapore (n.d.)

oblivious that tour buses spew them out above a gargantuan stormwater reservoir. Hidden beneath the Singapore Botanic Gardens' new coach park is a very large concrete box that extends 28 m into the ground, big enough to temporarily detain 38,000 tons of rainwater from upstream—equivalent to 15 Olympic pools—and insuring against a repeat of the ruinous floods of 2010 and 2011 that inundated Orchard Road, Singapore's famous shopping street.

Even when pathways require enlarging, it can often be achieved in new and ingenious ways. The re-naturalisation of Singapore's Kallang River in the Bishan-Ang Mo Kio Park (Fig. 2.3), previously an ugly concrete canal, is illustrative. Although most people would look at it as parkland beautification, it was, first and foremost, drainage improvement. The Kallang River is Singapore's longest river, connects Lower Pierce Reservoir to Marina Reservoir, and is a crucial stormwater pathway. With her own floodplain inside of Bishan-Ang Mo Kio Park, that stretch of the Kallang River is now able to provide more flood protection than ever before.

At the receptor end, stormwater mitigation usually comes in the form of more "brute-force" applications, like raising an entire road or the platform of a new building. The aim here is to give flood waters no chance of affecting transport or dwelling. At the municipal level, this is perhaps best demonstrated by the Marina Barrage in Singapore. Built to seal off a lagoon reclaimed from the sea, Marina Barrage was conceived of and functions as a flood alleviation scheme—to eliminate the influence of high tides on drainage in low-lying areas in the city's centre and to release excess stormwater from its catchment.

2.6 Scarcity and Fickle Rains

Climate change threatens to bring unprecedented inundation, from both encroaching seas and from extreme rain. It will also threaten water supply, by exacerbating scarcity. When the rains do not come as anticipated, water stocks deplete, reservoirs dry up, waterworks lose input, and taps eventually go dry. This has happened with alarming regularity in recent years, afflicting large systems in different parts of the world.

After good rains in 2013 and 2014, the city of Cape Town experienced three consecutive years of drought. Without winter rains to replenish its reservoirs, on which the city depends entirely for its water supply, dam levels dropped from 72% in 2015 to 61% in 2016, to 39% in 2017 (City of Cape Town and National Department of Water and Sanitation 2019). Four million Cape Towners were resigned to "Day Zero" sometime in April 2018 when the city's water system would be shut down, and water strictly rationed and distributed from centralised locations. Fortunately, the rains returned, "Day Zero" was never declared, and Cape Town dodged a bullet.

Chennai in India must have wished it had better luck. The city of 10 million declared its own "Day Zero" on 19 June 2019 when there was no water left in its

Fig. 2.3 Kallang River at Bishan-Ang Mo Kio project. *Source* PUB Singapore

four reservoirs (CNN 2019). Two years of failed monsoons had led to this day. Thankfully, Chennai's water crisis would not be repeated in 2020 as its dams got restocked, desalination plants came online, and recycled sewage now supplies industry.

Singapore's fate is, in reality, no better. In spite of plentiful rain, Singapore is severely water-stressed. In its 2015 report, the World Resources Institutes declared Singapore to be at the highest risk of becoming severely water stressed by the year 2040, out of the 167 territories it surveyed (Water Resources Institute 2015). The lack of land, rather than of rain, fundamentally constrains Singapore from achieving water sufficiency in a conventional way. There is just not enough room in Singapore to capture and then to husband all of the water that it needs. As early as 1903, municipal engineer Robert Peirce assessed that even if all potential water resources in Singapore were to be fully developed and utilised, water supply would still not be able to keep up with growing population into the 1920s (Yeoh 2003).

Pierce's appraisal of Singapore's fundamental water scarcity could not have been more prescient. Scarcity of water supply was, is, and will forever remain a preoccupation for Singapore's decision makers and water planners. Unquestionably, it was this realisation that led Singapore to ceaselessly search for means to overcome water scarcity, and to make its next innovation. The water system in present-day Singapore and the way it is managed is the culmination of a decades-long quest by Singaporean water engineers for alternative water sources. The Singapore method is distinguished by three qualities: integration, circularity and reuse (Ng and Teo 2019).

When potable reuse technology became viable, at the turn of the millennium, PUB adopted it with great enthusiasm. And it has been producing for close to 20 years, on an industrial scale, NEWater—its own brand of ultra-high-quality recycled water. Singapore collects every drop of its sewage, and turns much of it into potable water again. Today, recycled used water can meet in excess of a third of total demand.

In order to make potable reuse possible, it was necessary to bring the water supply, the sewerage and sanitation, and the drainage systems, previously functioning independently, together and be operated as one integrated whole. When seawater desalination subsequently became affordable, Singaporean desalination plants were incorporated into the cycle, so that every drop of desalted water can also be endlessly reused.

Closing the "water loop" in this manner may be simple enough. However, in most of the rest of the world, water systems still tend to be run in discrete pieces. In most places, the water economy is still a linear one. Good water is taken from nature, and used just once by humans, and then the used water simply discarded. In many places, the water department is separate from the sewerage department, which is separate again from the drainage department. Invariably, all three will work at cross purposes.

Singapore's closing of the water circle was thus a stroke of administrative and operational genius. Its secret for a sustainable water supply is to manage the entire water system as one whole. In this way, the Singapore water economy also turned into a fully circular one, long before it was even fashionable to speak of circularity. In the shadow of looming climate change, this circularity has proven to be fortuitous. Because of early and deliberate diversification of sources, water supply in Singapore is largely resilient to fickle rains. Recycled NEWater and desalted seawater are manufactured, and therefore immune to the vagaries of climate.

2.7 A Smaller Footprint

Rain is free, and making it drinkable is fairly cheap. Seawater desalination and the production of NEWater are, in contrast, far dearer ways of producing potable water. While only 0.2 kWh of energy is enough to treat 1 m^3 of rainwater, close to 1 kWh/m^3 is needed to turn sewage effluent into NEWater. Desalination may be weather-resistant, but even state-of-the-art reverse osmosis, at about 3.5 kWh/m^3, is still positively energy guzzling.

In a climate-changed future, desalination's carbon footprint poses an upsetting concern. In the range of 0.4–6.7 kg CO_2eq/m^3 (Cornejo et al. 2014), meeting Singapore's entire daily water demand of about 2,000,000 m^3 through desalination would mean releasing more than 2800 tons of CO_2eq into the air every day, or the equivalent of running 600 automobiles. Clearly, sustainable desalination and, to a lesser extent, potable reuse, both of which otherwise possess significant sustainability benefits, demand that their energy intensities be meaningfully reduced.

For resource-poor Singapore, which imports almost all of its energy, options for mitigation are understandably limited, but not non-existent. There are two ways: one is obviously to find means to recycle and desalt far more efficiently; the other is to swap out some of the fossil fuel presently powering treatment plants for renewables.

Working with collaborators, Singapore's PUB is demonstrating electrodeionisation as a superiorly efficient way of seawater desalination, aiming to at least halve the current industry standard of 3.5 kWh/m^3 in the process. But why stop there? Mother Nature, as always, does it best. Mangrove and fish in the sea need fresh water too and are capable of removing excess salt with minimal effort. Biomimicry and biomimetics offer great promise and are areas of research that PUB devotes considerable resources to.

On the second count, one of the largest integrated floating solar panel arrays in the world was recently installed on a reservoir in the west of Singapore. Commissioned by PUB and operated by a subsidiary of Sembcorp Industries, a private company, the Tengeh solar farm has a rated capacity of 60 MW-peak. When up and running in 2021, it will generate enough to satisfy 7% of PUB's entire power requirement, and sufficient to simultaneously power every one of its conventional treatment plants. The satisfactory operation of the Tengeh project shall no doubt encourage PUB to make additional and larger investments in renewable energy. At the project's groundbreaking in August 2020, your author had positively proclaimed: *'With this floating solar power plant, which we believe to be one of the largest in the world, PUB takes a big step towards enduring energy sustainability in water treatment. Solar energy is plentiful, clean and green, and is key to reducing PUB's and also Singapore's carbon footprint'* (CNA 2020).

2.8   Conscientious Consumption

The typical system response to climate change tends to be grand and expansive. Emission targets, carbon pricing, massive infrastructure, extensive construction, international agreements, national pledges, corporate promises, and so on, come to mind. But the world is made up of people. Ultimately, every act of adaptation and mitigation must aim to save life and increase well-being. The fight to adapt, contain and maybe even reverse climate change cannot exclude individual effort.

To the individual, what is asked must surely be to use water in the most efficient and productive fashion. Singaporean household water consumption (PUB Singapore 2020)—at 141 L/person/day—although already low by international standards, is still deemed to be profligate. Singapore's populace is perennially challenged to be even more frugal and to use water ever more judiciously. Effective demand management, especially for the individual and the household, requires intervention at various levels. Children in kindergarten and grade school is a good place to start. Everyone can be taught, from infancy, that water is precious—even if affordable—and must be treasured. This is precisely the motivation behind the decades-long inclusion of water conservation in the curriculum of Singapore's schools.

Ultimately, a litre of water that is not required, and therefore not used, is a litre not produced. The consequent savings in energy, chemicals, labour and material, plant and equipment are not just expenditure avoided, but also go towards reducing the collective carbon footprint. Producer and consumer both do the Earth a favour by not having to make or use up that marginal litre of water.

Certainly, everyone in a developed jurisdiction like Singapore can and should use less water. Doing so is entirely feasible and should not be bothersome or inconvenient. It also does not require giving up modern-day comforts, or compromising personal and public hygiene. But it does require behavioural change. Even here, the plethora of smart meters and gadgets now coming into common usage, which are enabled by a ubiquitous internet, and empowered by cloud computing and artificial intelligence, is conspiring to drive positive change in ways not previously possible. Digitalisation will eventually help to make conscientious consumption second nature, allowing the end-consumer to become a smarter user of water, who can do better sense-making every time he decides to flip that tap.

2.9 Conclusion

Climate change is real and will most likely get worse. Its effects are ominous. If ignored, devastation is guaranteed. Therefore, water utilities, regulators and system operators disregard the weather effects to come at certain peril. To avoid tragedy, action is required and must be taken well before disaster arrives. However, climate change is neither a death sentence nor even the end of civilisation. Practicable strategies for adaptation can be conceived and implemented, which will mediate and moderate the worst consequences. Possibilities for mitigation also present themselves, many of which are already readily realisable.

There *is* a way.

Singapore's way is to find the best means of holding back the rising seas, coping with flooding that accompanies extreme and unpredictable rainfall, and further reinforcing an already weather resilient water system. With strong will, clarity of vision and determination, it is entirely possible to adapt well and to mitigate profitably. In the process, it may well again prove the old cliché true, that there is always opportunity in crisis. The opportunities are many: to create new land, to return nature to the city, to pivot to renewables, to advance the state-of-the-art, to digitalise, to transform, and to become smarter.

Singapore seeks to guarantee its own water security in the face of climate change the same way it has dealt with the crippling scarcity, pervasive floods and dreadful lack of sanitation of the past. And this is through remaining highly pragmatic, taking the long view, planning well in advance and executing relentlessly. As always, the aim is to ensure that Singapore's water system remains adequate, resilient and sustainable, all at once. This strategy will, once again, allow tiny Singapore to turn disadvantage into strength, and what seemed an insurmountable vulnerability into endless opportunity.

References

Bereiter B, Eggleston S, Schmitt J, Nehrbass-Ahles C, Stocker TF, Fischer H, Kipfstuhl S, Chappellaz JA (2015) Revision of the EPICA Dome C CO_2 record from 800 to 600 kyr before present. Geophys Res Lett. https://www.ncdc.noaa.gov/paleo-search/study/17975. Accessed 30 Nov 2020

CCRS (Centre for Climate Research Singapore) and Met Office Hadley Centre (2015) Singapore's second national climate change study: climate projections to 2100

City of Cape Town and National Department of Water and Sanitation (2019) City of Cape Town: dam levels report. City of Cape Town Isixeko Sasekapa Stad Kappstad

CNA (2020) Construction begins on Tengeh Reservoir floating solar farm, touted as one of world's largest. https://www.channelnewsasia.com/news/singapore/floating-solar-farm-tengeh-largestclimate-change-carbon-13029904. Accessed 30 Nov 2020

CNN (2019) India's sixth biggest city is almost entirely out of water. https://edition.cnn.com/2019/06/india/chennai-water-crisis-intl-hnk/index.html. Accessed 30 Nov 2020

Cornejo P, Santana M, Hokanson D, Mihelcic J, Zhang Q (2014) Carbon footprint of water reuse and desalination: a review of greenhouse gas emissions and estimation tools. J Water Reuse Desalin 4(4):238–252

Housing and Development Board, Singapore (n.d.) Tengah. https://www.hdb.gov.sg/cs/infoweb/about-us/history/hdb-towns-your-home/tengah. Accessed 5 Dec 2020

National Climate Change Secretariat (n.d.) Coastal protection. https://www.nccs.gov.sg/singapores-climate-action/coastal-protection. Accessed 5 Dec 2021

Ng P, Teo C (2019) Singapore's water challenges past to present. Int J Water Resour Dev 36(2–3): 269–277

Prime Minister's Office Singapore (2019) National day rally 2019. https://www.pmo.gov.sg/Newsroom/National-Day-Rally-2019. Accessed 15 Dec 2020

PUB Singapore (2020) Make every drop count: continuing Singapore's water success. https://www.pub.gov.sg/news/pressreleases/MakeEveryDropCountContinuingSingaporesWaterSuccess. Accessed 15 Dec 2020

Spalding M, McIvor A, Tonneijck F, Tol S, van Eijk P (2014) Mangroves for coastal defence. In: Guidelines for coastal managers & policy makers. Published by Wetlands International and The Nature Conservancy. https://www.researchgate.net/publication/272791554_Mangroves_for_Coastal_Defece_Guidelines_for_Coastal_Managers_Policy_Makers. Accessed 15 Dec 2020

Tans P, Keeling R (n.d.) Global monitoring laboratory—carbon cycle greenhouse gases. https://www.esrl.noaa.gov/gmd/ccgg/trends/data.html. Accessed 22 Dec 2020

The Straits Times (2016) Pulau Tekong to get extra land the size of two Toa Payoh towns using new reclamation method. https://www.straitstimes.com/singapore/singapore-to-use-new-land-reclamationmethod-in-pulau-tekong. Accessed 22 Dec 2020

Water Resources Institute (2015) Ranking the world's most water-stressed countries in 2040. https://www.wri.org/blog/2015/08/ranking-world-s-mostwater-stressed-countries-2040. Accessed 22 Dec 2020

Yeoh BSA (2003) Contesting space in colonial Singapore: power relations and the urban built environment. NUS Press, Singapore

Chapter 3
Consequences of Declining Resources on Water Services: The Risks if We Do not Act!

Diane D'Arras

Abstract The impact of climate change on water resources is a crucial issue. Many people have dealt with it and will deal with it in the future. To complement those contributions, I would like to address the consequences of reduced water resources on water supply services, the need to adapt to these consequences, the importance of involving citizens in this policy area, and the role of the International Water Association in this enormous challenge.

Keywords Water services · Climate change · Adaptation measures · Water demand management · Non-conventional sources of water

3.1 Consequences of Climate Change on Water Distribution Services

It is clear and indisputable that the great water cycle is already impacted and will be even more impacted by climate change, regardless of the debates that arise around the importance of the phenomenon. The current balances between the location of populations and their water resources will be clearly and strongly impacted by climate change. The situation is already difficult in arid countries and will in most cases worsen. In more temperate countries, we also already face and will face lack of water resources, which means difficulties for supplying drinking water, particularly during the summer. Two degrees more for the planet is an annual global average. This means that some regions will be more affected than others (the geographic average) and that certain times of the year will be more impacted than others (the annual average).

D. D'Arras (✉)
International Water Association, London, UK
e-mail: diane.darras@bunzi.international

© The Author(s), under exclusive license to Springer Nature Singapore Pte Ltd. 2022   33
A. K. Biswas and C. Tortajada (eds.), *Water Security Under Climate Change*,
Water Resources Development and Management,
https://doi.org/10.1007/978-981-16-5493-0_3

If water resources decrease in certain places or at certain times, in the absence of alternative solutions (multi-year storage of raw water, desalination, recycling), *the water services will have to manage shortages by deploying restrictive measures on water distribution in order to adapt consumption to production.*

This situation is far from hypothetical. This is already the case today around the world, even in the so-called developed countries. Such consumption restrictions are sometimes only incentive measures, such as bans on watering gardens or on washing cars. This approach works only if the restrictions which are imposed are limited and if the population as a whole is ready to accept them.

If the frequency or severity of the restrictions increase, the result can be intermittent supply. *Such intermittent distribution can be seen as rationing—voluntary, or involuntary—that must be managed or suffered.*

It is a situation in drinking water services similar to the one linked to food restrictions observed both in wartime (example of the 1940 war) or in peacetime in countries such as the Sahel.

This situation, rightly criticised by professionals, is unfortunately commonplace in the world, since around two billion people are now supplied with water in this way. The current cause of intermittent distribution is most of time a combination of different causes: poor maintenance of networks due to a lack of financial resources or lack of competence; wastage of water by customers, with insensitivity to the value of water (price too low for all, even for those who could perfectly afford to pay for the service, or lack of meters); or lack of water resources. *If we do not take preventive measures to be able to adapt to the future level of our water resources, it is certain that the number of people supplied intermittently, and therefore chaotically, will increase dramatically.*

This intermittent water distribution is indeed a chaotic mode of distribution with many undesirable effects:

- Degraded water quality, because networks lack protection when they are no longer under pressure: runoff water can enter the network, and the presence of air in the water promotes bacterial development.
- Unjust distribution: the water, unlike car drivers, does not respect any green light, red light or prohibited direction. It flows where gravity takes it. It is therefore the areas near the resources or those downstream that will be served first, leaving the other areas aside.

And intermittent distribution is unfortunately a downward spiral:

- Unhappy customers do not feel like paying their bills.
- Technical management becomes a real puzzle, with the classic water service indicators (technical performance, volumes, pressures) being inconsistent, as the water meters which are supposed to give important information on volumes count mainly air passing through the network. It becomes incredibly difficult for the utility maintenance teams to understand what is really happening.
- And, above all, the return of water after a period without power regularly causes a water hammer which can lead to breakage of the pipes. The network becomes

more and more leaky, losing 50% of the water—in some cases more—An infernal circle!

It is instructive to read or reread the book *Dealing with the Complex Interrelation of Intermittent Supply and Water Losses* (Charalambous and Laspidou 2017) which deals with all the misdeeds of intermittent distribution, to understand how important it is to maintain a permanent distribution for those fortunate enough to benefit from it every day of the year!

3.2 Implementing Adaptation Measures: Desalination or Reuse (a Little), Reduce Consumption (a Lot)

It is necessary to organise how to adapt to the new climatic conditions of today and tomorrow and their impacts on water resources. Where resources are set to decrease permanently or episodically, the demand side must absolutely be adapted permanently or on an ad hoc basis. This notion is a little new in the drinking water community. In the twentieth century, we were rightly proud of having developed water distribution services to improve sanitary conditions and comfort by bringing water to inhabitants, but the need to properly recover wastewater was largely forgotten, and we are fighting today to upgrade our concepts and services in this area.

The water sector is also not exempt from the need to develop mitigation measures to try to limit climate change impacts, and less energy-consuming technologies are possible. But we will mainly focus on understanding how to reduce the demand for water. Adapting to the possibility of supply reduction strategy is the best mitigation strategy.

The volumes consumed by manufacturers, agriculture or individuals can be hugely different from one country to another:

- The irrigation volumes needed are influenced by local rainfall or temperature and by the nature of the goods cultivated.
- Depending on the industrial products manufactured and the services developed, the volumes of water necessary for processing can be hugely different. In addition, they may have little impact on the final cost of the product: the motivation for reducing volumes will obviously depend on this.
- The use of water by individuals can vary greatly depending on the country, from less than 100 L per day per capita in some European countries to 500 L in some Latin American countries—a fivefold difference. The volumes represented by this use can however remain relatively small compared to industrial or agricultural uses. But unlike the other two uses, any interruption of service has dramatic health effects. It is therefore essential to manage the demand (consumption) according to the available resources!

3.2.1 Adaptation Through New Resources

Before discussing the reduction in consumption, I will recall that technical solutions have been found to make more resources with less: they are water desalination and reuse/recycling.

Desalination is an expensive technique per m^3 and is energy intensive even if the progress in reverse osmosis has brought this technology to an "acceptable" level for sanitary or industrial uses (but in no case for agriculture purposes). It is economically feasible if you do not have to transport water over long distances, with transport of water being hugely energy intensive. Water desalination is therefore possible only in the immediate vicinity of seas or oceans. Of course, it becomes necessary for climate change mitigation reasons to use renewable energy for that purpose!

Recycling of water (reuse) is another alternative solution that is being deployed. It is possible to multiply the use locally from the same resource by a cascade effect: urban wastewater can be reused in industry or agriculture after appropriate treatment. Treatment is necessary to restore the water to sufficient purity to allow the new use. A smart technology—but like desalination it has an economic cost and consumes energy—can be applied wherever there is a mix of usages able to share a common resource. Its application therefore involves redeveloping common sense among consumers and politics. Not always easy.

3.2.2 Much Better: Adapt by Reducing Consumption

The best global and long-term strategy remains one of adapting consumption to available resources. Of course, the needs may be different from one country to another depending on habits and different climatic conditions: we can imagine that in the heat of the day we are more inclined to take a second shower! But nothing justifies such differences in consumption (from one to 5)!

Change can come from three types of actors involved: the water utilities, the consumers, and the regulators.

Changing mentality of the water utilities

Factors that fail to lead to a culture of reasonable use include the following:

- The absence of meters, which does not allow consumers to understand the mechanisms of their consumption. And if they have meters, most of the time the bills speak clearly about money but not consumption. Most consumers therefore have no idea on how much water they use!
- A price of the service that is too low. A low price is not an incentive, and it is certainly counterproductive if this price is lower than the real cost of the service; it is a shame to lower the price of water for everyone when a professionally

managed selective strategy can be smarter and fairer. It is obvious that we must not deprive the poorest populations of water, but selective rules can be implemented!

– No short-term interest by water utilities in reducing their consumers bills. The costs of water services are predominantly fixed; therefore, any marginal revenue is obviously fundamental and welcome, as it is pure margin and has a great impact on the economic equilibrium of the services! It is very often only when the crisis arrives that the awareness of management occurs. Marginally, each cubic metre sold brings 80% of its price in more benefits, because the variable costs in energy and treatment products rarely exceed 20% of the average price. The utility therefore does not necessarily push consumers to reduce their consumption (even if it is pure water leaks!). If the price is reasonably high, consumers may be encouraged to repair their leaks, but most of the time have little incentive to shorten their showers or water their gardens. In the long run, the calculation of the water service can turn out to be wrong. In fact, if new resources are required, such as desalination, recycling or the construction of new facilities, average costs may increase when water prices remain stable. The extra m^3 can be extremely expensive! like the third child who makes a bigger house necessary!

– A great technical maturity is needed and above all a political maturity to be able to enter a positive circle of reduction in consumption and/or increase in prices so that the service remains balanced on a lower but more virtuous basis of consumption. As I have said, awareness has only developed in utilities that have been forced to fight for alternative resources. They are now leading the movement by deploying remarkable awareness policy among their consumers (for example the PUB in Singapore or the Water Corporation in Perth, Australia).

Developing consumer awareness

Consumers must be given the means to understand their consumption in order to be actors of change.

The necessity of metering water properly is fortunately no longer a discussion among water professionals. The relative cost of the technology (even the traditional ones) can still be considered as expensive in developing countries compared to local economic means. But it is impossible to implore consumers without giving them the information; awareness needs metering, even if the bill is not based on volumes!

Educational tools and information are key. Most people do not understand what a cubic meter of water is. Anglo-Saxons sometimes use the term "Megaliter" which I have always had trouble understanding, educated in the "cubic meter". But with hindsight, litres are perhaps easier to understand for the user, more visual! 1000 L speak more than one m^3! If possible, information should be given per day, per person, so that everyone can follow.

The city of Perth has distributed to its consumers a small stopwatch that strikes after four minutes as a reminder when you use the shower. A shower can be comfortable and efficient without spending hours in it. For that reason, I will mention the Hydrao shower head, which has been designed so that the user can take a pleasant shower while limiting water consumption to 6.6 L per minute, when a standard shower head consumes an average of 12 L per minute. The Hydrao shower head is equipped with a self-powered turbine which adds a fun element and intuitive pedagogy by using colour indications on the volumes of water consumed. This type of innovation is doubly interesting in a long-term vision because it certainly reduces water consumption, but better too the consumption of hot water, a source of energy that is often derived from fossil fuels. Hydrao has been tested in different places in the world with good effect. Being a very good innovation with climate change mitigation and adaptation effect, it was honoured by an IWA Innovation Prize in 2019.

Smart metering should also help raise consumer awareness, as many applications can be developed around the information given by on-line metering, such as night consumption, winter and summer consumption, leaks, or peak consumption.

Regulators as game-changing actors

Regulators can be fundamental in implementing a long-term strategy. As mentioned above, a good dose of economic and political maturity is needed for a water utility to embark on a policy of consumption reduction. The presence and action of regulators can be decisive in implementing such a strategy, as water price and the invoicing scheme are key. The Italian regulator for example was reflecting on implementing progressive prices with a relatively low price up to 40 L per day per person, and then higher beyond that. As a reminder, the average consumption can be 100 to 500 L per day per person depending on the country. It is open to discussion as to what the first level of consumption should be (probably linked to a minimum necessary), but tariff-setting will be important for changing the game. On-line meters can be useful as prices could be modified depending on the available resources, pushing users to adapt their consumption when weather conditions dictate.

Appropriate and adaptable pricing over time is therefore a highly interesting avenue for adapting to the increasing variability of the situations that we will encounter. It must obviously be accompanied and explained.

If all water utilities, operating 24 h a day, start to develop new strategies in a more reasoned vision, we can hope to avoid temporary or regular water cuts, escaping the infernal spiral of intermittent distribution. But what to do where this type of distribution is the current mode of operation?

3.3   Mobilise and Support the Water Services Who Must Manage Intermittent Water Supply

I have been discussing how to avoid the future scenario of falling into the dramatic situation of intermittent supply. But as I have said, it is today unfortunately a common situation in many countries today.

Nearly two billion people receive water from piped networks that experience intermittency. Furthermore, nearly 41% of piped water systems in lower- and middle-income countries operate intermittently. The twin parameters of low pressure and the time for which taps do or do not flow define the experience of intermittent supply from the consumer's perspective.

This experience varies widely. For example, in a study conducted by the Asian Development Bank looking at 18 utilities in India, the duration of water supply varied from 20 min to 12 h per day, while the average pressures at consumer connections ranged from 0 to 10 m (0–14 psi). Furthermore, unreliability of water supply is also a frequent issue among those already impacted by limited availability. The time when water is available can vary on an almost daily basis, leaving many paralysed by an inability to forward plan or to allocate their (limited) resources effectively. The only reliable thing about their supply is its unreliability!

If we do not move forward, climate change will increase the number of disastrous situations.

There is a great need for action and mobilisation in order to improve rapidly the situation and be more resilient!

The very nature of the impacts places this demand for action right at the heart of IWA's global community. This is the space in which IWA's Specialist Group on Intermittent Water Supply is ideally placed to lead a transformative agenda on an issue that blights the lives of many.

Knowledge is lacking for us to move beyond the realities of where we are today. Certainly, solutions need to be built around the experiences and the research work of experts in the regions most affected, particularly Africa and Asia. In this regard, we can learn from the diversity of case studies of working with intermittent supply and so build appropriate and context-specific solutions that can support a transition from intermittent towards continuous supply.

Even with an ambition for continuous water supply, the reality for many utilities around the world is that they will need to work for a while—which can mean five or ten years—with the practical challenges of delivering supplies on an intermittent basis. This is a particular area of responsibility for the IWA Specialist Group to contribute, facilitating exchanges on the many dimensions of Intermittent Water Supply. It is a space in which they can look to apply all the latest thinking and technologies that the sector has at its disposal.

With the piped water supply expanding rapidly across the globe, we need research that investigates some of the unique characteristics of Intermittent Water Supply, such as: de-pressurisation and pressurisation; secondary network modelling and the associated systems function with respect to pressure deviations and outflows; and links between sanitation provision, pipe condition and microbial contamination and propagation, etc. Such research will have a fundamental part to play in understanding how best to improve equity, reliability, and water quality in intermittent systems.

Awareness raising and capacity building will be especially important, particularly in the case of operators. Armed with greater data, metrics and insight on the issues of Intermittent Water Supply, including water quality, they will be better placed to deliver improvements where they are most needed.

It is therefore most important that we can improve rapidly the situation, as we already know that with decreasing resources due to climate change, the challenge will be even greater!

Our collective aim is to create a water-wise world. Intermittent water supply represents today one of the most glaring gaps in this ambition. If we do not take actions in those cases but also where the situation is for the moment under control, climate change will have a dramatic effect on water services! IWA's efforts can play a key role in better understanding the current global situation, identifying challenges on the ground, and in turn shaping and delivering potential solutions.

Reference

Charalambous B, Laspidou C (2017) Dealing with the complex interrelation of intermitent supply and water losses. IWA Publishing, London. https://www.iwapublishing.com/books/9781780407067/dealing-complex-interrelation-intermittent-supply-and-water-losses. Accessed 20 Feb 2021

Chapter 4
Resilience Through Systems Thinking for Water Infrastructure

Cindy Wallis-Lage and Zeynep Kisoglu Erdal

Abstract The civil infrastructure of our communities is an outcome of past practices. These practices are being tested frequently under today's realities, which include climate extremes as a major factor, and they catastrophically fail at an increasing rate. Infrastructure problems are often approached through siloed solutions to address a singular service: water supply, pollution control and stormwater diversion. While appropriate for the time, the changes in external and connected pressures require a new roadmap. To achieve the infrastructure resilience needed to manage and mitigate the impacts of climate change, population growth, economic and societal movements, we must envision a future that builds on today, yet provides a trajectory that is unlike our past. Our resilience is predicated on systems thinking, and our ability to do long-range planning for our water infrastructure considering phased and integrated implementation of innovative engineering and nature-based approaches. For that, we must start the resiliency journey of our water infrastructure by acting now and acting with urgency.

Keywords Climate change · Resilience · Water stress · Droughts · Reuse · Community-based partnerships

Our infrastructure is in peril

Today's infrastructure represents a history focused on siloed solutions to address a singular service: water supply, pollution control and stormwater diversion. While appropriate for the time, the changes in external and connected pressures require a new roadmap. Climate change response has increasingly centred on resilience, and the needed modifications to infrastructure design practices, investment analysis processes, and policy decisions regarding financing and disaster risk management.

C. Wallis-Lage (✉)
Black & Veatch, 8400 Ward Parkway, Kansas City, MO 64114, USA
e-mail: wallis-lagecl@bv.com

Z. K. Erdal
Black & Veatch, 5 Peters Canyon Road, Suite 300, Irvine, CA 92606, USA
e-mail: ErdalZ@bv.com

Fig. 4.1 A piece by street artist Banksy near the Oval bridge in Camden, north London in view of the UN Climate Summit in Copenhagen in 2009. *Source* UN WATER (2019)

Fundamentally, resilience is "the capacity of any entity such as an individual, a community, an organisation, or a natural system to prepare for disruptions, to recover from shocks and stresses, and to adapt and grow from a disruptive experience" (Rodin 2014) (Fig. 4.1).

The modern concept of resilience builds on insights from three fields: engineering, ecology, and systems thinking. The engineering application of resilience focuses on combining strength with flexibility or redundancy. The ecological resilience considers occurrence of large changes, through which the system can absorb large shocks without collapse. Systems thinking then considers the interconnection of engineered and natural ecosystems, and how they can operate together to respond to the change event.

Water utilities, cities and communities interconnected through built environments, infrastructure and nature spaces need to consider the steps to provide the necessary resilience of their entire water systems (clean water supply, used water management, stormwater management, natural water bodies) against extreme weather events—one of the primary ways in which the effects of climate change are impacting society now and into the future.

4.1   State of the Water Sector Under Climate Pressure

Water stress already affects every continent (Fig. 4.2). Physical water scarcity is often a seasonal phenomenon in these regions, and climate change is likely to cause shifts in seasonal water availability throughout the year in several places (IPCC 2014). About four billion people live under conditions of severe physical water scarcity for at least one month per year (Mekonnen and Hoekstra 2016). More than two billion people live in countries experiencing high water stress. The situation will likely worsen as populations and the demand for water grow, and as the effects of climate change intensify (UN Water 2018).

Climate change is projected to increase the number of water-stressed regions and exacerbate shortages in already water-stressed regions. Climate change impacts are most felt through changing hydrological conditions including changes in snow and ice dynamics. By 2050, the number of people at risk of floods will increase from its current level of 1.2 billion to 1.6 billion. In the early to mid-2010s, 1.9 billion people, or 27% of the global population, lived in potentially severely water-scarce areas. In 2050, this number will increase from 2.7 to 3.2 billion people. As of 2019, 12% of the world population drinks water from unimproved and unsafe sources. More than 30% of the world population, or 2.4 billion people, live without any form of sanitation (WMO 2020a).

World Meteorological Organization (WMO)'s *United in Science Report* (WMO 2020a) findings show that greenhouse gas concentrations are at their highest levels in three million years and continue to rise, reaching new record highs by the end of 2020. Meanwhile, large swathes of Siberia have seen a prolonged and remarkable

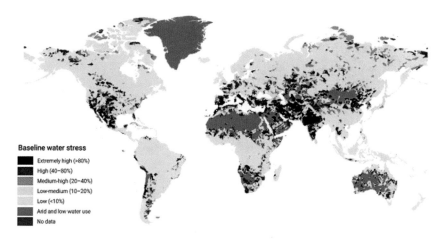

Baseline water stress

- ■ Extremely high (>80%)
- ■ High (40–80%)
- ■ Medium-high (20–40%)
- ■ Low-medium (10–20%)
- □ Low (<10%)
- ■ Arid and low water use
- ■ No data

Note: Baseline water stress measures the ratio of total water withdrawals to available renewable water supplies. Water withdrawals include domestic, industrial, irrigation and livestock consumptive and non-consumptive uses. Available renewable water supplies include surface and groundwater supplies and considers the impact of upstream consumptive water users and large dams on downstream water availability. Higher values indicate more competition among users.

Fig. 4.2   Global water stress map. *Source* WRI (2019)

heatwave during the first half of 2020, and 2016–2020 is set to be the warmest five-year period on record. Resulting picture of the global state of conditions indicates worsening of impacts through cascading actions of interconnected parts of our planet (Fig. 4.3).

The SARS-CoV-2-driven COVID-19 pandemic has highlighted the critical role water plays in managing the spread of disease and the need for flexibility and adaptability of our infrastructure. The pandemic has resulted in global lockdowns and changes in water use patterns and community movements and emphasised the need for strengthening the resilience of water systems as well as the broader human infrastructure systems. Never has this been more apparent than in 2021 as the world embarks on the United Nations Climate Change Conference 2021 (COP26) journey.

As was also originally recognised as part of the United Nations Sustainable Development Goals (SDGs) and further recognised as a result of the COVID-19 pandemic, water is truly central to our survival. However, a viable water sector requires many other essential elements of societies such as future-proofed energy, transportation and communication infrastructure. Without the broader infrastructure

Fig. 4.3 Current global events and their impacts on SDGs. *Source* WMO (2020b)

system, a community's public and environmental health, food and overall quality of life is threatened. Therefore, ensuring the resilience of all aspects of and adjacencies to the water sector is not only essential but also imperative for a community to survive.

Immediate impacts of climate change are apparent from evaluation of events that are currently experienced in communities, cities, and utilities. Increasing temperatures, extreme storm events and change in precipitation patterns are creating devastating outcomes across the globe. Figure 4.4 illustrates the projected changes in seasonal precipitation patterns through 2070–2099 (RCP8.5), indicating precipitation decreases in the South-western USA, Mexico, and the Caribbean, as well as movement of increased precipitation events to the northern parts of North America. Several studies have projected increases of precipitation rates within hurricanes over ocean regions, particularly for the Atlantic basins. Increased air temperatures are shown to result in increased water vapour which in turn results in the high-intensity precipitation and variability of events (Fig. 4.5). Since hurricanes are responsible for many of the most extreme precipitation events especially in South-eastern USA, such events are likely to be even heavier in the future resulting in greater catastrophic damage similar to Hurricane Katrina of 2005, Hurricane Sandy of 2012, Hurricane Harvey of 2017 and others that have followed. Similarly, typhoons in Asia–Pacific region have been observed to exhibit greater intensity.

Additionally, the land area impacted by extreme events has expanded as illustrated in Fig. 4.6. Scientists have evaluated the data records from the early 1900s through current time and found that with the expansion of storm sizes and intensity of events, the effect of one-day storms has substantially increased the area of land impacted, ultimately resulting in greater stress on the agricultural and urban areas. Combined with sea level rise, the resultant storm surges and flooding have increased as the intensities of these weather events have increased. Vulnerable low-lying areas such as the Florida and Gulf Coast of US (Fig. 4.7), as well as Pacific Islands have experienced large life and economic losses over the last couple of decades.

An important additional aspect of climate change is the potential for compounding extreme events. These can be events that occur at the same time or in sequence (such as consecutive floods in the same region) or at multiple locations within a given country or around the world (such as the 2009 Australian floods and wildfires). They may consist of multiple extreme events or of events that by themselves may not be extreme but together produce a multi-event occurrence (such as a heat wave accompanied by drought or unexpected deep freeze conditions as experienced in Texas in February 2021). Some areas are susceptible to multiple types of extreme events that can occur simultaneously. For example, certain regions are susceptible to both flooding from coastal storms and riverine flooding from snow melt, and a compounding event would be the occurrence of both simultaneously. Another compound event frequently discussed in the literature is the increase in wildfire risk resulting from the combined effects of high precipitation variability (wet seasons followed by dry), elevated temperature, and low humidity.

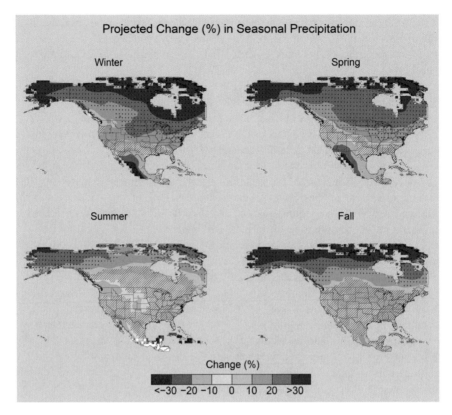

Fig. 4.4 Projected change (%) in total seasonal precipitation from CMIP5 simulations for 2070–2099. *Source* USGCRP (2017)

If followed by heavy rain, wildfires can in turn increase the risk of landslides and erosion.

The other type of risk due to compounding effects is when known types of compound events recur, but are stronger, longer-lasting, and/or more widespread than those experienced previously or projected by model simulations. One example would be simultaneous drought events in different urban and agricultural regions around the world, both challenging the food security as well as water supply sources used for agriculture versus other needs. More concerning is the unexpected ways the compounded effects can come together, similar to what happened during Hurricane Sandy where sea level rise, anomalously high ocean temperatures, and high tides combined to strengthen both the storm and the magnitude of the associated storm surge, resulting in the extreme level of inundation and infrastructure damage experienced especially in the mid-Atlantic region of US.

From a resultant risk perspective, the primary concern is the need to manage compounding extremes with additive or even multiplicative effects, and the predictability (or lack thereof) of such compounding effects. This requires the

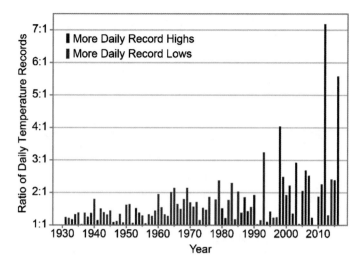

Fig. 4.5 Observed changes in the occurrence of record-setting daily temperatures in the contiguous USA. *Source* USGCRP (2017)

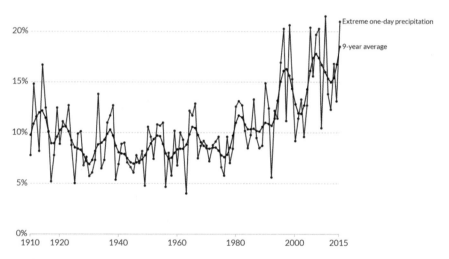

Fig. 4.6 Share of land area (expressed as % values on y-axis) experiencing extreme one-day precipitation. *Source* Our World in Data (2019)

development of potential operating scenarios and consideration of longer-term planning horizons, which may not be as desirable depending on the specific needs and dynamics of individual communities/utilities. It also requires clarity on the fiscal, political, and cultural views on the acceptable level of resiliency.

Fig. 4.7 Close to forty Houston, TX wastewater treatment plants flooded during Hurricane Harvey in 2017 (Wright 2018)

4.2 What Future Do We Seek?

The challenges are many, thus it will take strategic thinking and planning to understand the impacts of various events on the water sector. Critical to that thinking is understanding the overlay of events as well as clarity on the targeted level of resilience. Under these scenarios, mitigation opportunities can be identified, and a roadmap established to deliver a resilient water sector. Table 4.1 highlights the impacts and mitigation opportunities for various climate change events.

When we study the actions of leading utilities of today, a picture becomes clear: the mission and vision of these utilities, municipalities and other local governments, and their stated goals are not made up of disparate actions or siloed functions to meet today's demands of supplying clean and safe water, without integrating the needs of the whole community. The vision is tailored for them, and goals built on a continually self-assessing framework that allows innovation and technological advancements. The vision ultimately must serve the communities for the higher goal of stewarding the safety and health of the total environment in which the communities they serve are located, including the water systems.

One example of a utility innovatively transforming itself to address climate change, system resiliency and operational efficiency is Aarhus Vand, a Danish public limited company owned by Aarhus Municipality. According to its mission statement, it "exists to serve the purpose of creating health through the supply of clean water—to the population and the planet. The vision is to create a national platform as a driver for local and global solutions for a healthier water cycle." They produce and distribute more than 15 million m^3 per year drinking water, convey and

Table 4.1 Climate change-driven key events and opportunities to mitigate systemwide impacts

Events	Impacts on water sector	Mitigation opportunities
Extended droughts	• Uncertainty of supply reliability • Reduced revenues for water utilities due to lower flows • Competition for same sources • Overdrawn groundwater basins • Land subsidence over basins • Watershed and ecosystems stress • Salinity increases in water supply • Inefficient wastewater infrastructure due to low flows yet same organic/ nutrient load • Economic losses • Food supply disruption and food insecurity	• Long-term, deep uncertainty scenario planning • Permanent and transient water conservation measures • New financial models for water pricing • Water recycling for maintaining stressed streams and watersheds • Water reuse to supplement water supply • Recycled water use to replenish groundwater and prevent subsidence • Recycled water for agriculture, irrigation, industrial uses, to preserve potable sources • Water-efficient farming, vertical farming
Extreme precipitation events Extreme weather events (hurricanes, typhoons, polar vortex, etc.) More frequent precipitation events	• Uncertainty of the extent and durations of events • Short-term peak stress on watersheds and waterways • Property damage from erosion in community waterways • Coastal and riverine flooding with inundation areas covering wider perimeters under longer-sustained events • Non-operation of key facilities under longer-term flooding • Contamination of water supply sources and infrastructure • Overflows from sanitary sewers, agricultural and livestock operations • Economic losses • Catastrophic infrastructure failures (e.g. Oroville, New Orleans, Texas) that exacerbate correlated resilience concerns like water supply	• Long-term, deep uncertainty scenario planning • Nature-based solutions to impede and buffer watersheds, wetlands, marshes • Low-impact development and stormwater capture in urban and rural areas • Storage basins to recede flow impacts • Updated design standards • Digitisation to plan and monitor systems • City planning and emergency response planning to include hardened critical water and power infrastructure • Flood and impact proofing key facilities to sustain functionality, and physical integrity

(continued)

Table 4.1 (continued)

Events	Impacts on water sector	Mitigation opportunities
Warming air and sea temperatures Extreme and/or sudden changes in temperatures	• Reduced snowpack and permanent shrinking of glaciers • Changing runoff and hydrologic conditions that impact water supplies • Extreme events fuelled by increased water vapour, higher-intensity storms • Shifts and reversals in natural ocean and atmospheric cycles, impacting ecological and economic patterns, even resulting in deep freeze in Gulf of Mexico	• Efficient and renewable energy systems for water and city operations • Carbon footprint reduction and sequestration in water operations • Adaptive farming for food security • Vertical or indoor farming that can produce high yields and consumes less soil area and water, NextGen Ag • Natural treatment systems
Sea level rise	• Coastal communities' inundation • Compounding impacts of extreme weather events, greater storm surge and wave-run-up • Economic losses • Significant population displacement and correlating political instability (i.e. climate refugees)	• Long-term, deep uncertainty scenario planning • Coastal communities hardening, sea walls, updated levees • Carbon sequestration to reduce emissions • Designed/built inundation zones • Coastal ecosystem protection and restoration
Compounding and recurring occurrence of events	• Uncertainty of event sizes and durations, impacts, life loss • Exacerbation of possible impacts at the same, or at greater intensity • Loss of system elasticity, watershed buffering capacity • Repetitive infrastructure damage, rapid ageing of built systems	• Long-term, deep uncertainty scenario planning • Robust asset management plans • Scenario analysis considering long horizon adaptive planning to identify possible/ unforeseen worst-case conditions • Hardened facilities designed to operate in island mode and sustain communities • Decentralised yet interconnected systems
Resource limitations forming as a result of climate impacts (e.g. water, soil, nutrients, energy, food)	• Continued nonpoint pollution of watersheds, bays, inland water bodies	• Resource recovery at sewage treatment facilities designed to be net energy positive

(continued)

Table 4.1 (continued)

Events	Impacts on water sector	Mitigation opportunities
	• Impacts on phosphorus reserves and "peak phosphorus" limitations to be experienced after 2035 (Nedelciu et al. 2020; Alewell et al. 2020) • Lost/damaged agricultural land unable to buffer peak flows • Growing energy demand for water and ecosystems management	• Application of new efficient treatment technologies such as anammox • Water recycling at satellite facilities • Urine separation, ammonia and phosphorus recovery • Changing farming practices (i.e. NextGen Ag)

purify more than 30 million m³ of wastewater every year, maintain and operate eight water and four wastewater treatment plants, manage rainwater, and implement climate adaptation projects as well as safeguard a balanced and healthy water cycle (Fig. 4.8).

Aarhus Vand has the benefit of being part of a governance structure that understands and proactively wants to manage climate change impacts guided by science. In addition to the regulation of the broader water sector, Danish government's climate action plan and vision for transition to green energy set the framework for Aarhus Vand to support the goal of reducing greenhouse gas emissions by 70% in 2030. In order to accomplish these goals, Aarhus Vand collaborates with the neighbouring municipalities, other water utilities from neighbouring countries, and globally even with partner utilities such as Metropolitan Water Reclamation District of Greater Chicago in the US, as well as partnerships with the local authorities in India, South Africa and Ghana, Dubai, Sweden and Portugal (Àguas de Portugal), and Australia.

As part of their strategy to transform their business model and align their actions with UN SDGs, they are also reinventing their wastewater treatment facility. Aarhus ReWater is master planned to become the world's most resource efficient wastewater treatment plant and serve as a hub for innovation as part of the Danish Water Cluster, initiated under the leadership of Aarhus Vand. As part of their broader transformational strategy, the utility is also transforming its operations and approach to optimisation, efficiency and asset management programmes through digital transformation. Having collaborated with Aarhus Vand on development of the Digital Blueprint for Aarhus ReWater, one of the authors of this chapter were able to integrate resiliency to planning, design and in the future operations of the ReWater facility.

The key factors underlying the success of this utility can be informative for others in their journey to establish a future state of their resilient water systems:

Fig. 4.8 Carefully planned digital framework is allowing Aarhus Vand to integrate all ReWater functions on one platform, embedding resilience through systems thinking (BV 2020)

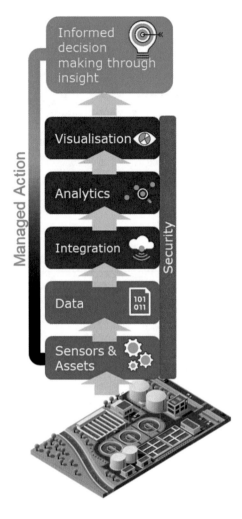

- Alignment of goals between governance units, engineers, utility and community and a unified vision for future state of the utility and its benefits to the communities.
- Partnership models driving collaborative actions towards meeting climate change and UN SDGs.
- Continuous assessment of policies, financial direction, new information and new technologies for integration into the transformation journey.
- Digitisation of functions and embedded digital solutions across the utility.
- Open-minded and planned innovative thinking embedded into the key strategies driving change.

- Engineered and nature-based resource recovery, water management, recycling, energy management, and safe drinking water solutions.
- Existing systems hardened, and new systems constructed to meet the assessed needs for resiliency.

4.3 Immediate Actions for Mitigation

As discussed previously, climate change is adversely impacting where, when and the quantity of water in a given community and the ability of the current water sector infrastructure to adapt is limited without investment to mitigate against these changes. One of the key needs for mitigation is "hardening" of the existing infrastructure. Hardening in this context can be:

- Supply hardening — An example of supply hardening is the City of San Diego's Pure Water Potable Reuse Program planned and implemented to ensure supply security for a large metropolitan area that in most cases lacks groundwater sources. Purified wastewater stored in reservoirs will provide long-term water supply resilience. Similarly, Metropolitan Water District of Southern California, City of Los Angeles and others such as the Public Utilities Board of Singapore are following the same path to self-reliance and complete regional supply resilience.
- Infrastructure hardening — Critical infrastructure hardening takes many forms from relocation to elevating critical facilities/units to strategic integration of grey/green infrastructure for maximum benefits to build the environment as well as the ecology of the surrounding area. Globally, gross domestic product (GDP) exposed to river floods alone is estimated to be $96 billion per year (Browder et al. 2019). Coastal wetlands in the USA are estimated to provide $23.2 billion/ year in storm protection services alone. To deal with extensive urban flooding, in 2005 DC Water had to develop a plan to reduce combined sewer overflows (CSOs) by investing in sewer, wastewater and green infrastructure projects including bioretention or rain gardens, permeable pavements, and three deep stormwater runoff tunnels (USEPA 2015). DC Water's plan also incorporated an environmental impact bond structure that is the first of its kind linking financial pay-outs with environmental performance (USEPA 2017).
- Sea level rise mitigation and sea walls/levees — Cities like San Francisco are implementing protection of its residents and building infrastructure from sea level rise and extreme weather events by building seawalls. San Francisco's seawall is intended to also provide protection from earthquake induced damage, improve safety, resilient infrastructure upgrades, and protect the historic waterfront of the City. In Singapore, the government has committed S$100B ($72B) to protect the nation from adverse impacts of climate change through a blend of coastal protection measures and hardening of existing infrastructure.
- Earthquakes, wildfires and physical hardening — In some parts of the world, including the Western USA, physical hardening of infrastructure against

earthquakes is also part of the equation. Although not a result of climate change, a compounding event in the case of an earthquake that can coincide with a mild to extreme weather event can be catastrophic, especially for coastal communities. Building codes need to be uniformly updated and structures reinforced or rebuilt. The extensive and long-lasting wildfires experienced in California, USA and in Australia in 2020 also showed that power infrastructure along with water infrastructure can be rapidly compromised if systems are not hardened to withstand these extreme, sudden events.

- Operational systems and basic utility infrastructure hardening — As experienced following the February 2021 polar vortex event in the USA, power systems can be challenged and taken out of service with rapid cascading impacts on water and sanitation systems, as well as access to basic necessities. Hardening against colder than original design conditions can be as important as hardening against extreme heat and flooding events. Interconnectedness of water, power and telecommunication infrastructure put Texas, US, at the centre of an unplanned and far-reaching stress test that could be a case study for utility planners and designers.

In spite of these clear needs and examples of utilities leading by taking action, a number of institutional and regulatory, system planning, and engineering and design challenges hinder most utilities from implementing resilience measures. For example, most utilities do not have incentives to integrate resilience measures in their planning processes due to lack of financial and technical tools/human resources to plan and implement successful resilient measures. Similarly, scenario analyses and risk assessments are often not uniformly used in planning of infrastructure systems with limited funds being prioritised for maintaining the status quo and aging systems, using methods familiar to them.

In many cases, master planning efforts, if not regulated or driven by standards, have commonly been separated from resilience and risk management efforts. In addition, design codes and standards are seldom updated to reflect changing average temperatures and precipitation rates, extreme conditions, or natural hazards and, therefore, do not reflect the shifting needs and urgency beyond the needs of day-to-day operations. There is an increased risk of issues that arise due to incompatible build out, i.e. when various investments are timed between utilities. An inherent fear of failure in combination with the vast risk and complexity of the issues can drive an inability to make a decision, especially when the decision makers are tied to election cycles.

An integrated methodology to assess water systems includes a stepwise evaluation of existing system elements against the risks and climate change scenarios likely to occur independently and concurrently under different operational scenarios (Fig. 4.7). One application of such methodologies is being implemented by USEPA Region 9 who has oversight over California water agencies and water systems, and provides an example of top-down regulatory driven integration of climate resilience features for water systems specifically targeting vulnerabilities. For example, Los Angeles Regional Water Quality Control Board is beginning to require climate change vulnerability assessments and mitigation plans as part of the permitting process for wastewater utilities. Through the permit renewal cycle, the utilities are asked to assess and then develop plans to eliminate/mitigate climate change vulnerabilities to ensure continued operation of the sanitation facilities safely and reliably. Though encouraging, institutional hurdles and priorities to solve more pressing imminent issues can have primacy even in the presence of such plans if not enforced or if there is not alignment between all stakeholders, political leaders and leadership of the utility. The question then arises "Is there a set of options that reasonably satisfies all objectives and criteria?" and for which communities does it apply especially where affordability is a key element of decisions. The issue of water equity is not new and will come into play even more under this new paradigm of climate change resilience (Fig. 4.9).

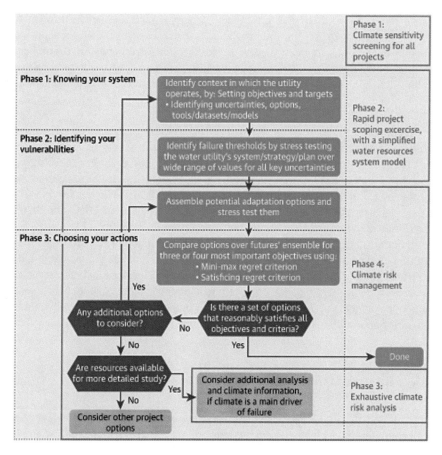

Fig. 4.9 Basic framework for assessing climate change vulnerabilities and identifying actions. *Source* World Bank (2018)

In the face of these challenges, digital transformation can support data-driven approaches by providing quick, reliable insight into quantity and quality of resources, utility operations, performance efficiency and asset life, as well support automation of systems with greater confidence. It can even support vulnerability assessments and provide transparent information to garner public and regulator confidence. The role of digital tools such as operations information visualisations, financial information dashboards, and even visualisation of physical assets through use of digital twins can play a key role in building trust and efficient systems, that can in turn drive policy decisions and investment.

Especially considering the importance of making climate change-based policy decisions at state, national, and continental levels, well-designed digital tools can increase understanding of base conditions and anticipated risks, and support efforts for unifying standards to use as the basis for these policy decisions. Different parts of the globe experience different conditions when working towards implementing policies. For example, countries like Singapore or Denmark have the ability to move faster due to their size, governance structures, and the general trust of the public in leaders of key institutions. In cases such as in the USA, federal- and state-level decisions are made to meet the needs of the lowest common denominator or politically expedient, short-term approaches, rather than the needs for a resilient future. Even in these conditions, though, there are examples of communities and public opinion shaping the policy decisions through political and social events, albeit at times the influence is short lived.

The impact of the extreme weather events goes beyond the immediate asset damage. Long-term impacts have developed as the insurability of assets is under risk or is not possible in some parts of the USA. For example, the insurance industry is not able to cope with the increasing frequency and/or concurrence of certain types of extreme events especially after the floods experienced in 2017–2018 in Eastern/South-eastern USA. Similarly, the wildfires of 2019 and 2020 in California also resulted in a similar outcome of non-insurability.

Impacts of climate change-driven events on communities, public and private utilities are undeniable. To make policy decisions take root and garner support, and to render the policy frameworks a bridge instead of a barrier to advancement and progress, policies that balance economic sustainability and resilience with cost of action, encourage public–private partnerships, and generate new jobs will yield the greatest opportunities for change.

4.4 Intermediate Future Actions for Mitigation

Considering the initial phase actions are taken including the hardening against vulnerabilities under event scenarios customised for the specific utility and its locale, intermediate future actions can then be built on the analysis of these events. Means for the utility to modify, expand, or upgrade its current or proposed infrastructure and policies to reduce its vulnerabilities can then be implemented in a prioritised fashion. The decision-making and prioritisation process chosen must answer questions like:

- Are there low- or no-regret options that help to achieve objectives no matter what future occurs?
- Are any combinations of options robust over all plausible futures?
- What are the trade-offs among the implementation options?
- Can the utility defer some actions and implement only if conditions warrant?
- Can the utility make its plans more robust by monitoring and adjusting over time?

Implementation of actions specific to water systems rest on the water infrastructure pillars—drinking water, used water/wastewater, water resources, waterways, and urban and natural water systems—can take different forms for different communities. However, integration of specific actions with broader systemwide strategies and elements is what can make a whole system infrastructure more resilient compared to those that take disparate actions. For example, integration of water and energy infrastructure vulnerabilities with forward-thinking approaches can create resources that can be reused, better used, saved to preserve depleting resources, or transition to more efficient resources—nutrients, freshwater sources, fossil fuels.

Using one of the pillars of water system resiliency as an example, we can see how specific ways to add resiliency to existing systems could work. To better manage used water including sanitary sewage (wastewater) in combination or separate from stormwater flows depending on the design of the infrastructure can have a central place in our water infrastructure considering the essential role these facilities play to protect our environment and provide resources that can be reused. With urbanisation, our communities have invested heavily in infrastructure for both conveyance and treatment of wastewater. Traditional approaches have focused on centralised treatment that created large collection systems and treatment facilities. Decentralised systems are being considered for future growth; however, financial limitations will most likely drive the need to utilise existing assets as part of a resource recovery solution and allow a stepwise transition to a renewable energy, sustainable agriculture, one water paradigm.

A potential solution for many of these issues is the development of "Resource Recovery Facilities" (Fig. 4.10) as an enhancement or replacement to existing wastewater treatment plants (Wallis-Lage et al. 2011). These facilities can provide ample opportunities to address the growing water, energy, and nutrient needs associated with increasing population and standard of living expectations as illustrated on Fig. 4.11. While the technologies needed to recover water, energy, and nutrients are commercially available, the challenge is to identify how resource recovery can be implemented within existing facilities as part of driving resilience, making the best use of asset investments and managing through limited financial funding, and creating positive business models generating a positive ROI for the utility.

While it would be truly ideal to find the perfect synergy to recover all resources available, the reality is that one resource will typically dominate and some resources

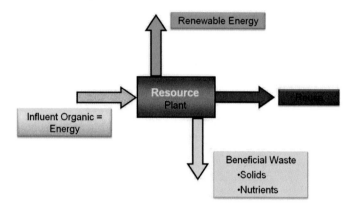

Fig. 4.10 Resource plants recover energy, water, solids, and nutrients

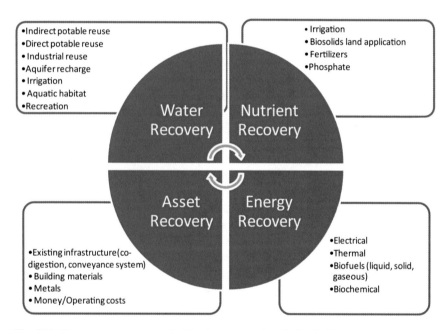

Fig. 4.11 Resource recovery opportunities for water and sanitation facilities

are sacrificed. Examples from across the world show that there are ways to implement integrated solutions that result in resource recovery while meeting multiple climate resiliency goals.

Brisbane, Queensland, Australia: When the water level in Wivenhoe Dam was dropping rapidly in Brisbane Australia (Fig. 4.12) and the long-term water supply appeared to be less than 18 months, finding an additional water supply was the single driver in implementing reuse. The decision to develop three advanced water treatment plants (AWTPs) using microfiltration, reverse osmosis and advanced oxidation that would augment the City's water supply was made despite a concomitant power requirement increase. However, it is also important to note that the primary end users for the reclaimed water are power stations. With the increasingly strict water restrictions with respect to the potable water supply, the power stations were in danger of losing their cooling water source. The implementation of the AWTPs provided both an increased water supply and a guaranteed energy supply. Thus, while energy was not recovered at the actual treatment facility, the implementation of the advanced water treatment plants recovered energy for the community. This example illustrates the close relationship of water and energy. Even though nutrients had to be removed to meet water quality requirements, limited consideration was given to recovery as a combined focus on water, energy, and nutrients resulted in competing objectives that could not be satisfied.

Irvine, California, USA: In Irvine California, energy production from biosolids was a key driver in the decision to develop a new biosolids treatment system for the Michelson WRP. In California, high energy prices, in combination with renewable energy grants, encourage solutions that maximise biogas production and on-site power generation using fuel cells. This set of circumstances led to an evaluation of WAS pre-treatment and FOG addition to an anaerobic digestion process to increase biogas production. While the FOG facilities were included in the final design, the decision was to leave space for the option of WAS pre-treatment in the future. The biogas produced will be used to meet a portion of the on-site energy demands. In addition to producing energy from biogas, digested solids will be thermally dried to create a pellet fertiliser or fuel product that can be used by a local cement kiln (Fig. 4.13). Water scarcity is also a driver, which led to the use of an MBR process to produce high-quality water that can be used to meet growing reuse water demands despite its higher energy requirements. The implementation of energy recovery from

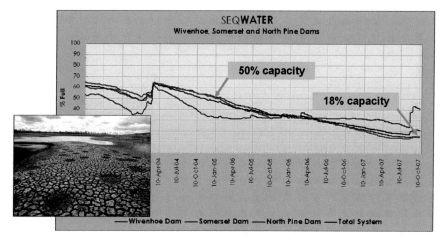

Fig. 4.12 Wivenhoe Dam water level in 2006–2007

the biosolids helps offset some of the increased energy for the liquid stream. While phosphorus recovery was not considered a viable solution at this time, space is available within the solids handling facility footprint for future implementation.

Integrating with this new paradigm, new system designs with climate-based design standards that consider new design storm conditions, and resilient engineering features are needed. Even in certain instances relocating the existing facilities to prevent repeated damage and risk to public safety can be more viable financially, as was the case for Iowa City's North Wastewater Treatment Facility that was inundated multiple times, and ultimately relocated after repeated extreme flooding (Fig. 4.14).

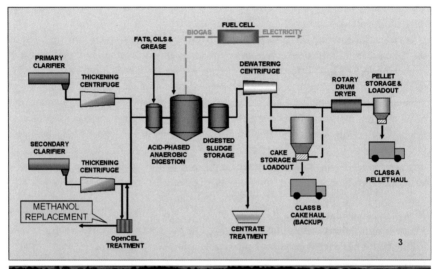

Fig. 4.13 IRWD biosolids handling facility. The final design excluded the fuel cell and the OpenCEL treatment

(a)

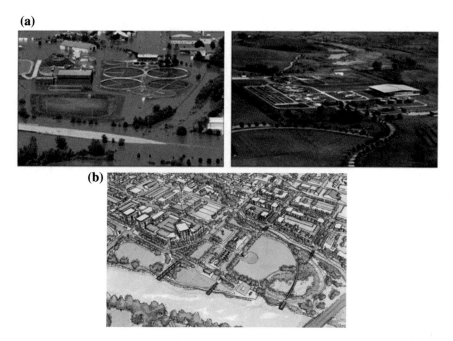

(b)

Fig. 4.14 Iowa City's north wastewater treatment facility relocated after extreme flooding. **a** Flooded facility and the new facility after relocation. *Source* GAO-20-24 (2020). **b** Illustration of the riverfront restoration after removal of wastewater facility. *Source* USEPA (2021)

Similarly, the city of Anacortes, WA, water treatment plant located along the Skagit River was vulnerable to current floods and future climate risks. Because moving the facility out of the floodplain was cost prohibitive, the city rebuilt on the existing site taking climate projections into account (Fig. 4.15).

Vulnerabilities against more frequent and intense storms, saltwater intrusion, and increased sedimentation levels were mitigated by using the design storm conditions and flood projections through the 2080s. These vulnerabilities included an expanded 100-year floodplain, an estimated 350% increase in peak suspended sediment load in winter, and anticipated upstream migration of the saltwater wedge due to the effects of sea level rise.

One innovative solution that is spurring public and private partnerships is community-based partnerships (CBPs). Under this model, a local government or public utility aggregates multiple stormwater improvement projects into a single, integrated procurement, creating one point of private sector accountability. It uses a performance-based contract, linking the partner's payments to specific, measurable goals. The private partner assumes both short- and long-term budget and schedule risks, incentivising best value and a whole-life compliance solution. The performance contract also requires the partner to deliver a system that improves environmental sustainability and resilience. The community can also require its partner to achieve specific key performance indicators (KPIs) related to community equity,

Fig. 4.15 Anacortes, Washington, rebuilt a water treatment plant to mitigate potential impacts of climate change on the facility and safety of water supply. *Source* USEPA (2020)

job creation, minority- and woman-owned enterprises (MBEs/WBEs) engagement, create public green spaces and multi-benefit facilities and achieve other community selected goals.

In 2014, Prince George's County, Md. (Fig. 4.16), was faced with an enormous regulatory challenge. Compliance with its current NPDES permit required retrofitting 6,000 impervious acres with green infrastructure—with the future requirement for up to 15,000 acres of untreated impervious area by 2025—at an estimated cost of $1.2 billion. The county could have utilised traditional project delivery methodologies and procurement; however, given the magnitude of the challenge, county leaders sought an alternative solution. Prince George's County formed a CBP known as the Clean Water Partnership—a 30-year stormwater programme partnership between the county and Corvias—to tackle its stormwater compliance projects, starting with a $100 million pilot. When the project met specific delivery and community performance metrics, that triggered payments to the private sector partner. The county recently completed the initial pilot, retrofitting 2,000 acres and saving more than 40% compared to traditional, non-bundled procurements (Fig. 4.16). The county also achieved regulatory compliance requirements and used greater than 87% local, minority-owned businesses.

4.5 Resilient Water Infrastructure and Systems is a Journey

Resilient infrastructure is already in planning or on a path to be in place according to a roadmap developed for some of the forward-thinking utilities as was discussed previously. They know that resilience planning is an ongoing effort that

Green Infrastructure
installation at
Francis Scott Key
Elementary School,
Prince George's
County, Md.

Fig. 4.16 Green infrastructure as part of community-based partnership

characterises known and emerging threats/vulnerabilities and utilises adaptive planning and implementation strategies to move towards a more resilient future. For others, reaching a resilient future may have to be after moving through immediate and intermediate strategic thinking and planning steps. Following the initial planning and hardening of systems, the utility should keep their focus on resiliency by using milestones to measure progress. To best accomplish this, key performance indicators (KPIs) need to be established at the beginning of the journey. Having adequate KPIs and setting the right milestones on the road to resilience can clearly illustrate that the utility is making progress (Fig. 4.17). Some example KPIs that align with climate change resilience and mitigation actions include:

- Reuse of water (% of received, mgd, m^3/yr).
- Beneficial use of organic and other resource end products (lbs/yr, tons/yr, kW/yr).
- Progress towards SDGs (SDG Tracker 2018).
- Renewable energy generation and GHG reduction.
- Carbon and water footprint.
- Asset management programme progress towards goals (% complete).

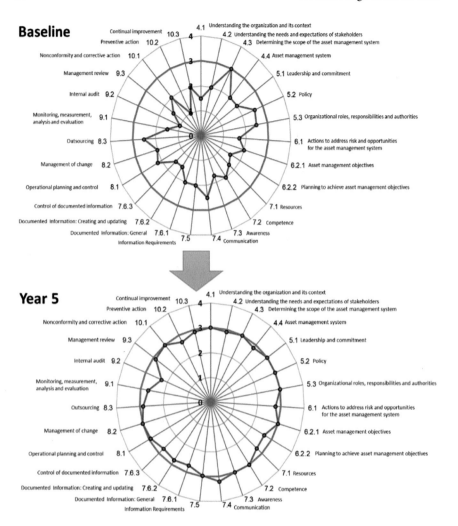

Fig. 4.17 Monitoring progress towards goals and KPIs is critical to success of resilience programmes (BV 2021)

4.6   The Time to Act is Now

The fragility of the water sector infrastructure to manage and adapt to the challenges of today and tomorrow is clear, as a result we must develop a forward-looking roadmap focused on mitigating the impacts of climate change, population growth and economic and societal movements. We must envision a future that builds on today while it provides a diametrically opposing diagram of possibilities that we can collectively move towards to address the changing climatic conditions we live in. Our centuries-long ways of living which started with the great human migration across the globe beginning 80,000 years ago (Gugliotta 2008) and marked with the key events such as the discovery of the Americas, plagues and famines, resulted in settling in areas where we had access to water—the key ingredient of life. Like the stories of the cities of Anatolian civilisations, one built over another going back to 12,000 BC, or Amesbury of England going back to 10,000 BC, our cities and city infrastructure of today are still informed by our concept of communal living, society and managing life sustaining resources (Balter 2005). How do we now transform our thinking to insure a climate resilient future for our cities and our critical infrastructure? History is also filled with examples of humans having to relocate due to climate change/disasters/droughts similar to the examples of Egyptian drought and famine, Anasazi collapse, and others. We have the tools now to better understand and plan for changes to change our old ways, we must learn the lessons from history, invest in resilience or, we too, are highly susceptible to a more rapidly, changing climate.

We must start the resiliency journey of our water infrastructure by recognising the factors that hinder advancement so we can address and separate our decisions from historical thinking. Factors like political will, affordability, water equity, risk tolerance, and long-term investment commitments have interrupted progress. Without overcoming these factors, the resilience required will not be accomplished. To start the journey, communities/utilities need to establish a strong foundation. Key elements include:

- Defined community expectations for risk management and willingness to support investment.
- Clarity on societal/economic benefits of resilient infrastructure and a commitment to changing how infrastructure is planned and built.
- Defined key milestones to act as guidepost on the path to achieve resiliency.

Now is the time to build on the current global conditions and leaning into the immediate opportunities:

1. Turn the recovery from the pandemic into an opportunity to build a better future.
2. Incorporate consistent and solid science, backed by the strong collaboration of scientific institutions and academia, to underpin policy decisions.
3. Keep the forward momentum, even if the movement is in steps.

References

Alewell C, Ringeval B, Ballabio C (2020) Global phosphorus shortage will be aggravated by soil erosion. Nat Commun 11:4546

Balter M (2005) The seeds of civilization. Why did humans first turn from nomadic wandering to villages and togetherness? The answer may lie in a 9,500-year-old settlement in central Turkey. Smithsonian Magazine. https://www.smithsonianmag.com/history/the-seeds-of-civilization-78015429/. Accessed 15 Feb 2021

Browder G, Ozment S, Rehberger I, Gartner T, Lange GM (2019) Integrating green and gray. Creating next generation infrastructure. WRI and World Bank. https://files.wri.org/s3fs-public/integrating-green-gray_0.pdf. Accessed 15 Feb 2021

BV (2020) Digital blueprint for Aarhus Vand ReWater. Technical Memorandum

BV (2021) Milwaukee metropolitan sewerage district asset management analysis. Technical Memorandum

GAO-20-24 (2020) Water infrastructure report to US congressional requesters

Gugliotta G (2008) The great human migration. Why humans left their African homeland 80,000 years ago to colonize the world. Smithsonian Magazine. https://www.smithsonianmag.com/history/the-great-human-migration-13561/. Accessed 15 Feb 2021

IPCC (2014) Climate Change 2014: impacts, adaptation, and vulnerability. Part A: global and sectoral aspects. Contribution of working Group II to the fifth assessment report of the intergovernmental panel on climate change. Cambridge University Press, Cambridge/New York, United Kingdom/USA. www.ipcc.ch/site/assets/uploads/2018/02/WGIIAR5-PartA_FINAL.pdf. Accessed 15 Feb 2021

Mekonnen MM, Hoekstra AY (2016) Four billion people facing severe water scarcity. Sci Adv 2 (2):e1500323

Nedelciu CE, Ragnarsdottir KV, Schlyter P, Stjernquist I (2020) Global phosphorus supply chain dynamics: assessing regional impact to 2050. Global Food Secur 26:100426

Our World in Data (2019) All our charts in natural disasters. Share of land area which experienced extreme one-day precipitation, USA. https://ourworldindata.org/natural-disasters. Accessed 15 Feb 2021

Rodin J (2014) The resilience dividend: being strong in a world where things go wrong. Public Affairs, New York

SDG Tracker (2018) Measuring progress towards the sustainable development goals. https://sdg-tracker.org/. Accessed 15 Feb 2021

UN Water (2018) SDG 6 synthesis report 2018 on water and sanitation. United Nations. https://www.unwater.org/publications/highlights-sdg-6-synthesis-report-2018-on-water-and-sanitation-2/. Accessed 15 Feb 2021

UN Water (2019) Water and climate change. United Nations. https://www.unwater.org/water-facts/climate-change/. Accessed 15 Feb 2021

USEPA (United States Environmental Protection Agency) (2015) District of Columbia Water and Sewer Authority, District of Columbia Clean Water Settlement. Overviews and Factsheets, May 18

USEPA (United States Environmental Protection Agency) (2017) DC Water's environmental impact bond: a first of its kind. USEPA Water Infrastructure and Resiliency Finance Center. https://www.epa.gov/sites/production/files/2017-04/documents/dc_waters_environmental_impact_bond_a_first_of_its_kind_final2.pdf. Accessed 15 Feb

USEPA (United States Environmental Protection Agency) (2020) Anacortes, Washington rebuilds water treatment plant for climate change. https://www.epa.gov/arc-x/anacortes-washington-rebuilds-water-treatment-plant-climate-change. Accessed 15 Feb 2021

USEPA (United States Environmental Protection Agency) (2021) Iowa City, Iowa closes vulnerable wastewater facility. https://www.epa.gov/arc-x/iowa-city-iowa-closes-vulnerable-wastewater-facility. Accessed 13 Apr 2021

USGCRP (US Global Change Research Program) (2017) Climate science special report: fourth national climate assessment, vol I (Wuebbles DJ, Fahey DW, Hibbard KA, Dokken DJ, Stewart BC, Maycock TK (eds)]. U.S. Global Change Research Program, Washington, DC, USA, 470 pp. https://doi.org/10.7930/J0J964J6

Wallis-Lage C, Scanlan P, de Barbadillo C, Barnard J, Shaw A, Tarallo S (2011) The paradigm shift: wastewater plants to resource plants. Proc Water Environ Federat WEFTEC 14:2680–2692

WMO (World Meteorological Organization) (2020a) United in Science 2020: a multi-organization high-level compilation of the latest climate science information. https://public.wmo.int/en/resources/united_in_science. Accessed 15 Feb 2021

WMO (World Meteorological Organization) (2020b) Key messages. United in Science 2020: a multi-organization high-level compilation of the latest climate science information. https://public.wmo.int/en/resources/united_in_science. Accessed 15 Feb 2021

World Bank (2018) Building the resilience of WSS utilities to climate change and other threats. A roadmap. The World Bank, Washington, DC

WRI (World Resources Institute) (2019) WRI Aqueduct website. www.wri.org/aqueduct. Accessed 15 Feb 2021

Wright P (2018) Sea level rise will inundate coastal sewage plants, study says. The Weather Channel. https://weather.com/science/environment/news/2018-04-03-sea-level-rise-wastewater-treatment-sewage-plants. Accessed 13 Apr 2021

Chapter 5
Water Security and Climate Change: Hydropower Reservoir Greenhouse Gas Emissions

María Ubierna, Cristina Díez Santos, and Sara Mercier-Blais

Abstract Water storage is a driver for economic growth and often mentioned as a proxy for water security. Hydropower storage projects deliver multiple benefits contributing to water and energy security; however, the reservoir creation raises concerns about greenhouse gas (GHG) emissions and puts in doubt how clean hydropower generation is. As storage becomes more relevant under climate change, adequate assessment is necessary to ensure projects' sustainability. This study quantifies hydropower global median lifecycle greenhouse emissions at 23 gCO_2e/kWh using the G-res Tool to estimate the net emission for 480 hydropower storage projects. This result is aligned with the IPCC estimates.

Keywords Hydropower · Greenhouse gas emissions · Water security · Climate change · Climate resilience and adaptation · Hydropower carbon footprint

5.1 Introduction

Hydropower is the leading player on the world's renewable energy stage, responsible for around 16% of global electricity generation (IEA 2019). If sustainably managed, its infrastructure provides vital freshwater services, such as water supply and storage, irrigation, flood control and drought prevention.

Hydropower contributes directly to meet several 2030 Sustainable Development Goals (SDG) like SDG 7 Affordable and Clean Energy and SDG 13 Climate

M. Ubierna (✉)
International Hydropower Association (IHA), London, UK
e-mail: mariaubierna@gmail.com

C. D. Santos
International Hydropower Association (IHA), London, UK
e-mail: crisdiezsantos@gmail.com

S. Mercier-Blais
UNESCO Chair in Global Environmental Change, University of Quebec at Montreal (UQAM), Montreal, Canada
e-mail: saramercierblais@gmail.com

Action. It can also help meet many others like SDG 6 Clean Water and Sanitation. The ability to store freshwater in reservoirs can provide several services such as irrigation, water supply, navigation, fisheries and recreation. The water storage, to be redistributed in space and time, provides a higher systemic resilience and capacity to adapt to climate change as it procures flood control and drought mitigation. Its storage and flexibility also balance and enhance the variable renewable energy sources—such as wind and solar.

Hydropower on its role in renewable energy systems and freshwater management can support the efforts to tackle and adapt to climate change, bringing security for energy and water supply.

IRENA estimates the need for 850 GW of newly hydropower installed capacity over the next 30 years, roughly similar to the entire power system capacity of the European Union in 2020, to meet global climate change targets. Global estimations by FAO (2017) expect an increase in water and food needs in the range from 50 to 100% for the next 30 years, and higher impacts on water availability and water-related ecosystems and socio-economic systems. Climate disruption will heighten the situation, increasing the importance of water storage and water conservation.

5.1.1 Hydropower and Water Security

Current progress on the targets of Sustainable Development Goal (SDG) 6 on water and sanitation is alarmingly off-track. UN data statistics suggests that achieving universal access to basic sanitation by 2030 demands doubling the current annual rate of progress.

Water, climate and energy are inextricably linked. Yet too often decision-makers have taken the approach of managing policies and markets within separate silos. Integrated strategies are required to handle the use of these resources sustainably. Integrated Water Resources Management (IWRM) concept as a means of integrating water-related management components at the river basin scale is covered by the UN Target 6.5. Yet, it also encompasses direct contributions to achieving other SDGs such as ending poverty, providing clean and affordable energy, achieving gender equality and protecting ecosystems (Benson et al. 2020). The intertwined relation becomes even more significant as climate change impacts the global hydrological system, increasing extreme weather events.

Hydropower plays an essential role in the climate change adaptation of water resources availability (Berga 2016), and IWRM is a crucial aspect when developing hydropower projects. Several studies (Xiaocheng et al. 2008; Thoradeniya et al. 2007) have raised concerns over the cumulative impacts of large deployments of small-scale hydropower projects, which can match or outweigh those of a larger-scale hydropower project providing an equivalent energy output. Mayor et al. (2017) study assessed the hydropower's impacts in the Spanish Duero Basin and suggested that large-scale hydropower projects contributed more to energy and water security. In contrast, small-scale hydropower projects showed more significant repercussions due to the cumulative cascading effects.

In terms of water footprint, the global perception is that smaller-scale hydropower projects are more beneficial. Larger-scale hydropower projects that usually correspond to reservoirs with ample storage can consume a lot of water due to evaporation from the reservoir surface. The studies from Scherer and Pfister (2016) and Pfister et al. (2020) examined about 50% of the total hydropower generation and found that about half of the hydropower plants assessed had negative water scarcity footprints, suggesting that they alleviate rather than worsen water scarcity. These studies on cumulative impacts and water footprint underline the need for IWRM principles to avoid discriminating a technology based on the development's size.

Water storage is often cited as a proxy to water security (Sanctuary et al. 2007) and becomes even more relevant under climate change scenarios. Pokhrel et al. (2021) report that the global land area and population in extreme terrestrial water storage drought would be more than double by the late twenty-first century. Moreover, water storage is a driver for economic growth and is of particular importance to smooth intra-annual and special variations in rainfall that otherwise have significant impacts on economic growth (DFID 2009). Brown and Lall (2006) analysed the water storage needs for countries with significant hydrological variability and revealed a statistically significant relationship between hydrological variability and lower per capita GDP.

Multipurpose dams can generate direct and indirect economic benefits. In, Hogeboom et al. (2018), the reservoir services produce an economic benefit estimated at US$265 billion globally, mainly due to the electricity generation and water supply. Although hydropower has the largest water footprint given its high economic value, the findings show that this activity is not even the second purpose in water-scarce basins.

Single-purpose and multipurpose hydropower projects deliver a range of power and non-power benefits to society and the environment. Over and above electricity generation, power-related benefits include flexible generation and storage, as well as reduced dependence on fossil fuels and avoidance of pollutants. Wenjie et al. (2020) report that the China Three Gorges Project provides multiple benefits, particularly flood control and navigational capabilities that significantly support the region's socio-economic development. The waterway shipping developed at the largest hydropower plant in the world has reduced road and air transportation substantially. Thus, further contributing to the reduction of GHG emissions from those types of transports.

According to the International Commission on Large Dams (ICOLD 2021), most of the world's existing large dams are single purpose. Irrigation accounts for the most common single purpose with 30% followed by hydropower with 10%. As improved water storage capacity and water security will be particularly required in climate zones characterised by low rainfall and major rainfall variability, multi-purpose reservoirs will be able to play a major role. While multipurpose reservoirs can provide many investment benefits (from a macroeconomic perspective), they entail an added complexity in terms of planning and dealing with multiple stakeholders. Nonetheless, a hydropower dam designed for single purpose often becomes multipurpose by practice. In many cases, this evolution does not allow the optimum realisation of the project benefits, neither allows the reflect the positive benefits of the services provided by the reservoir.

5.1.2 Resilience and Adaptation

Hydropower resilience to climate change is essential to provide energy and water services safely. Hydropower projects are susceptible to climate change impacts due to their dependency on precipitation and runoff and exposure to extreme weather events. Both hydropower operations and infrastructure need to be resilient to overcome the increasing climate variability and provide adaptation services. This is particularly important for storage dams that have the capacity for flood control and drought mitigation.

If not designed and managed correctly, hydropower projects could exacerbate climate change impacts on local communities and the environment. Failure to adequately consider climate risks may lead to shortcomings in technical and financial performance, safety aspects and environmental functions. Furthermore, by not assessing climate change-related opportunities, investment decisions may not adequately recognise the role of hydropower infrastructure in providing climate-related services.

Hydropower systems are characterised by their longevity and are traditionally designed based on long-term historical hydrological data and forecasts. However, a rapidly changing climate means historical data may no longer represent the current climate state. Academics described it as the death of stationarity (Milly et al. 2008).

In response, novel approaches and methods to guide through Decision-Making under Deep Uncertainty (DMDU) are appearing (Hallegatte et al. 2012). Their characteristic is a bottom-up analysis, and they rely on robust planning. Local project stakeholders define the functions and metrics for study as opposed to top-down analysis developed to look at the generic case first and then into the details. And by robustness, these approaches value the ability to perform well across a wide range of unpredictable possible futures over the ability to optimise their performance in an expected future condition.

Therefore, planning hydropower systems from a long-term, climate-resilient perspective will guarantee that current and future generations inherit hydropower infrastructure that will not be compromised by climate change and will ensure water and energy services.

In 2019, IHA launched the Hydropower Sector Climate Resilience Guide (IHA 2019) to address the need for state-of-the-art science and international industry good practice on how to incorporate climate resilience into hydropower project planning, design and operations. Following the approaches mentioned above to address inherent climate change uncertainty, the guide provides practical guidance to identify, assess and manage climate risks resulting in more robust and resilient hydropower projects.

The hydropower sector has welcomed the guide as it is the first of its kind for renewable energy and is sector-specific. The guide's core methodology is founded on the World Bank's framework for confronting climate uncertainty in water resources planning (Ray and Brown 2015). During the pilot phase, seven projects tested the guide. In the following two years after its launch, fifteen projects have undertaken or are undertaking a climate risk assessment using the climate resilience guide. These applications around the world have demonstrated the guide's core principles of applicability to any project's type (storage, run-of-river, pumped storage) and scale, single and cascade schemes, greenfield and modernisation projects, and in different types of hydrological and geographical conditions with varying levels of data availability.

Multilateral development banks like the World Bank (2021) refer to the guide for hydropower project's climate risk assessments and recognise its value to design climate-resilient project, thus, supporting the access to climate finance aligned with their climate investment targets. Lenders also recommend the guide's use in modernisation projects which present a unique opportunity to adapt the project to the new hydrological and socio-economic conditions (Ubierna et al. 2020).

5.1.3 Sustainability Assessment Tools

Adequate assessment of environmental, social, governance and technical aspects is necessary to ensure that hydropower projects impacts do not overcome their benefits to water and energy security. Due to its long-life span and complex integration with the environment and local communities, hydropower projects had sometimes not been developed in an environmentally and socially sustainable manner.

Following the World Commission on Dams (2000) recommendations for a comprehensive assessment of risks implied by a project, a multistakeholder forum was formed to build consensus on what a sustainable project is. The forum included representatives of social, community and environmental organisations, governments, commercial and development banks and the hydropower sector.

Ten years later, on behalf of the forum, IHA (2010a) published the Hydropower Sustainability Assessment Protocol, an assessment tool to measure performance compared to defined basic good practice and proven best practice.

In 2020, the hydropower sector counts with a suit of assessment tools and guidance documents aimed at driving continuous improvement in hydropower development and operation. The Hydropower Sustainability Tools define and measure sustainability in the hydropower sector, providing a common language to all stakeholders involved in a project. The tools consist of the Hydropower Sustainability Guidelines on Good International Industry Practice (HGIIP), the Hydropower Sustainability Assessment Protocol (HSAP) and the Hydropower Sustainability ESG Gap Analysis Tool (HESG) (IHA 2018a, 2010a, 2020a).

As of the end of 2020, 32 hydropower projects had conducted an official assessment using the HSAP and 6 projects using the HESG. The projects range in scale from 3 MW to 14,000 MW all around the world, half of them in low- and middle-income countries. The tools have had a greater uptake in some countries like Colombia, Brazil and Iceland, but major hydropower markets like China or the USA had no assessments.

Nonetheless, the tools have passed through expectation for what they were designed to do. Sarawak's electricity company in Malaysia has embedded the HSAP in its internal processes for hydropower projects development (Sarawak 2021). The state-owned Icelandic electricity company has adapted the HSAP to assess their geothermal assets, thus, developing the new Geothermal Sustainability Assessment Protocol (Landsvirkjun 2017).

On the other hand, training and capacity building on the tools attracts a lot of interest by lenders for their internal staff or clients and development partners. For example, Swiss cooperation supports the development of How-to Guides to enhance practitioners' knowledge to meet good international industry practice as defined by the HGIIP. As of March 2021, four How-To Guides are available at hydrosustainability.org on the topics: Downstream Flow Regimes, Resettlement, Erosion and Sedimentation and Benefit Sharing.

5.1.4 Climate Mitigation

Mitigating climate change is one of the most significant challenges of our times, and it is vital to ensure human and natural systems health.

The Paris Agreement, adopted in 2015 by all countries, aimed to increase the global response by keeping a global temperature rise this century well below 2°C above pre-industrial levels. But the Intergovernmental Panel on Climate Change Special Report on Climate Change in 2018 (IPCC 2018) focuses on the impacts of global warming of 1.5°C above pre-industrial levels. The negative effects will have far-reaching consequences to human and natural systems and adaptation will be more difficult with higher levels.

According to the Emissions Gap Report 2019 by the United Nations (UNEP 2019), emissions must drop 7.6% every year from 2020 to hold the global temperature rise to below 1.5°C by 2030. This can only happen if countries increase their national commitments under the Paris Agreement more than fivefold.

In 2020, the COVID-19 pandemic forced strict measures worldwide with bans in travel and drop of industrial activities, which caused a sudden reduction of carbon dioxide emission in the first half of the year. However, once the countries lifted the measures and industry recovered, emissions sharply rebounded. By the end of the year, global emissions show only a reduction of 4.4% while they fell up to 12.5% in the USA, 9.8% in Brazil and 8.1% in India as reported by Carbon Monitor (2021).

This decline in global emissions due to large sections of the world's economy greatly impaired illustrates the climate change crisis's magnitude. Worldwide, international organisations, civil society and the renewable energy sector call on governments to take the current situation as an opportunity to address the climate emergency with an ambitious decarbonisation agenda. The UN Secretary General calls for recovery plans that trigger systemic shifts to reduce greenhouse gas emissions (UN 2021).

Financial stimulus packages must accelerate the transition toward cleaner and lower-carbon energy sectors supporting renewable energy deployment. As the world's largest renewable electricity generation source, hydropower contributes significantly to the avoidance of GHG emissions. If hydropower was replaced with burning coal for electricity generation, this would result in the emission of around 3.5–4.0 billion t of additional greenhouse gases annually. Global emissions from fossil fuels and industry would be about 10% higher as estimated by IHA (2020b).

Yet, hydropower projects are not free of greenhouse gas emissions. To maximise their role in mitigating climate change, hydropower projects need to be developed and operated sustainably, and that includes, limiting greenhouse gas emissions. Special consideration is for reservoir hydropower projects with a large, impounded surface.

5.1.5 Greenhouse Gas Footprint

Hydropower greenhouse gas footprint, especially biogenic emissions caused by the impoundment of a reservoir, has raised concerns about hydropower designation as clean energy.

In the past, lack of scientific consensus on quantifying this footprint has proved a significant obstacle for policy and decision-makers to justify hydropower financing, in particular the construction of large dams, as a clean low-carbon energy source.

In its Fifth Assessment Report in 2014, the IPCC noted a wide reported range of lifecycle greenhouse gas emissions for hydropower with a minimum of 1 gCO_2-eq/kWh and a maximum of 2200 gCO_2-eq/kWh. The panel cautioned that few studies had appraised the net emissions of freshwater reservoirs, subtracting pre-existing natural sources and sinks and unrelated human emission sources.

Addressing the need for international guidance, scientific partnership led by IHA and UNESCO Chair in Global Environmental Change developed the GHG Reservoir (G-res) Tool to accurately estimate the net change in greenhouse gas emissions attributable to the creation of a specific reservoir (Prairie et al. 2017).

Using the tool gives investors, regulators and local communities greater confidence in the carbon footprint of a reservoir. Assessing the net greenhouse emissions of a reservoir using tool such as the G-res Tool is now an expectation for sustainability reporting (European Commission 2020) and a requirement for financing a hydropower project through climate finance (Climate Bonds Initiative 2019; Patel et al. 2020).

5.2 Materials and Methods

This study aims to quantify hydropower GHG emissions taking a net approach and the project's lifecycle. The study uses a sample of hydropower projects which is representative of the world hydropower storage fleet. Then, the study compares the hydropower emissions to other sources of energy.

5.2.1 Approach

The creation of a reservoir alters the natural carbon cycle of the affected ecosystem. Impoundment causes a prolonged water residence time than in a flowing river and adds an important amount of carbon due to the new soil flooded. Also, it concentrates biological processes to happen in the reservoir that otherwise may not have occurred at all or may have occurred downstream in a pre-impoundment situation, such as relocation of sediment carbon pool (Mendonça et al. 2012). Therefore, Prairie et al. (2018) argue that parts of the emissions, mostly CO_2, occurring in the reservoir, are not new but displaced in space.

The impounded land is subject to new physical and chemical processes that are likely to increase the greenhouse gas production whose emissions will end eventually in the atmosphere. For example, the impound of shallow littoral areas can contribute to more CH_4 through bubbling (Maeck et al. 2014).

However, these processes' complexity makes both measurement and modelling a significant challenge due to the temporal and spatial variability between production and emissions and to site-specific variability.

A UNESCO and IHA's initiative proposes a consensus-based scientific approach on standard measurement techniques for field measurement campaigns worldwide (IHA 2010b). The objective is to build a credible and comparable dataset that can provide reliable information to develop predictive tools that can assess the net GHG emissions in planned and existing reservoirs.

Scientific research takes divergent approaches to document reservoirs emissions. Deemer et al. (2016) take a gross approach to compute global carbon budgets from existing reservoirs using average of field measurements. Fearnside and Pueyo (2012) study based on physical measurements conclude that reservoirs located between the tropics emit a significant amount of greenhouse gas, predominantly methane. While Chanudet et al. (2011) confirm that a 40-year-old subtropical reservoir is a carbon sink.

In contrast, the net approach allows counting for the greenhouse gas emissions prior to flooding. Retrofitted natural lakes illustrate well this concept. It would be ludicrous to account the greenhouse emissions of a natural water body to hydropower generation.

The net approach taken by the tool used for this study and published in Prairie et al. (2018) uses a unique framework to represent only the GHG emissions that are attributable to the introduction of the reservoir in a catchment. It first includes the explicit consideration of the GHG footprint of the affected landscape before impoundment (GHG_{pre}). It factors the climatic, geographic, edaphic and hydrologic settings of each reservoir site, the displaced emissions that would have occurred somewhere else regardless of the reservoir presence and the temporal evolution over the lifetime (GHG_{post}). It subtracts the unrelated anthropogenic emissions associated with any human activity in the catchment such as settlements, industry or agriculture ($GHG_{unrelated}$) and adds the emissions incurred during the construction (GHG_{const}). Therefore, the formula for the net GHG emissions calculation is:

$$Net\,GHG\,emissions = \left(GHG_{post}\right) - \left(GHG_{pre}\right) - \left(GHG_{unrelated}\right) + \left(GHG_{const}\right)$$

where
GHG_{post}: post-impoundment GHG balance of the reservoir site
GHG_{pre}: pre-impoundment GHG balance of the reservoir site
$GHG_{unrelated}$: GHG emissions from the reservoir due to unrelated anthropogenic sources
GHG_{const}: GHG emissions due to the construction of the dam.

5.2.2   Tool

Since 2017, the web-based G-res Tool is publicly available at g-res.hydropower.org for hydropower companies and researchers to estimate and report net greenhouse gas emissions from a reservoir. In the absence of site-specific temporal data to calculate the net greenhouse gas emissions, the use of G-res Tool is considered the most reliable and comprehensive approach according to Levasseur et al. (2021) that compares nine methods.

As explained in the Technical Documentation (Prairie et al. 2017), the tool can be calibrated for any geographical location using local parameters and the approach considers multiple types of emissions (Diffusive CH_4 and CO_2, Bubbling CH_4 and Degassing CH_4), as well as the specific characteristics of each reservoir. It considers the state of the land pre-impoundment, naturally occurring emissions and emissions related to other human activities over the lifetime of the reservoir. The lifetime is assumed to be 100 years.

It also offers a methodology for apportioning the net greenhouse gas emissions to the various freshwater services that a multipurpose reservoir can serve, such as water supply, irrigation, flood and drought management, hydropower, navigation, fisheries and recreation.

The expert committee continues further improvement of the tool with new advances in science and research findings. The net GHG footprint used for the present analysis was extracted from G-res version 3.0 (Prairie et al. 2021), which will be integrated to the online interface as soon as it is accepted in a peer-reviewed journal for publication. This third version slightly modifies the prediction following some improvements to the statistical process behind the models.

5.2.3 Methodology

Using the G-res Tool, the study estimates the net carbon footprint that corresponds to each hydropower reservoir. As output, the G-res Tool gives the estimated total GHG footprint in gCO_2-eq/m^2/yr over each reservoir's lifetime (100 years). These estimates were coupled with average annual hydropower generation data (GWh/yr) to obtain the GHG emissions intensity (gCO_2-eq/kWh) of hydropower operations at each reservoir. To represent the portion of total emissions corresponding to hydropower, we apply an allocation methodology for multipurpose reservoirs as described below.

Allocation Method

A reservoir can provide a range of different economic, social and environmental services such as hydropower, irrigation, water supply, navigation, flood control and fisheries. However, fishing or recreational activities usually are secondary purposes that emerge from the creation of a reservoir. We also find drought mitigation and food and water security among indirect services, which enable societal growth and further business opportunities.

It seems logical to allocate any footprint, whether water consumption or greenhouse gas emissions, to the different uses benefiting from the reservoir creation. However, the allocation methodology raises lots of discussion inside and outside academia. The hydropower industry demands a justifiable allocation technique for multipurpose reservoirs because of the negative attention received by studies that allocate water consumption or greenhouse emissions solely to this activity. For example, Mekonnen and Hoekstra (2012) allocated the full footprint to hydropower where it was the main reason for the construction of the reservoir. While Bakken et al. (2016) conclude that volume allocation to be the most robust approach, Hogeboom et al. (2018) prefer allocation by economic value and Zhao and Liu (2015) propose an approach based on the relation to the ecosystem services provided.

These approaches have some setbacks. The economic method is the most data-intensive, while the volumetric one poses difficulties to allocate footprint in terms of water volume utilised by services such as flood control, navigation, fisheries or recreational activities.

A third approach reflects on the reservoir operating rules from multipurpose reservoirs to maximise the benefits and prioritise services whenever there is a conflict of interest. The G-res Tool follows this prioritisation allocation methodology (Table 5.1) by weighting the uses depending on their classification. It allows the following eight services: (1) flood control, (2) fisheries, (3) irrigation, (4) navigation, (5) environmental flow, (6) recreation, (7) water supply and

Table 5.1 Allocation methodology used in the G-res Tool

Importance	Allocation (%)	Notes	
Primary	80	If there are more than one service in the level, allocation is split equally between them	If there are more than one service in each level, GHG emissions are split equally between these. There is a maximum of three services for each level
Secondary	15	Where there are no secondary services, the allocation (15%) is split between primary services	
Tertiary	5	Where there are no tertiary services, the apportionment (5%) is split between secondary services	

Table 5.2 Classification of purposes for multipurpose reservoirs followed in the G-res Tool

Importance	Explicit prioritisation	Operating rule curve
Primary	Ranked 1–3 in operation hierarchy	Operating rules are designed to maximise these services for part or all of the year
Secondary	Ranked lower than 3 in operational hierarchy, or place constraint on operations	The service places operational constraints on the operating level of the reservoir for part or whole of the year
Tertiary	Provides benefits but does not alter the operation of the reservoir	The service provides benefits but has little or no impact on the operation of the reservoir

(8) hydroelectricity ranked from primary to tertiary use. Table 5.2 shows how the classification relates to the operational regime of explicit prioritisation.

Given the scope, data availability and the use of the G-res Tool, we select the explicit prioritisation of the reservoir services as the allocation method. According to the operational regime, the dataset ranks the reservoir services in primary, secondary and tertiary purposes.

5.3 Data

The characteristics of the reservoirs and dams of hydropower plants were extracted from the Global Reservoir and Dam Database (GRanD) (Lehner et al. 2011), the scientific literature, the IHA Global Hydropower Station Database, the US Energy Information Administration and from individual surveys to hydropower operators. It includes information on the asset name, location, installed capacity, average annual electricity generation, reservoir surface area, and primary, secondary and tertiary purposes of the reservoir.

Fig. 5.1 Location map of the 480 hydropower dams considered in the study

The extraction of the information resulted in 480 hydropower plants that this study analyses. They are spread across the world, as shown in Fig. 5.1, with a greater representation in countries with higher hydropower installed capacity such as China, Brazil, USA, Canada, Turkey and South of Europe.

This study represents a small fraction, approximately 4%, of the total hydropower stations recorded in the IHA Global Hydropower Stations Database that counts with 12,315 hydropower operational stations in the world representing about 90% of the global hydropower installed capacity by 2020. In terms of installed capacity and generation, the 480 hydropower plants dataset, adding to 204 GW and 833 TWh, covers 16% of the global hydropower installed capacity and 20% of the total average electricity generated by hydropower in 2019 (IHA 2020b).

According to ICOLD (2021), globally, about 40% of the hydropower reservoirs are multipurpose and 60% single purpose. The 480 hydropower stations of this study, however, do not follow the same representation. About 64% of the hydropower stations analysed are multipurpose and supply on average two (35% of the multipurpose stations) and in cases up to eight different uses. The remaining ones, about 36%, served hydropower as their only purpose. Multipurpose reservoirs where hydropower is the primary use are dominant in the dataset, though, at 74%. Hydropower is the secondary use for 14% reservoirs, and only 14 reservoirs count hydropower as the tertiary purpose.

Due to the limited amount of data available, the study analyses how well the study dataset (480 hydropower plants) fits the distribution from the total hydropower stations recorded in the IHA Global Hydropower Stations Database and the GRanD database. The study uses the Chi-Square goodness of fit test to analyse the sample representativeness for installed capacity, climatic zones and surface area variables. The Chi-Square test was used to determine whether there is statistically significant difference between the sample corresponding to the 480 reservoirs and the IHA Global Hydropower Stations database (installed capacity) and the GRanD database (climatic zones and surface area). The obtained X^2 is considered not significant at a conventional p value of 0.05 for the three variables. The justified interpretation following the rejection of the null hypothesis would be to conclude that there is not statistically significant difference between the distributions of the study dataset and the IHA Global Hydropower Stations database, and GRanD database.

5.3.1 Installed Capacity

The size range covers from mini-scale plants (lower than 1 MW) to the largest hydropower plant (22,500 MW installed capacity) in the world. The average generation values range from 0.1 GWh/yr to 89,500 GWh/yr. More than half of the study's hydropower plants (56%) have an installed capacity lower than 100 MW, and 90% of the plants have less than 1000 MW. Both installed capacity and generation datasets follow a positive skewed distribution. Table 5.3 shows the basic statistical description of the dataset. The mean and median values for the installed capacity and generation show that most values are clustered around the left tail of the distribution. The high values in the standard deviation for both the installed capacity and generation are consistent with the dispersion of the dataset values relative to the mean.

Table 5.3 Descriptive statistics for installed capacity and generation of the study sample dataset of 480 hydropower stations

	Installed capacity (MW)	Generation (GWh/yr)
Minimum	0.013	0.100
Q1	19.4	51.1
Median	75.5	200.0
Q3	264.8	681.8
Maximum	22,500	89,500
Range	22,500	89,500
Total	203,556	843,686
Mean	424	1758
Standard Deviation	1492	6881

Due to the positively skewness distribution of the sample datasets, the installed capacity dataset was compared with the total dataset of the IHA Global Hydropower Stations Database, with a total of 12,315 operational stations. Figure 5.2 presents a visual representation of the installed capacity distribution of the study dataset and the IHA Global Stations database. While both datasets are positively skewed, the shape of the study dataset is less pronounced, with about 56% of the hydropower stations below 100 MW, while the global IHA Global Stations database shows about 84% of all the stations below 100 MW. Moreover, there are about 2% of the study dataset stations with higher operational capacity than 3000 MW while that only represents 0.2% in the global IHA Global Stations database.

The Chi-Square test was used to determine whether there is a statistically significant difference between the sample corresponding to the 480 reservoirs and the total IHA Global Station database in terms of their installed capacity distribution. The obtained $X^2 = 8E{-}87$ is not significant at a conventional p value of 0.05. In the annex, it included the distribution of the study dataset and IHA Global Stations database and the expected values to perform the Chi-Square test. The justified interpretation following the rejection of the null hypothesis would be to conclude that there is no statistically significant difference between the distributions of both datasets. While high levels of skewness can generate misleading results from statistical tests, the distribution of the 480 hydropower stations of this study is proven representative of the global operational hydropower fleet.

After performing a Chi-Square test between this study dataset (n = 480) and the IHA Global Hydropower Stations database (see Annex), we can conclude that there is not statistically significant difference between the distributions of both datasets ($X^2 = 7.9E{-}87$). While high levels of skewness can generate misleading results from statistical tests, the distribution of the 480 hydropower stations of this study is proven representative of the global operational hydropower fleet in terms of installed capacity.

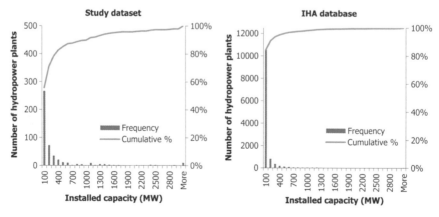

Fig. 5.2 Histogram of installed capacity for the study dataset and IHA Global Hydropower Stations database for operational hydropower plants

5.3.2 Climatic Zone

The 480 stations corresponding to the study dataset are distributed over the four climatic zones (boreal, temperate, subtropical and tropical). The geographical location of the study dataset was compared to the GRanD database. A total of 1528 reservoirs with hydropower as primary purpose had available information on the climate zone. The location of the 480 hydropower dams was compared with the location of the 1528 hydropower dams from the GRanD database using a Chi-Square test.

The Chi-Square test was used to determine whether the sample corresponding to the 480 reservoirs follows a different distribution than the GRanD database in terms of the climatic zone (Table 5.4). The obtained $X^2 = 5.5E-5$ allows us to confirm that there is no statistically significant difference between the study dataset and the GRanD database in terms of the climate zone (see Annex).

5.3.3 Surface Area

The surface area of the hydropower stations reservoirs ranges from 0.2 to 6988 km^2 with a median of 17.85 km^2. To assess if the surface area of the 480 stations is representative of the total hydropower stations reservoirs globally, the study considered all GRanD database reservoirs with available surface area data and with hydropower as the primary use. We considered 1528 reservoirs, with a minimum area of 0.1 km^2 a maximum 67,165 km^2 of and a median of 8.7 km^2.

Figure 5.3 presents a visual representation of the surface area distribution of the reservoirs of the study dataset and the GRanD database. While both datasets are positively skewed, the shape of the study is slightly less pronounced, with about 52% of the hydropower stations reservoirs being below 20 km^2, while the GRanD database shows that about 65% of all the hydropower reservoirs are below 20 km^2. Both datasets show a significant number of reservoirs with a higher surface area than 500 km^2, which in the study dataset represents about 7% and in the GRanD database about 5%.

The Chi-Square test was used to determine whether there is statistically significant difference between the sample corresponding to the 480 reservoirs from the

Table 5.4 Distribution of number of stations per climate zones of study dataset and GRanD database

	Study dataset	GRanD database
Climate	# Stations	# Stations
Boreal	69	291
Temperate	338	1056
Subtropical	30	46
Tropical	43	135
Total	480	1528

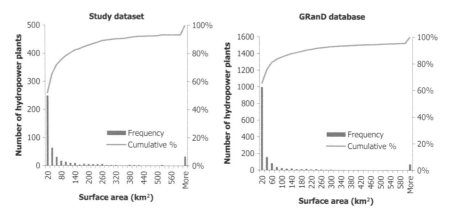

Fig. 5.3 Histogram of surface area for the study dataset and GRanD hydropower plants database

study dataset and the 1528 hydropower reservoirs from the GRanD database (see Annex). In terms of their surface area distribution, both datasets were not significantly different ($X^2 = 0.0007$), and this study dataset is thus considered representative of the total hydropower reservoirs GRanD database.

5.4 Results

The G-res Tool was used to estimate the GHG footprint of the 480 hydropower stations in the study dataset. These estimates were coupled with the average annual hydropower generation to obtain the GHG emissions intensity (gCO_2-eq/kWh) allocated to hydropower at each hydropower station reservoir. Table 5.5 shows that the evaluated GHG emissions from hydropower reservoirs cover a wide range with ~ eight orders of magnitude difference between minimum and maximum, which supports previous observations (IPCC 2014). We estimated the median lifecycle assessment for hydropower emissions is 23 gCO_2-eq/kWh. In comparison to the reservoir's total GHG emissions, the allocated GHG emissions to hydropower are about 50% lower.

As shown in Fig. 5.4, there appears no correlation between the average annual electricity generation and the GHG emissions per unit of generated electricity.

The findings were compared to the lifecycle intensity emission of other sources of energy reported in the IPCC Fifth Assessment Report published in 2014. For hydropower IPCC reported 24 gCO_2-eq/kWh median lifecycle emissions.

The study results align with those reported by IPCC although the spread is wider. Figure 5.5 shows the spread by comparing the statistics from this study results and those of the lifecycle emissions estimated in AR5 reported in IPCC. The greenhouse emissions associated with reservoirs range an order of magnitude at the global scale with minimum −922 gCO_2eq/kWh and maximum 4295 gCO_2eq/kWh.

Table 5.5 Comparison of the total median global average emissions and allocated to hydropower

	Net GHG emissions (gCO$_2$eq/m^2/yr)	Net GHG emissions intensity (gCO$_2$eq/kWh)	Net GHG emissions intensity allocated to hydro (gCO$_2$eq/kWh)
Minimum	−607.76	−921.52	−921.52
Q1	227.23	9.46	5.45
Median	334.43	43.09	22.72
Mean	617.34	277.36	170.03
Q3	605.25	185.22	98.71
Maximum	11,000.18	10,536.28	4294.54

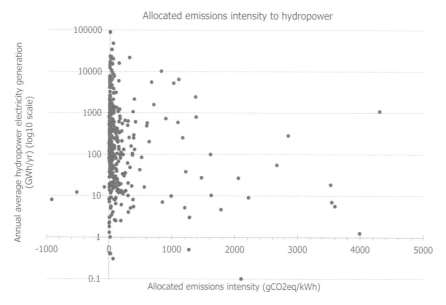

Fig. 5.4 Allocated emissions intensity to hydropower calculated for the 480 hydropower stations of the study dataset

The IPCC range is smaller (0–2200 gCO$_2$eq/kWh), but the upper quartile, which represents the 75th percentile, doubles (200 gCO$_2$eq/kWh) the one from this study (99 gCO$_2$eq/kWh).

Figure 5.6 compares the intensity emission of different sources of energy reported by IPCC and the findings from this study. As shown, hydropower median emissions are significant less than fossil fuels and comparable to other renewable energy sources. The median is one order of magnitude lower than those of fossil fuels and biomass. Among the renewable energy source, only wind, both onshore and offshore, has a median lifecycle emission lower than hydropower at 11 gCO$_2$-eq/kWh and 12 gCO$_2$-eq/kWh, respectively. Solar and geothermal power median emissions double the one from hydropower.

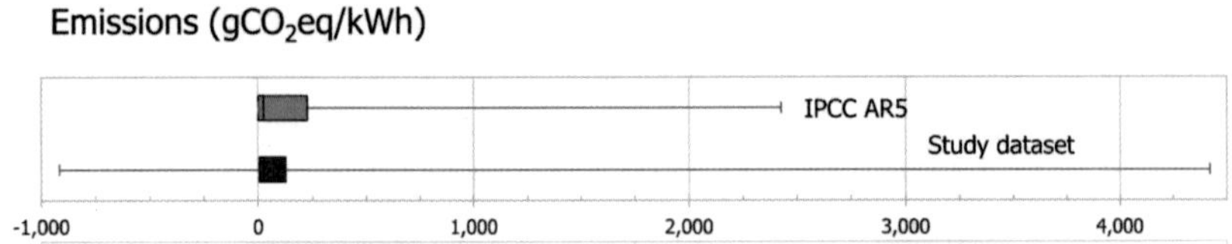

Emissions (gCO$_2$eq/kWh)

	Study dataset	IPCC AR5
Min	-922	0
Q1	5	1
Median	23	24
Q3	99	200
Max	4295	2200

Fig. 5.5 Lifecycle emissions (gCO$_2$-eq/kWh) for hydropower as estimated in the study dataset and (Purple) and the IPCC AR5 (Blue). Q1 represents the 25th percentile and Q3 the 75th percentile

5.5 Discussion

Hydropower plays an essential role in the climate change adaptation of water resources availability. It can provide multiple benefits to societal development and growth, especially in contributing to guarantee water and energy security.

While IRENA estimates the need for 850 GW of newly hydropower installed capacity over the next 30 years to meet global climate change targets, the creation of hydropower projects is still a source of controversy due to their environmental and social impacts. A rational treatment of this issue requires applying the commonly accepted climate change policy principles described in the United Nations Framework Convention on Climate Change (UNFCCC), as well as promoting participatory multiple water use management plans through IWRM.

Planning hydropower systems from a long-term, climate-resilient perspective will guarantee that current and future generations inherit hydropower infrastructure that will not be compromised by climate change and will ensure water and energy services. Moreover, hydropower assets have now available the Hydropower Sector Climate Resilience Guide to assess climate change-related opportunities to help investment decisions recognise the role of hydropower infrastructure in providing climate-related services.

As climate disruption will heighten the situation, the importance of water storage and water conservation will continue increasing. Hydropower reservoirs can provide multiple benefits to societal development and growth, especially in contributing to guarantee water and energy security. The water stored in storage hydropower projects, besides providing clean, reliable, sustainable energy, provides a higher systemic resilience and capacity to adapt to climate change as it procures flood and drought mitigation.

However, the storage hydropower projects have also raised concerns due to the biogenic emissions caused by the impoundment of a reservoir. In the past, lack of scientific consensus on quantifying this footprint has proven a significant obstacle

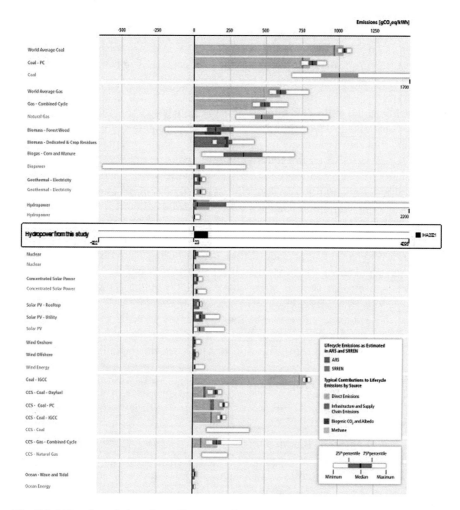

Fig. 5.6 Lifecycle emissions from all sources of energy (Modified from Table A.III.2 from the IPCC Fifth Assessment Report 2014)

for policy and decision-makers to justify hydropower financing, in particular the construction of large dams, as clean low-carbon energy sources.

This study estimates the median lifecycle assessment for hydropower emissions is 23 gCO$_2$-eq/kWh. The results are aligned with those reported by IPCC. The findings of this study show that hydropower median emissions are significantly less than fossil fuels and comparable to other renewable energy sources. Among the renewable energy sources, only wind has a median lifecycle emission lower than hydropower.

The study also underlines the significant wide spread of results, highlighting the importance of measuring and reporting the emissions from individual reservoirs to properly assess their lifecycle emissions. To maximise their role in mitigating climate change, hydropower projects need to be developed and operated sustainably, including limiting greenhouse gas emissions. Given the complexity of field measures and research based on the extrapolation, modelling tools such as the G-res Tool (g-res.hydropower.org) facilitate the estimation of greenhouse emissions during the pre-feasibility study, so these are not an obstacle to justify the creation of a reservoir.

Recognising the multiscale impacts of dams, policies and measures are needed at reservoir, catchment, national and global levels for dealing with GHG emissions. Moreover, adequate assessment of environmental, social, governance and technical aspects is necessary to ensure that hydropower projects impacts do not overcome their benefits to water and energy security. The hydropower sector counts with the Hydropower Sustainability Tools, a suite of assessment tools and guidance documents, providing a common language to all stakeholders involved in a hydropower project.

Also, in assessing GHG emission for hydropower projects to justify its development, developers and financier should take a holistic approach and take into account the avoided emissions not only from hydropower but also of other uses that the project can provide such as navigational capabilities that could substitute other means of goods transportation.

Lastly, these considerations will be important at the strategic level when national GHG policies are developed and at the operational level to make different actors assume responsibility and engage in mitigation efforts corresponding to different uses.

Acknowledgements This study was based on the previous work carried out by ex IHA staff members, Mathis Rogner and Emma Smith, under the funded UNESCO/IHA research project on the GHG status of freshwater reservoirs and published in the IHA Hydropower Status Report 2018 (IHA 2018b).

Declaration of Interest Statement The authors declare that they have no known competing financial interests or personal relationships that could have appeared to influence the work reported in this chapter.

Annex

Distribution of the installed capacity of study dataset and IHA Global Hydropower Station database and expected values to perform Chi-Square test

Installed capacity	Study dataset	IHA database		Study dataset expected values
	# Stations	# Stations	Distribution	# Stations
100	268	10,357	0.841006902	404
200	72	810	0.065773447	32
300	35	351	0.028501827	14
400	21	180	0.014616322	7
500	12	117	0.009500609	5

(continued)

(continued)

Installed capacity	Study dataset	IHA database		Study dataset expected values
	# Stations	# Stations	Distribution	# Stations
600	10	87	0.007064555	3
700	2	49	0.003978888	2
800	5	46	0.003735282	2
900	4	37	0.003004466	1
1000	1	41	0.003329273	2
1100	9	37	0.003004466	1
1200	2	41	0.003329273	2
1300	5	28	0.00227365	1
1400	5	14	0.001136825	1
1500	3	14	0.001136825	1
1600	2	9	0.000730816	0
1700	2	8	0.000649614	0
1800	1	8	0.000649614	0
1900	0	9	0.000730816	0
2000	0	7	0.000568413	0
2100	2	5	0.000406009	0
2200	1	4	0.000324807	0
2300	0	1	$8.12018E-05$	0
2400	3	8	0.000649614	0
2500	2	7	0.000568413	0
2600	0	3	0.000243605	0
2700	0	1	$8.12018E-05$	0
2800	1	2	0.000162404	0
2900	2	2	0.000162404	0
3000	0	4	0.000324807	0
>3000	10	28	0.00227365	1
Total	480	12,315		
Chi-Square test				$7.89169E-87$

Distribution of climate zones of study dataset and GRanD database and expected values to perform Chi-Square test

	Study dataset	GRanD database		Study dataset expected values
Climate	# Stations	# Stations	Distribution	# Stations
Boreal	69	291	0.190445	91.4136
Temperate	338	1056	0.691099	331.728
Subtropical	30	46	0.030105	14.4503

(continued)

(continued)

	Study dataset	GRanD database		Study dataset expected values
Climate	# Stations	# Stations	Distribution	# Stations
Tropical	43	135	0.088351	42.4084
Total	480	1528		
Chi-Square test p value				5.5E−05

Distribution of surface area of study dataset and GRanD database and expected values to perform Chi-Square test

	Study dataset	GRanD database		Study dataset expected values
Surface area	# Stations	# Stations	Distribution	# Stations
20	249	997	0.652486911	313.1937173
40	64	156	0.102094241	49.0052356
60	31	82	0.053664921	25.7591623
80	17	37	0.02421466	11.62303665
100	14	25	0.016361257	7.853403141
120	10	19	0.012434555	5.968586387
140	10	20	0.013089005	6.282722513
160	4	12	0.007853403	3.769633508
180	7	13	0.008507853	4.083769634
200	5	15	0.009816754	4.712041885
220	5	9	0.005890052	2.827225131
240	5	11	0.007198953	3.455497382
260	6	8	0.005235602	2.513089005
280	2	6	0.003926702	1.884816754
300	2	7	0.004581152	2.19895288
320	2	4	0.002617801	1.256544503
340	1	4	0.002617801	1.256544503
360	1	2	0.001308901	0.628272251
380	3	3	0.001963351	0.942408377
400	2	3	0.001963351	0.942408377
420	2	3	0.001963351	0.942408377
440	0	3	0.001963351	0.942408377
460	1	1	0.00065445	0.314136126
480	0	2	0.001308901	0.628272251
500	1	2	0.001308901	0.628272251
520	3	4	0.002617801	1.256544503
540	0	2	0.001308901	0.628272251
560	0	2	0.001308901	0.628272251
580	0	2	0.001308901	0.628272251

(continued)

(continued)

	Study dataset	GRanD database		Study dataset expected values
Surface area	# Stations	# Stations	Distribution	# Stations
600	0	2	0.001308901	0.628272251
>600	33	72	0.047120419	22.61780105
Total	480	1528		
Chi-Square test				0.000708537

References

Bakken TH, Modahl IS, Raadal HL, Bustos AA, Arnøy S (2016) Allocation of water consumption in multipurpose reservoirs. Water Policy 18(4):932–947

Benson D, Gain AK, Giupponi C (2020) Moving beyond water centricity? Conceptualizing integrated water resources management for implementing sustainable development goals. Sustain Sci 15(2):671–681

Berga L (2016) The role of hydropower in climate change mitigation and adaptation: a review. Engineering 2(3):313–318

Brown C, Lall U (2006) Water and economic development: the role of variability and a framework for resilience. In: Nat Resour Forum 30(4):306–317 (Oxford: Blackwell Publishing Ltd.)

Carbon Monitor (2021). https://carbonmonitor.org/. Accessed 6 Feb 2021

Chanudet V, Descloux S, Harby A, Sundt H, Hansen BH, Brakstad O, Serça D, Guerin F (2011) Gross CO2 and CH4 emissions from the Nam Ngum and Nam Leuk sub-tropical reservoirs in Lao PDR. Sci Total Environ 409:5382–5391

Climate Bonds Initiative (2019) Open for public consultation. The Hydropower Criteria. Climate Bonds Standard. https://www.climatebonds.net/hydropower. Accessed 6 Feb

Deemer BR, Harrison JA, Li S, Beaulieu JJ, DelSontro T, Barros N et al (2016) Greenhouse gas emissions from reservoir water surfaces: a new global synthesis. BioScience 66(11):949–964

DFID (Department of International Development) (2009) Water storage and hydropower: supporting growth, resilience, and low carbon development (A DFID evidence-into-action paper). Policy Booklet

European Commission (2020) Financing a Sustainable European Economy—taxonomy report: technical annex. https://ec.europa.eu/info/publications/sustainable-finance-teg-taxonomy_en Accessed 6 Feb 2021

FAO (2017) The future of food and agriculture: trends and challenges. FAO, Rome

Fearnside PM, Pueyo S (2012) Greenhouse-gas emissions from tropical dams. Nat Clim Chang 2:382–384

Hallegatte S, Shah A, Lempert C, Brown C, Gill S (2012) Investment decision making under deep uncertainty: application to climate change. Policy Research Working Paper 6193. Washington, DC: World Bank

Hogeboom RJ, Knook L, Hoekstra AY (2018) The blue water footprint of the world's artificial reservoirs for hydroelectricity, irrigation, residential and industrial water supply, flood protection, fishing and recreation. Adv Water Resour 113:285–294

ICOLD (International Commission on Large Dams) (2021) World register of dams. General synthesis. https://www.icold-cigb.org/GB/world_register/general_synthesis.asp. Accessed 6 Feb 2021

IEA (2019) World energy outlook 2019. https://www.iea.org/reports/world-energy-outlook-2019. Accessed 6 Feb 2021

IHA (International Hydropower Association) (2010) GHG measurement guidelines for freshwater reservoirs: derived from: the UNESCO/IHA greenhouse gas emissions from freshwater reservoirs research project. International Hydropower Association, London

IHA (International Hydropower Association) (2010) Hydropower sustainability assessment protocol. International Hydropower Association, London

IHA (International Hydropower Association) (2018) Hydropower status report 2018: sector trends and insights. International Hydropower Association, London

IHA (International Hydropower Association) (2018) Hydropower sustainability guidelines. International Hydropower Association, London

IHA (International Hydropower Association) (2019) Hydropower sector climate resilience guide. International Hydropower Association, London

IHA (International Hydropower Association) (2020) Hydropower status report 2020. International Hydropower Association, London

IHA (International Hydropower Association) (2020) Hydropower sustainability environmental, social and governance gap analysis tool. International Hydropower Association, London

IPCC (2014) Climate change 2014: mitigation of climate change. In: Edenhofer O, Pichs-Madruga R, Sokona Y, Farahani E, Kadner S, Seyboth K, Adler A, Baum I, Brunner S, Eickemeier P, Kriemann B, Savolainen J, Schlömer S, von Stechow C, Zwickel T, Minx JC (eds) Contribution of working group III to the fifth assessment report of the intergovernmental panel on climate change. Cambridge University Press, Cambridge, United Kingdom and New York, NY, USA

IPCC (2018) Summary for policymakers. In: Masson-Delmotte V, Zhai P, Pörtner H-O, Roberts D, Skea J, Shukla PR, Pirani A, Moufouma-Okia W, Péan C, Pidcock R, Connors S, Matthews JBR, Chen Y, Zhou X, Gomis MI, Lonnoy E, Maycock T, Tignor M, Waterfield T (eds) Global Warming of 1.5°C. An IPCC Special Report on the impacts of global warming of 1.5°C above pre-industrial levels and related global greenhouse gas emission pathways, in the context of strengthening the global response to the threat of climate change, sustainable development, and efforts to eradicate poverty. World Meteorological Organization, Geneva, Switzerland, 32 pp

Landsvirkjun (2017) Theistareykir first geothermal power plant to undergo Geothermal Sustainability Assessment Protocol. https://www.landsvirkjun.com/company/mediacentre/news/news-read/theistareykir-first-geothermal-power-plant-to-undergo-gsap-sustainability-assessment. Accessed 6 Feb 2021

Lehner B, Reidy Liermann C, Revenga C, Vorosmarty C, Fekete B, Crouzet P, Doll P, Endejan M, Frenken K, Magome J, Nilsson C, Robertson JC, Rodel R, Sindorf N, Wisser D (2011) Global Reservoir and Dam Database, Version 1 (GRanDv1): Dams, Revision 01. Palisades, NY: NASA Socioeconomic Data and Applications Center (SEDAC). https://doi.org/10.7927/H4N877QK. Accessed 6 Feb

Levasseur A, Mercier-Blais S, Prairie Y, Tremblay A, Turpin C (2021) Improving the accuracy of electricity carbon footprint: estimation of hydroelectric reservoir greenhouse gas emissions. Renewable Sustain Energy Rev 136:110433

Maeck A, Hofmann H, Lorke A (2014) Pumping methane out of aquatic sediments: ebullition forcing mechanisms in an impounded river. Biogeosciences 11:2925–2938

Mayor B, Rodríguez-Muñoz I, Villarroya F, Montero E, López-Gunn E (2017) The role of large and small scale hydropower for energy and water security in the Spanish Duero Basin. Sustainability 9(10):1807

Mekonnen MM, Hoekstra AY (2012) The blue water footprint of electricity from hydropower. Hydrol Earth Syst Sci 16(1):179–187

Mendonça R, Kosten S, Sobek S, Barros N, Cole JJ, Tranvik L, Roland F (2012) Hydroelectric carbon sequestration. Nat Geosci 5:838–840

Milly PCD, Betancourt J, Falkenmark M, Hirsch RM, Kundzewicz ZW, Lettenmaier DP, Stouffer RJ (2008) Stationarity is dead: whither water management? Science 319:573–574

Patel S, Shakya C, Rai N (2020) Climate finance for hydropower: incentivising the low-carbon transition. http://pubs.iied.org/10203IIED. Accessed 10 Feb 2021

Pfister S, Scherer L, Buxmann K (2020) Water scarcity footprint of hydropower based on a seasonal approach-Global assessment with sensitivities of model assumptions tested on specific cases. Sci Total Environ 724:138188

Pokhrel Y, Felfelani F, Satoh Y, Boulange J, Burek P, Gädeke A, Wada Y (2021) Global terrestrial water storage and drought severity under climate change. Nat Clim Change 1–8. https://doi.org/10.1038/s41558-020-00972-w

Prairie YT, Alm J, Beaulieu J, Barros N, Battin T, Cole JJ, del Giorgio PA, DelSontro T, Guérin F, Harby A, Harrison J, Mercier-Blais S, Serça D, Sobek S, Vachon D (2018) Greenhouse gas emissions from freshwater reservoirs: what does the atmosphere see? Ecosystems 21:1058–1071. https://doi.org/10.1007/s10021-017-0198-9

Prairie Y, Alm J, Harby A, Mercier-Blais S, Nahas R (2017) The GHG Reservoir Tool (G-res) Technical documentation, UNESCO/IHA research project on the GHG status of freshwater reservoirs. Joint publication of the UNESCO Chair in Global Environmental Change and the International Hydropower Association

Prairie YT, Mercier-Blais S, Harrison JA, Soued C, del Giorgio PA, Harby A, Alm J, Chanudret V, Nahas R (2021) A new modelling framework to assess biogenic GHG emissions from reservoirs: the G-res Tool. Environ Model Softw 143(105117):1–16. https://doi.org/10.1016/j.envsoft.2021.105117

Ray PA, Brown CM (2015) Confronting climate uncertainty in water resources planning and project design: the decision tree framework. World Bank, Washington, DC

Sarawak (2021) Sustainability and CSR. https://www.sarawakenergy.com/what-we-do/sustainability-csr. Accessed 10 Feb 2021

Sanctuary M, Tropp H, Haller L (2007) Making water a part of economic development: the economic benefits of improved water management and services. Stockholm International Water Institute (SIWI), Sweden

Scherer L, Pfister S (2016) Global water footprint assessment of hydropower. Renew Energy 99:711–720

Thoradeniya B, Ranasinghe M, Wijesekera NTS (2007) Social and environmental impacts of a mini-hydro project on the Ma Oya Basin in Sri Lanka. In: International conference on small hydropower. Hydro Sri Lanka 22:24

Ubierna M, Alarcón A, Alberti J (2020) Modernización de centrales hidroeléctricas en América Latina y el Caribe: Identificación y priorización de necesidades de inversión. Nota Técnica. Banco Interamericano de Desarrollo, Washington, DC

UN (United Nations) (2021) Goal 13: take urgent action to combat climate change and its impacts. COVID-19 response. https://www.un.org/sustainabledevelopment/climate-change/. Accessed 10 Feb 2021

UNEP (United Nations Environment Programme) (2019) Emissions Gap Report 2019. https://www.unenvironment.org/resources/emissions-gap-report-2019. Accessed 10 Feb 2021

Wenjie L, Dawei W, Shengfa Y, Wei Y (2020) Three Gorges Project: benefits and challenges for shipping development in the upper Yangtze River. Int J Water Resour Dev. https://doi.org/10.1080/07900627.2019.1698411

World Bank (2021) Resilience rating system: a methodology for building and tracking resilience to climate change. https://openknowledge.worldbank.org/handle/10986/35039. Accessed 10 Feb 2021

World Commission on Dams (2000) Dams and development. A new framework for decision-making. The Report of the World Commission on Dams. London: Earthscan

Xiaocheng F, Tao T, Wanxiang J, Fengqing L, Naicheng W, Shuchan Z, Qinghua C (2008) Impacts of small hydropower plants on macroinvertebrate communities. Acta Ecol Sin 28(1): 45–52

Zhao D, Liu J (2015) A new approach to assessing the water footprint of hydroelectric power based on allocation of water footprints among reservoir ecosystem services. Phys Chem Earth 79:40–46

Chapter 6
Climate Change and Its Implications for Irrigation, Drainage and Flood Management

Ashwin B. Pandya, Sahdev Singh, and Prachi Sharma

Abstract Climate change is transforming the hydrological regime globally, and one of the largest components is variations in the availability and distribution of water. Consequently, the impacts of climate change on water resources and agriculture, and subsequently food and water security, are increasingly dominating the sustainability debate. Essentially, these effects are a result of an imbalance between the availability of the resources versus the demands on them and our failure to put in alternative policies and measures for the restoration of such a sustainable balance. Hence, the climate change problem calls for climate-friendly solutions in the form of risk mitigation, coping mechanisms and other adaptive strategies to ensure the sustainability of our agricultural systems. Keeping this in mind, this chapter discusses climate change and its repercussions on agricultural water management, correlation with flood management, the role that data plays in determining these impacts and accordingly how creative adaptation strategies may be adopted to cushion the impacts of climate change on agriculture and water. Additionally, this chapter also discusses the role of the International Commission on Irrigation and Drainage plays in contributing to the fight against climate change and global warming.

Keywords Climate change · Agricultural water management · Irrigation · Drainage · Water security · Food security · Flood management

A. B. Pandya (✉) · S. Singh · P. Sharma
International Commission on Irrigation and Drainage (ICID), New Delhi, India
e-mail: sec-gen@icid.org

S. Singh
e-mail: sahdevsingh@icid.org

P. Sharma
e-mail: prachi@icid.org

© The Author(s), under exclusive license to Springer Nature Singapore Pte Ltd. 2022
A. K. Biswas and C. Tortajada (eds.), *Water Security Under Climate Change*,
Water Resources Development and Management,
https://doi.org/10.1007/978-981-16-5493-0_6

6.1 Introduction

Climate change is not a new phenomenon; throughout history, the earth has witnessed changes in the climate mostly due to minor variations in the earth's orbit resulting in a varied amount of solar energy received by the earth surface. However, for nearly 11,000 years, the climate has been consistent with minor variations. But with the rise of industrialisation since the mid-twentieth century, we have seen a significant rise in greenhouse gas emissions, particularly skyrocketing in the last few decades. This increased emission of greenhouse gases has accelerated climate change and resulted in increased global warming. According to the National Oceanic and Atmospheric Administration of the USA, the average temperature of the earth's surface (averaged across land and ocean) in 2020 has increased by 1.19°C since the pre-industrial era (Lindsey and Dahlman 2021).

During the Cold War, we were worried about nuclear Armageddon, but we were silently building another Armageddon of climate change with ever-increasing demands on the resources and exploitation of resources without regard to their sustainability. The race to industrialisation, considered directly proportional to the economy and growth of the country, has propelled the increased use of fossil fuels and thus increased emissions of greenhouse gases from industries and other pathways such as overuse of water for agriculture and industrial growth, deforestation, plastic pollution and burning of plastics, to name a few, all of which, directly or indirectly, have contributed to global warming and changing of the climate across the world. And since climate change is ubiquitous and does not recognise administrative or political boundaries drawn by the countries, its catastrophic effects, similarly, will not respect these borders.

For the agricultural sector, the marginal rise in temperature and concentration of CO_2 in the air are both favourable to most crop plants, but their impact on pest and soil microbial populations are yet to be fully researched and understood. Other accompanying consequences of climate change are the unanswered ambiguous questions. For example, gradual temperature rise would most certainly make more water available due to faster melting of glaciers in the short run, but adequate replenishment of freshwater would be a cause of concern in the medium to long term due to changing patterns of precipitation and erratic behaviours of extreme events such as flood and drought. Desertification of river basins and rise of sea level in the coastal parts will certainly add to our difficulties by limiting the supply of both land area and freshwater, the two of the three essential physical natural resources for agriculture, besides air.

6.2 Climate Change and Water

Water is an indestructible resource; the availability of which is governed by the hydrological cycle being part of global circulation patterns of the atmosphere and its interaction with the land. The impacts of climate change are visible mostly through the medium of water such as changes in the hydrological cycle, frequent and extreme floods and droughts, increased snowmelt, rising of sea levels and so forth. Other impacts of climate change on hydrology are apparent via changes in atmospheric circulation patterns, altered base flows, higher sediment flows, changes in ecosystems as well as changes in the biological and physical properties of soil (Muir et al. 2018). The effects of disturbance of thermal and water cycle are manifold. Changing areal distribution coupled with a change in temporal patterns can significantly influence the agricultural and industrial cycles of the society. It also has impacts on the disasters especially the floods which can increase their inundation spreads and also change their frequencies leading to increasing vulnerabilities of related infrastructure and also on habitations and domestic economic losses due to changed inundation patterns.

Changes in distribution patterns also unsettle the inter-societal arrangements for the distribution and consumption of water and can lead to tensions resulting from poor results from real-time transboundary water management practices. While impacts are discussed at great length, the quantification of impacts is generally lacking. Unless we quantify these future impacts, the solutions cannot be found. The quantifications are done at macro and micro-scales, where the macro-impacts and opportunities have to be quantified and adaptations at micro-levels have to be planned based on local conditions but the adaptations have to aggregate to the macro level provisions.

The water sector, especially in agriculture, has to migrate from the Cassandra approach and move towards the "*Bhagirath*" approach, i.e. not stopping at merely conjectural problem announcement but also bringing out implementable solutions. Implementable solutions have to be based on the present state-of-the-art technologies and quantitative science.

6.3 Impact and Opportunities for Agriculture Water Management

Water is principally used for agricultural operations which are the prime vehicles for providing food and fibre for mankind and domestic livestock. FAO estimates that about 70% of available freshwater resources are consumed by the agricultural sector. In certain countries, where agriculture dominates the economy, this percentage is even higher.

While climate change is putting pressure on freshwater resources and subsequently agriculture, non-climatic factors such as burgeoning population globally, urbanisation, global economic growth, competing demand for natural resources, agronomic management practices, modernisation, technological innovations and trade and food prices pose immediate impacts on water resources as compared to the climate change (FAO 2017). However, under the influence of climate change, the non-climatic characteristics are exacerbating and increasing water scarcity globally, meanwhile the demand for food security is rising. Thus, it is safe to conclude for most countries that the climatic determinants in conjunction with non-climatic determinants are forcing the farmers to grow more crops using lesser water.

Owing to this increased food demand, the agricultural water sector will get affected by the supply as well as demand side both. The changes in spatial and temporal availability and failure of storage and diversion infrastructure will affect the supply side. Whereas, the rise in temperatures and changes in climatic parameters like wind speed, sunshine and intense rainfall events will change the evapotranspiration demands. The life cycles of the crops may also change affecting the water distribution networks. These changes, especially in the form of floods or droughts, may not only affect at farm level but at the basin level as well.

Climate change may bring additional land areas for cultivation due to warming effects on the cooler climates. Such areas can need special monitoring and management to prevent undesirable side effects as well as economic sustainability. Vice versa is also true. The areas rendered unusable for specified cropping patterns may need alternate strategies for their continuance as the source of livelihoods. In addition, an increase in average temperature will lead to higher evapotranspiration rates, affect effective rainfall and altered river discharges. This, in turn, would require structural changes in the irrigation infrastructure inflicting additional financial burden.

Since agriculture is a prime livelihood source for least-developed countries and also for some developing countries, it is necessary to forecast the potential areas and take timely ameliorative actions. The crisis may be acute at the local level as the climate change effects will spread across national boundaries and the economic models of the individual countries will have to change. Least-developed countries need to be extra careful as they may not have adequate economic resilience for bringing more capital and also the subsistence level dependence on the agricultural sector leaving little surplus towards investment.

6.3.1 Availability and Consumption Patterns

Climate change is going to affect the temporal as well as the spatial distribution of precipitation as well as evapotranspiration, thereby affecting the availability of water resources (Konapala et al. 2020). Simultaneously, it is also driving the migration patterns, especially observed via the exodus of rural youth to urban centres for better employment opportunities creating a skewed pattern of water resources consumption. To remedy this, mitigation measures in terms of augmenting storage capacities need to be incorporated.

Climate change in coupling with increasing demands due to increasing population and economic uplift of the people has created the need for additional food and water demand. Since the irrigation sector accounts for the largest amount of global freshwater withdrawals, the supplementary demands for food and water are forcing the irrigation sector to adapt to newer parameters (IPCC 2007). Historical consumption patterns may not remain relevant due to changing water availability and increased competition from other sectors. This creates a double whammy effect of changed production patterns versus new demands of better nutrition and quality as well as increased quantities of output. Improvement in efficiencies at various levels for better utilisation of resources and also make the same resource last longer or spread over a larger area. This suggests that adaptation strategies need to be adopted not just for water security but food and fibre security as well in the context of climate change—scientific aspects, economic aspects and social behaviour areas.

6.3.2 Role of Irrigation in Achieving Global Food Security

Irrigation has been the prime mover of the green revolution and assurance of food security across the world. While irrigation is a prime source of agricultural sustainability, it is also the largest consumer of water. Since all sectors have to make adjustments, the burden of such adaptation on irrigation is equally high. Moreover, in all the projected scenarios of climate change and hydro-economic modelling, the overall demand for irrigation water is expected to increase; however, since the availability of freshwater resources is limited, the ability to expand irrigation will also be constrained.

Efficient water management can partly contribute to averting the difficulties faced by agriculture because of water scarcity in terms of crop failure and income loss. In that pursuit, so far, irrigation has merely been considered a tool for improving productivity; however, moving forward, it will also be required to be considered as a tool for climate change mitigation and adaptation.

As water security and food security are inherently interlinked, so will their management need to be based on the multi-criteria analysis inclusive of all stakeholders. To ensure that irrigation development and management copes up with the climate change impacts, key actors need to be identified, their roles need to be tailored according to the irrigation requirements and accordingly, climate-friendly strategies need to be adopted. Decision-making needs to rely on purely scientific knowledge in coherence with the traditional practices applicable for suitable microclimates, maintaining the socio-economic equilibrium.

Dependence on natural rainfall or precipitation does not assure a consistent output that can meet the growing needs of an expanding population. Thus, as an adaptation strategy, irrigation scheduling and management need to be incorporated to provide precise irrigation and on schedule to the newer varieties of crops. Initiatives such as integrated watershed management, implementation of on-farm water conservation techniques, irrigation based on storage, use of reclaimed water for irrigation may become a potential tool for coping with climate change. Many researchers have developed decision support systems for irrigation systems considering parameters such as air and soil humidity and temperature, plant evapotranspiration, precipitation intensity, wind direction and speed, and relative pressure, to optimise the use of water and energy resources in agriculture (Suciu et al. 2019). Additionally, institutional reforms in terms of policies and practices and concurrently enhancing the knowledge, skills and capacity of the irrigation and drainage service delivery practitioners will go a long way in achieving the goals of food security (Watanabe et al. 2017).

6.3.3 Land Drainage Requirements

As the methods of water application are changing, climate change may bring up different drainage requirements, in terms of structures of practices. Irrigation without adequate drainage may result in land degradation via salinisation. Newer areas opening up for agriculture may need fresh drainage arrangements in different groundwater regimes. Management of drainage and drained water may require extra treatment for making the reuse and recycling of the same feasible.

6.4 Flood Management

The frequent changes in the hydro-climatic regime brought about by global warming are becoming increasingly visible. Rising mean sea levels are expected to increase the likelihood of coastal flooding, whereas changes in the magnitude, intensity and frequency of rainfall and runoff would increase the riverine flooding (Burrel et al. 2007). The floods, thus, not only make the human settlements vulnerable but also pose a great threat to agriculture, endangering food security. To combat this, climate change adaptation strategies for flood protection and risk management need to be factored in and integrated for efficient flood management.

6.4.1 Close Coupling with Land Management

Flooding and land use are closely correlated. The rapidly increasing population and the proportionately developing urbanisation are depriving the land of its properties of natural drainage and increasing the risk of flooding, especially in the urban areas.

According to WMO, some of the potential impacts on the water resources triggered by the land-use change in systems are listed below (WMO 2009):

- Higher frequency of floods and droughts.
- Increased soil erosion and loss of fertile soil.
- Higher pollution load in rivers resulting in deterioration of the water quality.
- A decrease in low flows impacts the regulation of the reservoir which can distress the reservoir regulation capacity and subsequently affects the water supply, irrigation, navigation and hydropower at different stages.
- Salinity intrusion, especially in the coastal region due to erosion.
- Increased vulnerability of life and property in the flood-prone areas.

6.4.2 Structural and Non-structural Measures

Historically, structural measures to capture floods have always been preferred due to their physical presence providing a sense of security. Structural measures such as the building of dams, reservoirs, levees, embankments, floodwalls, sea walls or natural detention basins are generally adopted for flood protection. Other structural measures include channel improvement, drainage improvement or diversion of floodwaters. These measures help, to a great extent, in impounding floods, protecting vulnerable areas and regulating the flow downstream; however, they cannot eliminate the risk of flooding entirely. Additionally, the structural measures pose adverse hydrological, morphological and environmental impacts as well as possess the ability to influence socio-economic development (Hamburg University of Technology 2021). Thus, to support the functions of structural measures, non-structural measures are devised as complementary or independent measures for flood protection. Some of the non-structural measures against flood risk management include flood forecasting and warning, land-use regulations through flood-plain zoning, flood proofing and flood insurance (Das 2007). Other non-structural measures include emergency preparedness, response and recovery. With the advancement in remote sensing, GIS and other hydrological and climatic models, flood forecasting has become an important tool in flood risk protection by issuing pre-warnings for preventive measures to be taken beforehand. It also allows the respective authorities to respond with respect to dam operations.

Traditionally, flood forecasting has been implemented with pre-established observation networks and for such areas that have been facing the flood furies from historical times. However, it is being observed that apart from such areas, the newer areas are becoming vulnerable on an infrequent and random basis. The year 2017 saw many medium and small size basins in the Indian state of Gujarat experience heavy rainfall and sudden floods with the reservoirs which used to remain usually empty or half-filled, discharging large surplus volumes and desertic areas getting submerged and waterlogged. Such phenomena are being noticed very routinely. This poses a challenge to the forecaster. Usually, flood forecasting relies heavily on the flow patterns observed in past and correlates them with the key events like rainfall, levels in the upstream reaches and travel times. However, in the situations described above, the information is not available at very short notice. This requires flood modelling relying on the rainfall data and mathematical modelling using the topographical information of the river course. It has been observed that such measures are reasonably effective in pre-warning the disaster management establishments for getting into readiness mode to lessen the impact of the disaster.

Similar situations are noticed in the hilly areas especially the middle and higher Himalayas where the formation of landslide dams and their breaching creates large disasters in rather inaccessible areas with poor logistics. India has recently witnessed two such events like the Kedarnath disaster of 2013 and the recent Chamoli disaster of 2021. Apart from these, there have been relatively smaller level incidents from 2003–04 onwards where pre-emptive modelling of the flood waves was able to prevent large scale damages and disasters. However, these disasters have pointed out the need for better surveillance and forewarning mechanisms with appropriate awareness building in the local disaster management authorities. Better collaboration is also needed between the development agencies to establish specific collective disaster warning measures for pre-warning. The impact generated by such phenomena is getting accentuated as the development progresses under pressures of economy and population and the variability induced through creeping climate change in sensitive areas. In some cases, like cyclones on the east coast of India, advance warning mechanisms from a cyclone progress point of view are well established giving promising results but similar options are yet to be established for the hilly regions.

6.4.3 Safety and Sustainability of Water Infrastructure

Floods not only pose challenges to human lives and property but also severely impact the safety and well-being of costly water infrastructure like dams and conveyance works for irrigation water. The flood risk to the dams is one of the greatest challenges to the well-being of the structures. Most of the dam failures in India have occurred due to lack of spillway capacity or faulty operation of gates and consequent catastrophic failures of the dam (Kumar 2021). While the dams were being planned in the early 1950s extending to the 1980s, the design provisions were largely based on historical floods observed and return period estimation through an extreme value distribution-based approach was quite prevalent. At present, the dams have increased their economic and social values manyfold. However, the reassessment of flood risks taking into account newly observed phenomena of intense rainfall events and probability-based envelope flood assessment (Probable Maximum Flood) is generating demands for rehabilitation where costly structural measures coupled with non-structural measures are needed to keep the hazards at acceptable levels while safeguarding the benefits from such dams. The climate change effects are currently not directly involved in these efforts but they have the potential of requiring additional investments and thereby putting a burden on the economy.

6.4.4 Adaptive Flood Risk Management

While structural measures have so far been considered remedial for risk management, non-structural and adaptive flood risk management approaches have gained grounds in the recent decades. It responds to the pertinent challenges faced by the engineers such as managing increasing flood risks with limited resources; evading adverse hydrological and environmental consequences of flood control projects and dealing with the uncertainty.

"*Adaptive management is a structured, iterative process of optimal decision making in the face of uncertainty, with an aim to reducing uncertainty over time via system monitoring*" (Emami 2020). As shown in Fig. 6.1, Watanabe et al. (2017), the adaptive flood management strategies entail qualities such as adaptability, flexible decision-making, monitoring and evaluation, adopting resilient technologies and approaches, adaptive learning and stakeholders' participation.

6.5 Role of Real-Time Data and Forecasting

The hydrological processes are dependent upon the global climatic processes and are not yet fully understood or modelled in quantitative terms. Therefore, the rainfall process is considered a random process and very little specific information about spatial and temporal availability of the rainfall is available with a reliable

Fig. 6.1 Strategies of adaptive flood management (AFM). *Source* Watanabe et al. (2017)

forecast model on which quantitative decisions like cropping areas and possible inflows can be generated well in advance. With this constraint, the role of data becomes invaluable to compare the trends and derive planning decisions therefrom using various stochastic models. In the absence of real-time data, trends cannot be discerned and thereby the reliability of food and energy production cannot be ensured.

With the real-time data-based strategies, comes the need for intensive sampling. It is widely appreciated that the spatial distribution of rainfall even among the medium-sized catchments is also highly uneven. While the development of rainfall-runoff models, the data available has to be consolidated in terms of averages and standard deviations. However, these values are strongly influenced by the sampling frequencies. It is often necessary to model smaller units of the catchments to capture the variability. For these elemental hydrological response units to reflect their true behaviour, the precipitation processes must be sampled within their domain. Consequently, the sampling intensity per unit area has to perforce go up. In many parts of the world especially the developing and least-developed countries, the data sampling rates in terms of the spatial and temporal resolution range from very poor to non-existent. Many times, modelling and planning are based on the global models which may have a poor resolution as regards the individual planning units are concerned. At present, the tendency is to provide such solutions to relatively small community units, which are reliant upon the small resource base as well. For example, the tendency is to provide for community ponds and check dams having catchment areas of hardly a score square kilometres to as low as few hectares. The resilience of such interventions based on large scale models is accordingly unreliable in a crop cycle or watering interval. For these units to succeed, decisions based on real-time data are essential. Climate change is a creeping process and is not likely to arrive in one shot. This provides us with an opportunity to establish such infrastructure as a part of the climate change adaptation programme. Establishment of infrastructure in terms of hardware and corresponding processing software modules and making communities learn about their consumption in their day-to-day agriculture water management activities is a time-consuming process and is better started as an adaptive strategy in time. This will be a challenge for the developing and least-developed world with high dependence on agriculture and small farmers. With the landholdings averaging around 1 ha, the individual farmers will have to manage their own data acquisition and transmission so that problem specific advice can be provided. It will be necessary to devise systems that can cater to them in local languages and their gradual migration to informed decision-making processes rather than custom governed practices.

With climate change bringing unprecedented changes in the global hydrological regime, the availability of freshwater resources is expected to shrink, especially in regions that are already water-stressed. Consequently, the food security and economic growth of these regions also face the menacing impacts of climate change. In this regard, data science can play a vital role in the planning and decision-making processes of holistic water resources management to ensure water security and complementing food and energy security (Pandya and Sharma 2021). Thus, scientific research, policy-making and project implementation need to stay ahead of the curve to ensure viable agricultural water management.

Considering the major user of freshwater resources, the climate change adaptation strategies need to be adopted in the agriculture sector to maximise crop water productivity as well as water use efficiency, reduce the water losses and manage the drought-affected regions. Accordingly, the multi-disciplinary aspect of data science caters to simulating, determining and optimising the crop water demand under different scenarios and maximising the outputs. The characteristic data pertaining to irrigation development may include physical data of the catchment, hydrometric data, meteorological and climatic data, agricultural, potable and wastewater, industrial, navigation, hydroelectric power, environmental, demographic, institutional and economic data (Molden and Burton 2005).

6.6 Efforts at ICID

While dealing with issues as immense as climate change which has demonstrated global impact, the role of international institutions is as crucial as it is manifold. The International Commission on Irrigation and Drainage (ICID 2021a), similarly, provides a platform and a network to the global community working within the irrigation and drainage domain. Working with government institutions, international organisations, multilateral establishments, private firms and professional experts, the ICID network represents more than 95% of the global irrigated area. Through its various working groups, task forces, national committees and partners, ICID's main activity is to promote three core areas viz, irrigation, drainage and flood management by addressing their engineering, agronomic, environmental, social, financial and institutional aspects. For more than seventy years, ICID network has symbolised the share and exchange of knowledge and technology for agricultural water management (AWM). Under the limiting natural resources, climate change and rising conflicts, the task of the ICID network has become even more critical and daunting. The newly emerging and competing demands for water, coupled with the uncertainty of the impact of climate change on food productivity, have challenged the ICID stakeholders and partners to redouble their efforts (ICID 2017).

Right from its inception, ICID has taken up the integration of haves and have-nots in bringing about a change in agricultural water management policies. A study of the composition of the membership at inception and after 70 years at present reveal that ICID has a well-balanced mix of highly developed countries and regions like the USA, Canada and Australia and extending to developing countries like Somalia for example enabling the network to address problems at every level.

ICID consists of several working groups enveloping the core areas of irrigation, drainage, flood management, climate change and agricultural water management, rural development and sustainable on-farm irrigation development. Some of the efforts to support combating climate change directly or indirectly carried out at ICID are given in the sections below.

6.6.1 ICID Vision 2030

With a vision to create a water secure world free of poverty and hunger through sustainable rural development, in 2017, ICID released its publication: "A Roadmap to ICID Vision 2030". The six organisational goals are aligned along the lines of Sustainable Development Goals (SDG) of the United Nations, especially contributing to water and food security through sustainable agricultural water management in the era of climate change and global warming.

- Enable higher crop productivity with less water and energy.
- Be a catalyst for a change in policies and practices.
- Facilitate exchange of information, knowledge and technology.
- Enable cross-disciplinary and inter-sectoral engagement.
- Encourage research and support development of tools to extend innovation into field practices.
- Facilitate capacity development.

With climate change triggering complexities in the AWM, the interdisciplinary and multi-stakeholder aspect of AWM needs to be understood. Seven out of seventeen SDGs contribute to AWM (ICID 2017). Additionally, ICID Vision 2030 recognises the water-food-energy nexus and the need for stakeholders and users from these spheres to understand the synergies and trade-offs whilse planning and management of water resources. Keeping this in mind, the organisational goals laid out by ICID address the issue of AWM from a broader perspective, well-understanding the role of climate change and global warming.

6.6.2 Global Footprint

The National Committees (NCs), representing 78 countries and nearly 95% of the world irrigated area, constitute the core stakeholders of ICID. The NCs, hosted in the government departments or ministries, research institutions, universities, private sector companies, and in some cases farmers' groups, consisting of the ministries, public and private institutions with reigning expertise in water resources, irrigation, agriculture, rural development, hydropower, environment and flood management sectors, as well as finance and economics theme. Collaboratively with ICID and independently, the NCs play a key role in supporting the sustainable development agenda within their countries, in line with the ICID Vision and Mission. Collectively, the NCs aim to achieve the goals of food security, water security, poverty alleviation and rural development, especially in the face of climate change.

6.6.3 Promoting and Disseminating Knowledge on Water Saving Techniques

As a result of climate change and other anthropogenic activities, water scarcity has been observed in many parts of the world, and consequently, agriculture is facing severe pressure to increase water productivity in irrigated and rainfed agriculture, i.e. produce more food using less water per unit of output. To achieve this goal, appropriate water-saving technologies, management tools and policies need to be adopted. Thus, institutions across the world are encouraging workers and professionals involved in irrigation water management—policymakers, managers and farmers—to conserve water through appropriate policies and incentives. Similarly, by virtue of WatSave awards (ICID 2021b), ICID identifies, catalogues, awards and promotes cutting-edge research, management tools and conducive policies and practices to promote water saving in agriculture and to minimise wastages to mitigate negative environmental impacts. By design, the WatSave Awards also provide a unique opportunity to experts, innovators, young professionals and farmers, equally, to demonstrate myriad ways for climate change adaptation and water saving through modernisation, technological innovations and advanced management strategies. The awards given to individuals or a team of individuals are made in respect of actual realised savings and not for promising research results, plans and/or good ideas/intentions to save water (ICID 2021b).

The awards have been given to innovators working with numerous facets of agricultural water development and management such as micro and drip irrigation, controlled irrigation, development of new hardware and software techniques, simulation models and improvement in existing project infrastructure for water conservation. Additionally, practices to generate mass awareness and capacity building including Participatory Irrigation Management (PIM) are promoted and awarded through the awards towards the efforts of water saving in agriculture.

6.6.4 Understanding Effects of Climate Change

Since agriculture accounts for 70% of water use and up to 30% of greenhouse gas emissions, it contributes to and is threatened by climate change. Adapting water management policies and practices to the existing climate variability by building resilience is the best way to prepare for potential climate change. Improved water harvesting and storages (such as reservoirs, dams, pools, pits and retaining ridges, etc.), supplementing the water requirement for rainfed crops, highly efficient irrigation systems and best practices are fundamental for addressing the increasing variability of rainfalls and reducing the adverse impacts of extreme events of floods and droughts. ICID addresses these issues through its Working Group on Global Climate Change and Agricultural Water Management (WG-CLIMATE). The looming climate change and its likely impacts on water management for agriculture require cooperation cutting across institutional and disciplinary boundaries. It calls for intensification of data collection networks, research into methodologies to downscale the climate impacts on water and agriculture, review of the operation of storage systems, enhancing soil water storage with water harvesting structures, and sharing knowledge and information. WG-CLIMATE has the mandate to review the progression of and predictions for Global Climate Change (GCC) and climate variability and to explore and analyse the medium-term implications of climate change and climate variability for irrigation, drainage, and flood management. It stimulates discussion and raises awareness of water-related GCC issues within the ICID network and at national scales among scientists and policymakers. The working group collaborates with global partners like the UN System-wide Global Framework for Climate Services (GFCS) under the leadership of WMO.

6.6.5 Water Heritage and Sustainability

With the advent of more sophisticated technology and tools in the field of agri-culture and water management, modernisation is as essential as it is inevitable. However, while adopting such state-of-the-art innovative technologies and approaches is crucial, recognising the evolution of irrigation and water management throughout history holds a very significant role moving forward.

For centuries, food security by the virtue of sustainable agriculture has been the reason behind the rise and descent of numerous civilisations. The sustainability of these agricultural systems paved the way for the sustenance of the population and subsequent economic development of the region. ICID, through its programmes such as World Heritage Irrigation Structures (WHIS) (ICID 2021c) and World Water Systems Heritage (WSH) (ICID 2021d), has instituted mechanisms to honour and preserve the tangible and intangible water heritage, respectively. To trace the history and understand the evolution of irrigation in civilisations across the world, the WHIS programme recognises historic irrigation and drainage structures such as dams, barrages, water conveyance and storage structures, to name a few. Some of the recognised structures are as old as the 2nd Century BC, providing us with pearls of wisdom from that era on how agricultural water management was carried out and how the irrigation structures have proved their sustainability for several millennia. The WSH programme, on the other hand, focuses on the people-centred water management systems, organisations, regimes and rules considered to be of out-standing value to humanity that creates a coexistent social system for humanity and a sound environment and giving them recognition. In addition to this, ICID also has a Working Group on History of Irrigation, Drainage and Flood Control (WG-HIST) with a mandate to promote the interdisciplinary exchange of information, knowl-edge and experience, as well as networking on the topic (agricultural, political, socio-economic, climatologically and geographical, aspects) for proper under-standing of the technological developments on the subject via its National Committees (ICID 2021c, d).

The water heritage needs to be preserved and appreciated because of its contribution to the evolution of agriculture and humankind. The wisdom attained on sustainable agricultural practices which have survived through centuries has proved to be the foundation block of modern agriculture and water management.

6.7 Conclusions and Way Forward

While talking about the future, it is inevitable to exclude climate change from the dialogue, be it agriculture or any other sector of the economy. Irrespective of the localised sources of greenhouse gas emissions, the impacts of climate change are pervasive and visible everywhere. Moving forward, planning for economic growth needs to consider climate change impacts and appropriate coping mechanisms to ensure avenues to mitigate the impacts of carbon emissions globally.

Solving the increasingly pervasive and inextricable global warming crisis does not entail a one-step resolution, rather it involves an integrated approach involving systematic implementation of measures. This entails the adoption of climate-friendly technological solutions, using software and hardware tools available to manage the adverse impacts of climate change, especially to land and water resources, enriching knowledge, encouraging research and development and enhancing capacity development efforts across all institutions and generating mass awareness through climate education and activism.

As discussed before, in the era of rapidly changing climate, unprecedented changes are expected in the hydrological regime globally. As far as agricultural water management goes, the climate change effects will have to be integrated into the practices and solutions so that the outputs and outcomes do not get disrupted by the change. The key action here is to devise pre-emptive strategies to maintain the output levels and wherever possible, find the options which can enable communities to grow out of excessive dependence on increasingly skewed inputs in terms of water. Developments in associated fields of technology like better information and communication technologies (ICT) and developments in biological systems have potential which have to be nurtured by recognising them and propagating them to turn them into practices from being demonstrations. Newer challenges in disaster management and asset management due to extreme weather events randomly spread over space and time need continual reviews and fast responding approaches for managing the potential disasters.

References

Burrel B, Davar K, Hughes R (2007) A review of flood management considering the impacts of climate change. Water Int 32(3):342–359. https://doi.org/10.1080/02508060708692215
Das S, Gupta R, Varma H (2007) Flood and drought management through water resources development in India. WMO Bullet 56(3). https://public.wmo.int/en/bulletin/flood-and-drought-management-through-water-resources-development-india. Accessed 20 Apr 2021

Emami K (2020) Adaptive flood risk management. Irrig Drain 69(2):230–242. https://doi.org/10. 1002/ird.2411

FAO (Food and Agriculture Organization) (2017) Water management for climate-smart agriculture. Climate smart agriculture sourcebook. http://www.fao.org/climate-smart-agriculture-sourcebook/production-resources/module-b6-water/b6-overview/en/?type=111. Accessed 20 Apr 2021

Hamburg University of Technology (2021) Flood manager E-learning tutorials: integrated flood management (IFM). http://daad.wb.tu-harburg.de/tutorial/integrated-flood-management-ifm-policy-and-planning-aspects/. Accessed 20 Apr 2021

ICID (International Commission on Irrigation and Drainage) (2017) A roadmap to ICID vision 2030. https://www.icid.org/icid_vision2030.pdf. Accessed 20 Apr 2021

ICID (International Commission on Irrigation and Drainage) (2021a) http://icid-ciid.org/home. Accessed 20 Apr 2021

ICID (International Commission on Irrigation and Drainage) (2021b) Awards and recognition. WatSave awards. http://icid-ciid.org/award/watsave/43. Accessed 20 Apr 2021

ICID (International Commission on Irrigation and Drainage) (2021c) Awards and recognition. World heritage irrigation structures. http://icid-ciid.org/award/his/44. Accessed 20 Apr 2021

ICID (International Commission on Irrigation and Drainage) (2021d) World water system heritage programme. http://icid-ciid.org/inner_page/45. Accessed 20 Apr 2021

IPCC (Intergovernmental Panel on Climate Change) (2007) Working Group II: impacts, adaptation and vulnerability. IPCC fourth assessment report: climate change 2007. https:// archive.ipcc.ch/publications_and_data/ar4/wg2/en/ch3s3-5-1.html. Accessed 20 Apr 2021

Konapala G, Mishra AK, Wada Y, Mann ME (2020) Climate change will affect global water availability through compounding changes in seasonal precipitation and evaporation. Nat Commun 11:3044. https://doi.org/10.1038/s41467-020-16757-w

Kumar M (2021) Dam safety in India: dam rehabilitation and improvement project (DRIP). https:// damsafety.in/ecm-includes/PDFs/DRIP_II_Presentation/Dam%20Safety%20in%20India.pdf. Accessed 20 Apr 2021

Lindsey R, Dahlman LA (2021) Climate change: global temperature. https://www.climate.gov/ news-features/understanding-climate/climate-change-global-temperature#: ~ :text=According %20to%20NOAA's%202020%20Annual,more%20than%20twice%20that%20rate. Accessed 20 Apr 2021

Molden D, Burton M (2005) Making sound decisions: information needs for basin water management. In: Svendsen M (eds) Irrigation and river basin management: options for governance and institutions. International Water Management Institute (IWMI), Pelawatte, Sri Lanka, pp 51–74

Muir MJ, Luce CH, Gurrieri JT, Matyjasik M, Bruggink JL, Weems SL, Hurja JC, Marr DB, Leahy SD (2018) Effects of climate change on hydrology, water resources, and soil. In: Halofsky JE, Peterson DL, Ho JJ, Little NJ, Joyce LA (eds) Climate change vulnerability and adaptation in the intermountain region. Department of Agriculture, Forest Service, Rocky Mountain Research Station, Fort Collins, CO, U.S., pp 60–88

Pandya AB, Sharma P (2021) Importance of data in mitigating climate change. In: Pandey A, Kumar S, Kumar A (eds) Hydrological aspects of climate change. Springer, Singapore, pp 123–137

Suciu G, Uşurelu T, Bălăceanu CM, Anwar M (2019) Adaptation of irrigation systems to current climate changes. In: Abramowicz W, Paschke A (eds) Business information systems workshops. Springer, Cham, pp 534–549

Watanabe T, Cullmann J, Pathak C, Turunen M, Emami K, Ghinassi G, Siddiqi Y (2017) Management of climatic extremes with focus on floods and droughts in agriculture. Irrig Drain 67(1):29–42. https://doi.org/10.1002/ird.2204

WMO (World Meteorological Organization) (2009) Flood management in a changing climate: APFM technical document No. 14. Flood management tools series. https://library.wmo.int/ doc_num.php?explnum_id=7330#: ~ :text=CLIMATE%20CHANGE%20IMPACTS%20ON %20THE%20FLOOD%20FORMATION%20PROCESS,-20.&text=A%20variety%20of% 20climatic%20and,%2C%20volume%2C%20timing%20and%20phase. Accessed 20 Apr 2021

Chapter 7
Designing Research to Catalyse Climate Action

Bruce Currie-Alder and **Ken De Souza**

Abstract Climate action ahead of 2030 requires ambitious research that is fit for purpose: working across scale, creating synergy among cohorts of projects, and enabling capacity to pursue research uptake. Research needs to bridge local and national levels and provide evidence that informs decisions with decadal implications. To become more than the sum of its constituent activities, research programmes and consortia require learning frameworks and equitable partnership among participating organisations. Beyond scholarships and fellowships for training and independent study, exchanges and embedding in real-world settings practical experiences allow people to gain experience beyond academia in diverse host institutions. Greater emphasis needs to be given to the spectrum extending from research to its application, including co-production and knowledge brokering with local people and decision-makers.

Keywords Climate adaptation · Climate resilience · Capacity building · Scale · Collaboration · Research design

7.1 Introduction

The world has entered a decisive decade of climate action. Greater ambition is needed to cut carbon emissions in half by 2030 and achieve carbon neutrality by 2050. Current commitments are insufficient and on track to over 3°C warming. Even with ambitious action to hold the increase in the global average temperature to well below 2 °C, many places will surpass the threshold of 1.5°C warming. This necessitates adaptation now and more in the coming years. It is an age of imple-

B. Currie-Alder (✉)
International Development Research Centre, Ottawa, Canada
e-mail: bcurrie-alder@idrc.ca

K. De Souza
Foreign, Commonwealth & Development Office, London, UK
e-mail: ken.desouza@fcdo.gov.uk

© The Author(s) 2022
A. K. Biswas and C. Tortajada (eds.), *Water Security Under Climate Change*,
Water Resources Development and Management,
https://doi.org/10.1007/978-981-16-5493-0_7

mentation, from the global stocktake to the wise use of climate finance, building global climate resilience no longer waits upon rounds of global climate negotiations. And being resilient by 2030 is not enough, as people need to anticipate how to survive and thrive in the future climates by 2050. The choices made now can widen or constrain those opportunities moving forward.

How can research best catalyse climate action? The authors have wrestled with this question as our funding agencies learn from the past two decades and decide how to guide the next decade of research investment. This chapter shares insights arising from our reflection over 2018–20, including evaluation reports, scoping studies, learning reviews, conversations with principal investigators and contributions to the forward-looking strategies of our organisations. The impacts of climate change are, and will be, disproportionately experienced by marginalised and vulnerable groups. There is a moral imperative to support communities to adapt and be resilient by addressing gaps in knowledge, acting on what is already known, and enabling local capacity for further action. Research needs to go beyond assessing risks and identifying impacts, to find solutions that are user-centred and action-oriented. In short, the ways of organising research need to respond to the urgency and ambition of the climate crisis.

7.2   Recent Directions in Research

Scholarship and practice of water security and climate adaptation have evolved over the past two decades. Many water systems are now considered "non-stationary" such that they no longer reliably replicate past patterns. Historic observations are not a reliable guide to current and future patterns of precipitation, runoff and storms. Instead, the Anthropocene is characterised by altered landscapes and watercourses with changing risks of drought, floods, landslides and coastal erosion. Beyond the cascading effects of climate on water, this non-stationarity also stems from fragmenting mosaics of land use and the cumulative impact of the past century's efforts to store, divert and use water. The goal of water management has moved from optimising predictable and stable systems, to instead ensure robust and flexible responses to evolving complex systems. Approaches to decision-making and management must account for deep uncertainty and consider a range of possible futures (Smith et al. 2019).

Water security is created through relationships within society that enable people to enjoy water-related services. Residents do not value their municipal water plant or local water seller per se but rather the benefits they derive from using water for drinking, cooking and washing. Security is not simply a perimeter fence around such facilities but the results of efforts to administer municipal water, including adequate budgeting, operations and delivery. Security is embedded in the social, cultural and political relationships to water that enable people to lead a life that they have reason to value (Jepson et al. 2017). A similar turn can be seen within thinking and practice of climate adaptation, moving away from framing based on solutions and technical fixes, and towards appreciating adaptation as the lived experience of communities and as a means of addressing equality and inclusion (Nightingale et al. 2019; Pelling and Garschagen 2019).

Climate adaptation is a process of adjustment to actual or expected climate and its effects, how people and society seek to moderate harm or exploit beneficial opportunities (IPCC 2018). Adaptation efforts have strengthened over the past two decades under the United Nations' Framework Convention on Climate Change (UNFCCC). Some milestones include the emergence of the Least Developed Countries Expert Group to provide guidance on adaptation plans, establishing an Adaptation Knowledge Portal, and creation of the Adaptation Committee and the Green Climate Fund (UNFCCC 2019). Since 2019, the United Nations' Climate Action Summit and the Global Commission on Adaptation have elaborated the business case for investing in adaptation, highlighting opportunities spanning food and land use, cities and human settlements, water and nature, industry and infrastructure, disaster risk and local use of finance. Looking ahead, a global stocktake is expected to occur every five years, starting in 2023, which will assess the adequacy and effectiveness of adaptation and support provided for it, as well as review the overall progress towards the global goal on adaptation (Article 7 of the Paris Agreement).

Research topics and questions have evolved over time. The quantity and breadth of climate change research has exploded such that it is increasingly difficult for any one person or team to keep up with the frontiers of knowledge (Minx et al. 2017; Callaghan et al. 2020; Nalau and Verrall 2021). One response is the use of evidence maps and systematic reviews to map the evolving network of ideas and actors. While understanding climate risks and defining adaptation were major topics prior to 2010, the following decade emphasised assessing progress, enabling conditions and implementation (Klein et al. 2017). Under the World Climate Programme over thirty research priorities were identified for the previous decade, including: factors that support or hinder adaptation, improving the ways researchers and policymakers interact, learning from developing countries, understanding hotspots (such as coasts, semi-arid regions and mountains), and co-generating knowledge thru partnerships, and supporting collaboration across spatial scales (Rosenzweig and Horton 2013). The field has since matured in both theory and practice. Multiple authors have contemplated the future of research on climate adaptation, calling for greater opportunity for early career researchers and southern authors, embedding equity in research practice and improving upon stakeholder engagement (Mustelin et al. 2013). Desired changes include increased transparency and consultation in research design, demonstrating tangible impact for the lives and livelihoods of local people, and more effective knowledge brokering and learning (Jones et al. 2018). Research must be "more radical, bolder, more experimental" (Klein et al. 2017), engaging the roles of power and equality, the interplay of adaptation and climate justice (Newell et al. 2020), and connected to learning with marginalised people (Eriksen et al. 2021).

7.3   Designing Research for the Next Decade

This chapter identifies three features for research to be fit for the decade ahead to 2030, drawing on the experience of multiple programmes (Table 7.1). In a sentence, it is critical to embed research into climate action by working across scale in cohorts to enable capacity. In terms of cross-scale, bridging local and national-levels offers promising opportunities for research uptake, particularly addressing an evidence gap to inform decisions with decadal implications. Larger projects and programmes should articulate learning questions as the basis for monitoring and evaluation. In terms of cohorts, research programmes and consortia involve multiple activities and participating organisations. To become more than the sum of its constituent projects, such programmes require a framework that link these activities and must foster equitable partnership among participating organisations. In terms of capacity, beyond scholarships and fellowships for training and independent study, exchanges and placements in real-world settings allow people to gain practical experience beyond academia in diverse host organisations. Greater emphasis needs to be given to enabling capacity across a spectrum extending from climate science to services, including the ability to engage in co-production, knowledge brokering and research uptake with local people and decision-makers.

7.3.1   Cross-Scale

The next decade of research needs to be more intentional in addressing and bridging different levels, connecting across near- and medium-term time horizons, as well as the local-to-global expanses of geography. Country-level context remains key in determining domestic policy and action, as well as national contribution to global commitments. More research is needed for and with least developed and most vulnerable communities, especially in Africa and western Asia. Research also needs to address the evidence gap at the decadal time scale, to both enhance resilience by 2030 and navigate the climate futures towards 2050. There are ways to organise research to realise synergies between projects and programmes, by establishing common approaches to and platforms for data sharing, knowledge management and research uptake. Nested theories of change provide a focus for learning as well as describing how activities are intended to produce research results and contribute to society.

Scale describes the extent of a phenomenon across time and space. Weather events and environmental change occur over days, seasons, years and decades. Water systems and climate impacts extend from local communities, through larger watersheds, to continental and global systems. Beyond the natural world, spatial scale can also describe the levels used to organise society, ranging from municipalities or districts, through state or provincial boundaries, to the nation-state or regional grouping. "Scale" refers to any one of these dimensions—time, geography

Table 7.1 Comparison of select research programmes

Name	Description	Structure and budget	Main countries covered
Collaboration Adaptation Research Initiative in Africa and Asia (CARIAA)	Build resilience for people who live in high-risk climate hotspots	4 consortia 60 organisations £43 million/ 6 years	Bangladesh, Botswana, Burkina Faso, Ethiopia, Ghana, India, Kenya, Mali, Namibia, Nepal, Pakistan, Senegal
Conflict and Cooperation in Management of Climate Change (CCMCC)	Evidence on how climate change affects conflict or cooperation	7 projects 39 organisations £5 million/ 5 years	Bangladesh, Burkina Faso, Cambodia, Ghana, India, Kenya, Mexico, Myanmar, Nepal, Vietnam
Ecosystem Services for Poverty Alleviation (ESPA)	Evidence needed for sustainable ecosystem management	125 projects 922 researchers £44 million/ 9 years	53 countries
Future Climate for Africa (FCFA)	Generate new climate science and ensure it has an impact on human development	5 consortia plus knowledge exchange 72 organisations £19 million/ 7 years	Botswana, Burkina Faso, Kenya, Malawi, Mozambique, Namibia, Senegal, Tanzania, Uganda, Zambia, Zimbabwe
Science for Humanitarian Emergencies and Resilience (SHEAR)	Improved forecasting and decision-making for disaster preparedness and response	4 consortia plus knowledge broker 37 organisations £23 million/ 5 years	India, Kenya, Mozambique, Nepal, Uganda
Weather and climate Information and Services for Africa (WISER)	Quality, accessible and useable weather and climate information services for sustainable development	support for national meteorological agencies £34 million/ 6 years	Burundi, Ethiopia, Kenya, Rwanda, Tanzania, Uganda

or administration—while "level" refers to a unit of analysis within a given scale (Cash et al. 2006). For example, city and national governments span multiple levels of human geography, while the interplay of watersheds and water policy is cross-scale between the natural and human geography. Addressing climate action and environmental change often involve a "mismatch" between the scale of problems and how society distributes responsibility for them. For example, salmon fisheries extend from the upper reaches of watersheds out to the open ocean, involving actors from municipal to federal jurisdictions, and activities from forestry to transport sectors. This requires coordination among actors to overcome fragmentation of higher-level systems or contradictory actions emanating from different levels.

The national level remains key within climate adaptation and resilience. Parties to the UNFCCC are nation-states, which undertake national adaptation planning to organise their own response to a changing climate. They also identify Nationally Determined Contributions (NDCs) as their country's commitment toward the collective goals of the convention. Countries report on their progress through national communications, describing legislation and other means to reduce emissions, adapt to climate risks and assist other Parties. Climate Action Pathways encourage further commitments from non-party stakeholders such as cities, businesses and civil society. As research strives to be demand-led and action-oriented, an essential entry point is understanding who are the national-level actors responsible for NDCs: what are the decisions they are grappling with, what are the challenges faced in implementation and what forms of evidence and knowledge would be useful to them (Moosa et al. 2019). For example, research on climate impacts and how people respond helped inform Bangladesh's delta plan 2100, an ambitious national effort to guide investment to address flooding, land use, urban planning across the country over several decades.

Yet not all countries receive the same level of attention. Relatively less research has been conducted in smaller or more isolated countries and those with higher proportions of the population living in extreme levels of poverty. A review of the past decade of peer-reviewed articles in three prominent academic journals found relatively few articles on climate adaptation in large swathes of central, western and northern Africa or western Asia (Vincent and Cundill 2021). Certain countries feature prominently across multiple research programmes, while also leaving out large sections of the global south (Table 7.1). Over the next decade, there is a clear imperative to invest in research for and with the poorest and most vulnerable countries, especially those in the least developed countries group and Climate Vulnerable Forum (see LDC Climate change, n.d., and Climate Vulnerable Forum, n.d.). Such would be in keeping with the Paris Agreement mentions of developing countries that are particularly vulnerable to the adverse effects of climate change, and the 2030 Agenda pledge that "no one will be left behind… and to reach the furthest behind first".

Working across a range of scales and levels can reinforce each other. National-level actors may be initially reluctant to engage researchers simply to explore potential collaboration and discuss needs. Yet these same actors may be keenly interested to learn from research that provide insights from relevant local-level experience. For example, work in Botswana and Namibia drew upon practical actions at the district level to connect with the national policy processes on drought and vulnerability (Morchain et al. 2019). Local-level experience can also feed into global-level debates. For example, researchers have drawn on local empirical findings to present evidence to the United Nations' Commission on the Status of Women (ASSAR 2019). Such results were also compiled to show how environmental degradation affects women's agency across climate hotspots (Rao et al. 2019). Successful research teams not only integrate multiple disciplinary perspectives but share evidence from different locations, reinforcing learning from local-to-global levels. Larger-scale research efforts bridge global-level science of peer-reviewed literature and climate modelling, with local experience of communities striving to improve their lives and livelihoods in the face of flooding, drought and other extremes.

Yet there is a missing middle in the scale of evidence. The published literature and available datasets are strongest at extreme time horizons: the near-term opportunities for the next growing season and the risks of storms or drought over the coming months, as well as the long-term magnitude of climate change beyond 2080 under different emission scenarios. Yet the state of knowledge is relatively weaker over the medium-term period of multiple years and decades, precisely the time horizons that are most critical for assessing the consequences of policy and investments (Jarvie et al. 2020). More work is needed to address the uncertainties regarding the future climate from 2030 to 2050, and how these differ from each other. This is the scale in which people grapple with decisions over how to respond to climate risk, assessing the viability of infrastructure, economic activities and adaptation measures over time. Science is currently best able to forecast the weather in the coming months and the potential state of the world for future generations, yet the most useful scale lies between these two extremes.

Combined with greater attention to time horizons of 2030 and 2050, there is a similar "sweet spot" in the middle of spatial scales vital for understanding climate impacts and risks. The current literature speaks more to global-level projections and local-level case studies, offering relatively less insights into sub-national levels such as downscaled climate projection or how particular biomes and landscapes will alter and shift. There is a disconnect between what science can easily speak to and the context within which governments, investors and entrepreneurs decide how to act and invest in their future. Diversified and specialised data products and climate services can respond to different political and practical needs, making information available in formats and at scales that supports adaptation decision-making (Adaptation Committee 2020). Research must address the evidence gap to inform decisions with decadal implications. The mid-range is ultimately more useful for informing how society can transform over time and become more resilient to future states of the world.

A further understanding of scale concerns the size of research effort and the value expected to result from it. The size of research effort can range from an individual investigator working in a single location, through to large teams working across multiple labs and field sites. Understanding the impacts of climate change over a continent requires a large size of a research effort. The study of complex problems also requires multiple skills and disciplines, giving rise to inter- and trans-disciplinary approaches to research. The ambition of research funding has grown over the past decades to support sizable research efforts needed to address complex societal challenges. At the same time, this growth has raised expectations of generating greater value in the form of economic and social benefit, including a wider set of options and opportunities for the future. Funding agencies have articulated a desire for research projects to move beyond describing problems or testing pilot technologies, towards identifying and "scaling" solutions through implementation and replication.[1]

Achieving a sizable research effort can come through synergy among different projects. A research programme or consortium can permit different projects and activities to "speak" to each other by establishing common features in project design, articulating an overarching theory of change and fostering ongoing communication among participants. Research projects often originate as a response to a funding competition that identifies a topic of interest, eligibility and selection criteria, a range of acceptable expenses and budgets, and a deadline for submission. The competitive nature of such calls means that proposals are developed separately with little coordination among prospective applicants. One promising way to better catalyse climate action is for funding agencies to be more intentional in such future calls for proposals, providing guidance on shared features such as high-level research questions, methods or datasets.

A programme or consortium can also provide guidance on, and support for, such aspects as data sharing, stakeholder engagement and knowledge brokering. For example, CARIAA participants elaborated a shared framework for how to understand and pursue research uptake (Prakash et al. 2019). Articulating a "theory of change" identifies overall goals, assumptions regarding how research activities and outputs link to broader outcomes and impact, as well as how to monitor performance and assess success. Such a logic model provides guidance for projects and activities, identifying how they fit together and contribute toward a common goal, as well as inspiring more detailed theories specific to the project level. A theory of change also provides a basis for evaluation and learning, which can address the underpinning assumptions and the extent to which these are borne out by experience.

Communication is vital to bridge different levels of research effort and connect participants located in different organisations and locations. Some programmes established a knowledge manager or exchange unit to compile information coming from various projects and generating common outputs. CARIAA had a knowledge management platform as an online location for participants to share files, convene web conferences, coordinate schedules. Quarterly newsletters or weekly digests kept participants informed about recent events and publications, forthcoming meetings, celebrated achievements and foster the identity of being part of the programme. Given the dispersed nature of large-scale projects and programmes, such online platforms are vital for keeping participants engaged and motivated. Coordinating large-scale research efforts requires weaving together internal

---

[1] There remains a need to consider which scale of research effort is best suited to making meaningful progress on different problems of societal and scientific interest. There is some evidence that small teams can be more radical and challenge existing knowledge, while larger teams are better positioned to coordinate across countries and consolidate existing knowledge (Wu et al. 2019). Producing the IPCC sixth assessment report involved more than 700 experts, while most journal articles involve less than a dozen authors.

knowledge management across participants, monitoring-evaluation-learning efforts to track and reflect on progress, and external knowledge brokering to engage and reach stakeholders (Harvey et al. 2019a). These functions are vital if a project or programme is to become a cohesive whole and participants feel part of something larger. They also help set the stage for synthesis of research results, enabling participants to better articulate their collective contribution to global and regional agendas, such as the World Climate Research Programme, the Intergovernmental Panel on Climate Change or World Adaptation Science Programme.

7.3.2 Cohorts

A second facet of organising research concerns cohorts, gathering together different efforts to realise synergies. Research programmes and consortia involve multiple activities and participating organisations. To become more than the sum of its constituent parts, such programmes require a framework that link these activities and must foster equitable partnership among the organisations involved. Creating a cohort is more than simply synchronising the start-up and lifespan of a set of projects. It requires building opportunities for collaboration among projects as well as fostering relationships among diverse participants across these projects.

Programmes and consortia can achieve coherence among research efforts at different levels. At the project-level, funding is organised around specific activities which encourage diverse participants to come together to define and jointly manage a large budget. For example, the CARIAA call for proposals provided some guidance on how to structure research projects, specifying that each application needed to focus on semi-arid lands, river deltas or glacier-dependent basins. Yet prospective applicants were left ample latitude to define their own research activities and methods and geographic coverage. Consortia are large and complex projects which require strong leadership and coordination to set the research agenda, manage the contributions of participants, share resources and track and report on progress. The skill set and time required often surpass the ability of a single principal investigator. Successful consortia appoint a full-time manager or convener to work alongside the principal investigator and establish shared management structures that involve representatives from the partner organisations. For example, the Adaptation at Scale in Semi-Arid Regions (ASSAR) consortium was led from the University of Cape Town working jointly with START International, the University of East Anglia, Oxfam Great Britain and the Indian Institute of Human Settlements. Representatives from each of these organisations served a co-PIs, meeting at least monthly to jointly guide implementation.

The programme-level needs to provide some degree of coherence among different projects, a set of ideas or logic within which individual activities are situated. For example, the CARIAA was based on climate change hotspots or landscapes experiencing pronounced climate impacts and that are home to vulnerable, poor or marginalised people (De Souza et al. 2015). While each consortium tailored its own unique approach, they were interconnected by a programme-level theory of change and learning framework which established some overarching questions, as well as cross-consortium working groups which fostered collaboration among these teams. A share of the overall budget was then dedicated to integrating research in various ways, providing additional funding to incentivise participants to build on common interests or pursue novel collaboration across projects. For example, CARIAA supported additional efforts to contribute to the preparatory process for the Global Compact on Migration and to provide timely inputs into the IPCC Special Report on 1.5°C warming (O'Neill 2020). The programme utilised both intentional and emergent design, establishing collaborative spaces to bring participants together on foreseeable aspects while also maintaining an ability to respond as new opportunities arose (Cundill et al. 2018). To put it simply, programmes and consortia fit together their constituent projects and activities in ways that provide new insights and synergy.

Research programmes and consortia also need to provide guidance on the nature of partnership expected and nurture it over time. A partnership is the "ongoing, principles-based working relationships between diverse stakeholders, where solutions are co-designed and delivered, where each partner contributes a range of resources based on their strengths, commits to mutual accountability in return for mutual benefit, and where risks and benefits are shared" (Mundy 2020). Each project can be considered a partnership, whether as a small team of four individuals within a single faculty department at a university, or a large consortium involving more than a hundred individuals spread across several organisations. Whenever different people expect to cooperate on a joint endeavour, it can be useful to establish some common norms and practices. These become even more vital in situations when individuals work at distance from each other and do not regularly meet to sort out issues.

Where projects are identified through competitive calls for proposals, these have tended to privilege narrow considerations of scientific rigour and project management: assessing the academic merits of a proposal, the individual track records of participants in producing research, and a detailed work plan with a division of labour. Yet an emphasis on demonstrating why funding is needed and how it will be used can ignore how partners expect to work together. Future research needs to provide more explicit attention to partnership building during application and inception of new projects, considering matters related to project management, scope, research practices and communications (Martel 2020; Dodson 2017). Experience has shown that proposal development should also dedicate time and effort to clarifying expectations and ways of working among the people and organisations involved. Proponents can clarify norms and procedures for deciding on research design, methods and data management; access to funding and opportunities for travel; and the use of non-English languages in coordination and communication. Partnerships can only benefit if it is clear how decisions are made and who is involved.

Funding programmes can provide guidance on the nature of partnership and ensure partners establish and maintain a sense of fairness and equity. This starts with the research commissioning process. UKRI encourages research partnership to clearly articulate an equitable distribution of resources, responsibilities, efforts and benefits; recognise different inputs, different interests and different desired outcomes; and ensure the ethical sharing and use of data which is responsive to the identified needs of society (Mundy 2020). This is a good starting point yet could be supplemented with examples of successful past partnerships, creating conditions for partnerships to grow and evolve, and tools for creating and nurturing them. The experience of recent programmes underlines the need for research partnerships to understand the drivers and incentives that motivate different partners. Identifying and responding to what motivates each partner is essential to keeping them engaged over time, generating a range of outputs that satisfy diverse interests and needs.

Funding opportunities can unintentionally strain relationships by prescribing particularly organisational arrangements or size of budgets. For example, the call for proposals for CARIAA required applicants to identify five core partners and budgets equivalent to £7 million. These requirements displaced additional partners to a secondary status within projects and narrowed the pool of potential applicants to those able to receive and administer large budgets. In contrast, the research commissioning process can consider inviting a range of organisational arrangements and budgets. This would provide potential partnership with a degree of freedom to determine the size and structure needed to achieve its outcomes. Research partnerships can be encouraged to involve local and non-academic partners to help move knowledge from academic to real-world settings. They can also be encouraged to provide leadership roles for developing countries and non-research partners, including those from the private sector, government and civil society. Applicants should not only be assessed on their potential to contribute to cutting-edge knowledge on adaptation and resilience but their ability to fully understand and address the needs of real-world decisions and action. Applicants could also identify "gaps" or additional roles within their partnership, to be filled at a later date, providing space for the partnership to evolve during project implementation.

Partners can be encouraged to negotiate and agree to detailed ways of working, covering what each contributes and the added value to be realised from working together. Such a document serves to establish a shared perspective on overall purpose, means of coordinating the partnership, how to manage risks, protocols around decision-making and communication, as well as mechanisms to ensure mutual accountability and for recognising contributions, authorship and intellectual property. By agreeing to ways of working together, partners clarify expectations and establish a sense of fairness, equity and mutual accountability among themselves (Mundy 2020). Once underway, periodic reviews of the ways of working provide a routine "health check" on how the partnership is performing. Such reviews provide an opportunity to identify and address any issues and serve to induct new individuals and reaffirm commitment among existing ones. It can also recognise changes to the partnership over time, identifying the implications when a new organisation joins the project, an existing partner takes new responsibilities, or following the departure of an existing partner. The ways of working also set the stage for anticipating how partners act following a project: how they will use data, results and ideas beyond the project life, and what each partner contributes to preserving and making this work available into the future. A project is just one chapter in a longer story as people and organisations build their careers and pursue their mandates. A successful research project can benefit from established relationships and lay a foundation for future collaboration (Izzi 2018).

Research funding agencies also contribute to partnerships. Funders need to consider their own role, including the extent to which their systems, language and approaches facilitate or inhibit collaboration within the programme and engagement with principal investigators. Funding agencies describe a broad topic or high-level problem, providing the scientific community scope to identify and propose specific projects. A call for proposals specifies the selection criteria used to assess these proposals, with those that score higher being recommended for award. This approach to commissioning research tends to select each project on its individual merits, without necessarily considering how similar it may be to other proposals. Alternatively, funders can intentionally craft a portfolio in which different projects complement each other. For example, the CARIAA programme intentionally sought to fund activities in three distinct "hotspot" landscapes. Meanwhile, Future Climate for Africa selected projects that collectively covered western, eastern and southern parts of the continent as well as underpinning science to improve understanding of Africa's climate. Both these programmes brought participants together across projects to compile synthesis of diverse results across the portfolio and to reach audiences in policy and practice. As such programmes grew and evolved over time, they also benefited from adaptive management, shared between the funding agencies and principal investigators to jointly update budgeting, planning and procedures as needed (Currie-Alder et al. 2020).

Particularly when the source of research funding is official development assistance, or climate finance intended to assist developing countries, there is an expectation that the partnerships and programmes will have certain features and create some additional value. Beyond seeking geographic coverage or collaboration among projects, a portfolio approach to research can intentionally include projects that test alternative technologies, methods, or hypotheses. For example, within CARIAA, one project examined migration through household surveys carried out in communities that send and receive migrants, while another project conducted in depth life history interviews to uncover the lived experience of migrants (Singh et al. 2019). Privileging diversity of ideas and approaches within the portfolio can facilitate programme-level learning, permitting the comparison of project results to provide additional insights. Viewing the programme as a portfolio inspires thinking towards the unique value created by each project and the interaction among them, rather than simply generating more of the same types of results.

Funding competitions tend to discourage robust participation from least developed countries and work involving fragile and conflict-affected contexts states. Beyond having a relatively small domestic research community, many of the less-researched countries contain regions that are considered fragile or experiencing conflict. Without specific requirements for participants from—and activities in—these locations, the commissioning process can unintentionally simply favour collaboration among scientifically proficient peers across the global north and middle-income countries. There are many merits to encouraging projects that cover multiple countries, including the opportunity to compare results in multiple locations, foster peer learning across national borders and to hedge against risk. For example, the past decade witnessed punctual and longer-term security situations disrupt access to field locations in parts of Bangladesh, Egypt, Ethiopia, Mali, Mozambique and Tanzania. There is a wealth of guidance on how to conduct research in such contexts, covering the additional challenges in terms of planning, logistics and ethics; safety of participants and safeguarding against exploitation; and the potential negative consequences of research (Peters et al. 2020). The risks to any one project can be partially mitigated by spreading activities across multiple locations and partnering with local organisations and participants.

Many recent projects and programmes convened annual or periodic in-person gatherings to understand how different activities and work packages are organised, update each other on progress and findings and seize opportunities for combining insights or datasets. In-person gatherings were found to be good value for money in past climate research programmes. Annual project meetings served to ensure everyone clearly understood the project aims, jointly assess progress, reflect on what was working well (and what was not) and revise plans for the period ahead. In-person interaction over a number of days creates the space for having deeper conversations and critical debates on the ideas and design underpinning the research, confronting different understanding among team members and providing shared experiences that allow professional and interpersonal relationships to flourish. Such gatherings also served to orient participants who join later, allowing them to learn the stories and personalities behind project origins and implementation. In contrast, the regular routines of electronic mail, written documentation and web-based meetings do not serve as well to bring people together and form teams.

Virtual and distributed alternatives to these meetings, while still imperfect will continue to evolve, beyond the COVID pandemic and to make research more carbon–neutral. Compared to the past decade, future research programmes will be more limited in travel and mobility. This suggests that project design might shift towards more distributed models of organisation in which geographic nodes or sub-teams have a certain amount of autonomy and self-sufficiency. Yet whether virtual, hybrid, or in-person, periodic gatherings will remain vital to strengthening project- and programme-level coherence by identifying opportunities for learning and further research; conducting "health checks" that reflect on how each organisation contributes to and benefits from the partnership and further impact by engaging stakeholders to understand the context in which they operate, and their demands for insights and evidence.

7.3.3 Capacity

Our third facet of research design concerns capacity. Beyond offering scholarships and fellowships for training and independent study, exchanges of personnel between organisations in different locations and embedded experiences in real-world settings allow people to gain experience beyond academia in diverse host institutions. Greater emphasis needs to be given to enabling capacity across the spectrum from climate science to services, including the ability to engage in co-production, knowledge brokering and research uptake with local people and decision-makers. This section briefly reviews the understanding of capacity within climate action and considers recent insights to enable capacity through research.

Under UNFCCC, capacity building describes the ability of Parties to fulfil their obligations under the convention. Article 11 of the Paris Agreement describes capacity as the ability of developing countries to take effective climate change action, including adaptation and mitigation, as well as access to climate finance, education, training and information. The Agreement focuses on support to disadvantaged parties. There has been less attention to the capacity still needed even among developed countries or the efforts of developing countries to grow their own capacity.[2] It is clear that all societies require a capacity to understand and assess climate impacts, to plan and undertake climate action and to navigate among various potential futures. In this regard, there is promising literature on pathways and transformation, understanding how policies and investments function over time, opening and closing opportunities or "solution spaces" towards a more climate-resilient future (Haasnoot et al. 2020; Werners et al. 2021). Scholars have addressed the enablers and barriers to adaptation and the danger that potentially well-intended climate action can become maladaptive in future states of the world (ASSAR 2019; Gajjar et al. 2019).

The IPCC describes adaptive capacity as the ability of a system to adjust to climate change, whether households and societies choose to act and the extent to which do so. This notion borrows from ecology and the ability of natural systems to retain or modify their structure and function in response to shock and stress (Siders 2019). Like recent notions of waters security, understanding of adaptive capacity draws on the human capability approach which examines entitlements to material assets and social opportunities that permit people to exercise agency in deciding what to do or become (Mortreux and Barnett 2017). At its core, this understanding of capacity concerns the ability of human societies to rally cognition, resourceslingness in order to identify problems, understand their causality, assess potential solutions and act with purpose. In other words, how do societies overcome the "ingenuity gap" to detect and respond to risks and hazards in order to survive and thrive over time (Home, and wilr-Dixon 2000).

---

[2] There are promising signs that these issues are arising within the Paris Committee on Capacity Building, a voluntary network under UNFCCC to identify capacity gaps and needs, collect good practices and lessons learned and foster coordination and collaboration.

For research funding agencies, the sphere of control is investments intended to enhance or rally the ability to conceive, undertake, manage, share and use research and evidence. Such research can contribute to the capacity for action to implement UNFCCC and adaptive capacity to confront a changing climate. This ability to undertake and manage research is also referred to as the research capacity of an individual or an organisation. Over time, the notion of research capacity expanded from the skills and institutions required to solve problems, to consider how research products are used to bring about change as well as providing opportunities for peer learning from experiences elsewhere (Daniels and Dottridge 1993). Today research capacity is understood to include not only the skills and experience of an individual, but how individuals connect with others, to identify and analyse development challenges and to conceive, conduct, manage and communicate research that addresses those challenges (Neilson and Lusthaus 2007). In the UK, Vitae's research development framework includes 63 descriptors spanning the domains of knowledge and intellectual abilities, personal effectiveness, research governance and organisation, and engagement, influence and impact (VITAE n.d.). Individuals can be positioned along a timeline, ranging from early career to senior researchers. In the past, building one's research capacity was described in terms of gaining experience in different positions and increasing level of skill and responsibility over time. This can be achieved through an academic path from graduate student through postdoctoral awards, from initially contributing to work led by more established colleagues and collaborating with existing research teams, to gradually independence by progressing through faculty positions and gaining access to research grants.

Yet two factors have broadened the range of alternatives and complements to this traditional route. First is the rise of an "impact agenda" within research policy, predicating access to funding on demonstrating how the results of a project benefit society. Beyond merely addressing socially relevant problems and identifying some potential outcome, researchers are expected to undertake additional activities and engage with stakeholders in an explicit effort to bring about these outcomes. Second is the rising interest in large-scale collaborative models of research involving multiple organisations, multiple countries and multiple disciplines. As research funders seek greater levels of impact, there has been a rise in ambition whether under the assumption that larger-scale investments might generate greater returns, or that the breadth and complexity of societal problems requires proportionate broad and complex research efforts. Both these factors place a premium on skills not only to do and manage research but to collaborate with diverse actors beyond one's home organisation to interact and dialogue among scientists, policy makers and other actors, as well as joining and coordinating external partnerships (Araujo et al. 2020; Virji et al. 2012).

Various programmes have sought to enhance the capacity of individuals. The African Climate Change Fellowship programme (2007–2017) supported 120 early career awards across teaching and policy for doctoral and post-doc research. A subsequent Africa Climate Change Leadership programme provided small awards to 46 mid-careers to senior individuals in eastern and western Africa to orient research to development outcomes through different roles whether as thought leader, knowledge broker or research users (Meijerink and Stiller 2013). Climate Impacts Research Capacity and Leadership Enhancement (CIRCLE, 2014–2017) supported 97 individuals as visiting fellowships to undertake supervised placement in African host research institutions. A pilot training course seeks to strengthen knowledge on climate diplomacy and negotiation skills among African professionals (AGNES n.d.) while the IPCC Scholarships support Ph.D. students from developing countries for research that advances the understanding of climate impacts and options for adaptation and mitigation. Across these programmes, there is a shift from simply contributing to the careers of an individual, towards instead considering the functions needed within teams, organisations and partnerships to pursue and realise climate-resilient development in practice.

Many more capacity building efforts are embedded within research projects, rather than as stand-alone fellowships. For example, CARIAA supported over 260 people to benefit from graduate-level fellowships or internships, while 540 people participated in activities such as small grants, training or workshops. Capacity building within such programmes enables people to join the collaborative efforts involving large teams spread across multiple partner organisations and countries. The Future Climate for Africa programme found that early career researchers benefitted from being part of inter- and intra-consortium networks, accessing diverse resources that are not present in stand-alone or individual fellowship schemes (MacKay et al. 2020). Whereas pursuing graduate studies can be a solitary endeavour, capacity building within a consortium or project provides ready access to contacts in different countries (including potential peers and mentors), an opportunity to contribute to a larger research effort (including access to data, field sites, exposure to novel methods) and exposure to the collaborative skills as well as professional opportunities. Individuals also benefited from opportunities to grow as a professional within this community, increasing their level of responsibility and contributing in new ways. Similarly, the African Research Universities Alliance (ARUA) integrates training within its centres of excellence, including a network on climate and development among the universities of Ghana, Nairobi, and Cape Town. Ecosystems Services for Poverty Alleviation found it vital to give partners the opportunity to travel to each other's institutions throughout the project (Izzi 2018).

Future Climate for Africa embedded individual researchers within city governments. Researchers learned to act as a conduit for bringing scientific knowledge into local planning and decision-making. They were also exposed to real-world issues shaping the demand for that knowledge, appreciating how climate impacts interact with other governance challenges in cities and building relationships with non-academic audiences for research. The programme also supported exchanges among Harare, Lusaka, Windhoek and Durban to share insights on how different cities are responding to climate impacts on informal settlements, hydropower and water supply (Ndebele-Murisa et al. 2020). Similarly, the Climate and Development Knowledge Network (CDKN) has supported peer learning among professionals and practitioners in different countries. The emphasis is not to educate people on the findings from research but to provide exposure to how others are dealing with similar issues elsewhere. This involves connecting individuals across borders to share experiences and learning on delivering climate action on the ground. The overall message is not about transferring expertise from the global north, but sourcing capacity and insights within the global south.

Through these experiences, we are witnessing a shift beyond the capacity of researchers and organisations to do climate science and become leaders, to consider the capacity of diverse actors to access and use knowledge in the pursuit of climate action. As with the FCFA example, such approaches see research embedded in larger societal efforts to realise a more climate-resilient future, which necessarily means engaging and strengthening non-academic actors. One framing of this broader scope for capacity building considers the value created along chains and among actors that connect climate observation and information, to climate services and how they are used in decision-making. Considered in this way, the ultimate benefit or outcome depends not only on the existence of climate research, but the various links by which that knowledge is transformed and used across society. Weaknesses at any point along such chains can jeopardise the potential value of climate information. Funders need to recognise and support the interconnection among various actors, their outreach and interaction with decision-makers—whether politicians, firms or farms—and how their actions widen or constrain the opportunities available for such end users (Boulle et al. 2020).

Notions of research capacity have expanded. Traditionally research quality is viewed from the perspective of scientific peers in terms of its originality, relevance to academic audiences and integrity in terms of rigour and design. Yet research capacity is increasingly associated with the perspectives of practitioners and society: the extent to which research processes are seen as legitimate and pertinent in light of the concerns and values of stakeholders, as well as the extent to which the research outputs are seen as responding to practical needs and readily applicable to real-world contexts (McLean and Sen 2019; Clark et al. 2016). In particular, there is an expectation of co-design in research planning, co-production with stakeholders in the research process, as well as knowledge brokering to connect with users and their needs. The roots of co-production can be traced to traditions of participatory research and sustainability science (Miller and Wyborn 2020), involving local communities and interested stakeholders into the process of defining research questions, the process of gathering and assessing data and the creation of products based on that research. Co-production has been defined as bringing together different knowledge sources and experiences to jointly develop new and combined knowledge which is better able to support specific decision-making contexts.

Principles of co-production include transparency in the purpose of and methods used in research, tailoring the research process to specific context and decisions, timeliness in delivering results in keeping with those demands and communicating in ways that are accessible to diverse audiences (Vincent et al. 2021). High-quality knowledge co-production explicitly recognises multiple ways of knowing and doing, articulates clear goals that are shared among participants and allows for ongoing learning among actors through active engagement and frequent interactions (Norström et al. 2020). For example, the Pathways to Resilience in Semi-arid Economies consortium within CARIAA consulted with in-country decision-makers to understand their knowledge needs and priorities, as the basis for defining the research questions and study areas (Ludi et al. 2019). Building on such experiences, Harvey et al. (2019b) distinguish four approaches to co-production based on whether the process intends to produce usable knowledge or see it arise from interaction, and whether the process is brokered by a third party or involves purposefully engaging different perspectives. This means research must go beyond simply convening a dissemination workshop at the end of a research project. Instead, researchers should cultivate their relationships with such stakeholders over time and incorporate real-world needs into their research proposals, seeking to simultaneously advance academic knowledge and catalyse climate action.

7.4   Conclusion

How can research best catalyse climate action? The authors have wrestled with this question as our funding agencies learn from the past two decades and decide how to guide the next decade of research investment. The world has entered a decisive decade during which research must strengthen adaptation to the effects of climate change that cannot be prevented or reversed, supporting the most vulnerable worldwide. The urgency of this age of implementation cannot wait upon the customary linear process of research publications informing policy formulation, but rather requires more engaged models of research that are embedded in practice to foster real-time learning. Climate action ahead of 2030 requires ambition and design that is fit for purpose: working across scale, creating synergy among cohorts of projects and enabling capacity to pursue research uptake. By looking back at several large programmes over the past decade, this chapter identifies nine insights for designing the next decade of research investments (Fig. 7.1).

Research needs to cross scales to bridge local and national experience, address evidence gaps at decadal time scale and invest in programme-level learning. Connect community-level experience with how countries determine their national responses and contributions. Develop robust evidence across distinct locations, particularly for and with least developed countries and most vulnerable communities, especially in Africa and western Asia. Address the evidence gap between enhancing near-term resilience by 2030 and further climate impacts by 2050, as well as how adaptation choices taken now widen or constrain opportunities over time. Establish common approaches for data sharing, knowledge managemen and research uptake to allow different projects to "speak" to each other and assess how their activities both produce research results and contribute to society.

Cohorts link research efforts together to design for collaboration, foster partnership, and privilege diversity. Retain some flexibility in money and time, within the overall programme and in each project, to seize unexpected opportunities for

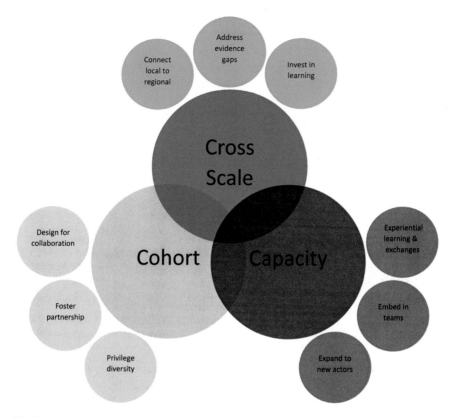

Fig. 7.1 Principles for organising research on climate adaptation and resilience

research impact and collaboration. Programmes can gather different projects and teams to share findings, assess progress and explore opportunities to work together. Within consortia, establish and refresh how partners work together including what each contributes and how they are accountable to each other. Involve partners and locations that complement each other, seeking autonomy and redundancy among in-country teams. Combine the strengths of multiple organisations, disciplines, countries and locations to generate new scientific knowledge, and enable actors to use evidence to realise more climate-resilient development.

Capacity is strengthened by designing research to enable experiential learning and exchanges, embed opportunities in larger efforts and expand to new actors and co-production. This constitutes a broader understanding of capacity beyond the mere conduct of climate science to encompasses a spectrum of skills connecting with its use in society. Consortia and large projects not only require principal investigators, but team members specialised in coordination, data management, gender equality and social inclusion, knowledge brokering and research uptake. Provide opportunities for practical experiences in real-world settings, such as researcher placements in city councils or working alongside practitioners. Being part of a consortium and programme provides early career researchers and professionals the benefits of working within a team, including mentoring and networking. Invest in capacity beyond academia to identify and address demand for climate knowledge. Provide practitioners and non-academic partners with leadership roles within projects to define research needs and realise impact.

In conclusion, designing research to catalyse climate action requires working across scale, creating synergy among cohorts of projects and enabling capacity to pursue research uptake. Together these features position research for impact, ensuring poor and vulnerable communities are more resilient to weather, climate change and related natural hazards in the near and longer term.

References

Adaptation Committee (2020) Technical paper on data for adaptation at different spatial and temporal scales. AC18/TP/7B. Bonn: UNFCCC. https://unfccc.int/sites/default/files/resource/ac18_7b_data.pdf. Accessed 15 Nov 2020

Araujo J, Harvey B, Huang YS (2020) A critical reflection on learning from the future climate for Africa programme. Cape Town: climate and development knowledge network

ASSAR (2019) Adaptation at scale in semi-arid regions (ASSAR): final report. http://hdl.handle.net/10625/58737. Accessed 24 Nov 2020

Boulle M, Scodanibbio L, Dane A et al (2020) Design scoping study for the capacity strengthening component of the CLARE programme. Change Pathways, Johannesburg. http://hdl.handle.net/10625/58680. Accessed 24 Nov 2020

Callaghan MW, Minx JC, Forster PM (2020) A topography of climate change research. Nat Clim Change 10(2):118–123. https://doi.org/10.1038/s41558-019-0684-5. Accessed 24 Nov 2019

Cash DW, Adger WN, Berkes F et al (2006) Scale and cross-scale dynamics: governance and information in a multilevel world. Ecol Soc 11(2):181–192

Clark WC, van Kerkhoff L, Lebel L et al (2016) Crafting useable knowledge for sustainable development. Proc Natl Acad Sci 113(17):4570–4578. https://doi.org/10.1073/pnas.1601266113

Cundill G, Harvey B, Tebboth M et al (2018) Large-scale transdisciplinary collaboration for adaptation research. Global Chall 3(4):1700132–1700132. https://doi.org/10.1002/gch2.201700132

Currie-Alder B, Cundill G, Scodanibbio L et al (2020) Managing collaborative research: insights from a multi-consortium programme on climate adaptation across Africa and South Asia. Reg Environ Change 20:117. https://doi.org/10.1007/s10113-020-01702-w

Daniels D, Dottridge T (1993) Managing agricultural research: views from a funding agency. Publ Adm Dev 13(3):202–215

De Souza K, Kituyi E, Harvey B et al (2015) Vulnerability to climate change in three hot spots in Africa and Asia. Reg Environ Change 15:747–753. https://doi.org/10.1007/s10113-015-0755-8

Dodson J (2017) The role of funders in equitable and effective international development collaborations. UKCDR, London. https://www.ukcdr.org.uk/wp-content/uploads/2017/11/Building-Partnerships-of-Equals_-REPORT-2.pdf. Accessed 25 Nov 2020

Eriksen SE, Schipper LF, Scoville-Simonds M, Vincent K et al (2021) Adaptation interventions and their effect on vulnerability in developing countries: help, hindrance or irrelevance? World Dev 141:105383. https://doi.org/10.1016/j.worlddev.2020.105383

Gajjar SP, Singh S, Deshpande T (2019) Tracing back to move ahead: a review of development pathways that constrain adaptation futures. Climate Dev 11(3):223–237. https://doi.org/10.1080/17565529.2018.1442793

Haasnoot M, Biesbroek R, Lawrence J et al (2020) Defining the solution space to accelerate climate change adaptation. Reg Environ Change 20(2):37. https://doi.org/10.1007/s10113-020-01623-8

Harvey B, Cochrane L, Jones L, Vincent K (2019a) Programme design for climate resilient development. IDRC, Ottawa. http://hdl.handle.net/10625/58293. Accessed 30 Nov 2020

Harvey B, Cochrane L, Van Epp M (2019) Charting knowledge co-production pathways in climate and development. Environ Policy Gov 29(2):107–117. https://doi.org/10.1002/eet.1834

Homer-Dixon T (2000) The ingenuity gap. Alfred A. Knopf, Toronto

IPCC (2018) Global warming of 1.5°C: an IPCC special report. www.ipcc.ch. Accessed 29 Nov 2020

Izzi V (2018) Research with development impact: lessons from the ecosystem services for poverty alleviation programme. University of Edinburgh. http://www.espa.ac.uk/publications/research-development-impact. Accessed 29 Nov 2020

Jarvie J, Vincent K, Bharwani S et al (2020) Enabling climate science use to better support resilience and adaptation practice. LTS International. http://hdl.handle.net/10625/58941. Accessed 29 Nov 2020

Jepson W, Budds J, Eichelberger L (2017) Advancing human capabilities for water security: a relational approach. Water Secur 1:46–52. https://doi.org/10.1016/j.wasec.2017.07.001

Jones L, Harvey B, Cochrane L et al (2018) Designing the next generation of climate adaptation research for development. Reg Environ Change 18(1):297–304. https://doi.org/10.1007/s10113-017-1254-x

Klein RJT, Adams KM, Dzebo A et al (2017) Advancing climate adaptation practices and solutions: Emerging research priorities. Stockholm Environment Institute. https://www.sei.org/wp-content/uploads/2017/05/klein-et-al-2017-adaptation-research-priorities.pdf. Accessed 29 Nov 2020

Ludi E, Nathe N, Gueye B et al (2019) Pathways to resilience in semi-arid economies: findings, recommendations, and learnings. IDRC, Canada. http://hdl.handle.net/10625/58343. Accessed 30 Nov 2020

MacKay B, Roux JP, Bouwer R (2020) Building research capacity in early career researchers. Future climate for Africa. https://futureclimateafrica.org/resource/building-research-capacity-in-early-career-researchers-insights-from-an-international-climate-research-programme/. Accessed 10 Dec 2020

Martel A (2020) Guide for research partnership agreements. Cooperation Canada, Ottawa. http://hdl.handle.net/10625/58810. Accessed 10 Dec 2020

McLean RKD, Sen K (2019) Making a difference in the real world? A meta-analysis of the quality of use-oriented research using the research quality plus approach. Res Eval 28(2):123–135. https://doi.org/10.1093/reseval/rvy026

Meijerink S, Stiller S (2013) What kind of leadership do we need for climate adaptation? Environ Plan 31(2):240–256. https://doi.org/10.1068/c11129

Miller CA, Wyborn C (2020) Co-production in global sustainability: histories and theories. Environ Sci Policy 113:88–95. https://doi.org/10.1016/j.envsci.2018.01.016

Minx JC, Callaghan M, Lamb WF et al (2017) Learning about climate change solutions in the IPCC and beyond. Environ Sci Policy 77:252–259. https://doi.org/10.1016/j.envsci.2017.05.014

Moosa S, Zhanje S, Ellis C et al (2019) Understanding African decision-makers needs for research and evidence. SouthSouthNorth and International Institute for Sustainable Development. http://hdl.handle.net/10625/58733. Accessed 10 Dec 2020

Morchain D, Spear D, Ziervogel G et al (2019) Building transformative capacity in southern Africa. Action Res 17(1):19–41. https://doi.org/10.1177/1476750319829205

Mortreux C, Barnett J (2017) Adaptive capacity: exploring the research frontier. Wires Clim Change 8:e467. https://doi.org/10.1002/wcc.467

Mundy J (2020) Commissioning research and improving the effectiveness of partnerships. Effective Collective, Melbourne. http://hdl.handle.net/10625/59616. Accessed 10 Dec 2020

Mustelin J, Kuruppu N, Kramer AM et al (2013) Climate adaptation research for the next generation. Climate Dev 5(3):189–193. https://doi.org/10.1080/17565529.2013.812953

Nalau J, Verrall B (2021) Mapping the evolution and current trends in climate change adaptation science. Climate Risk Manage 32:100290. https://doi.org/10.1016/j.crm.2021.100290

Ndebele-Murisa MR, Mubaya CP, Pretorius L et al (2020) City to city learning and knowledge exchange for climate resilience in Southern Africa. PLoS ONE 15(1):e0227915. https://doi.org/10.1371/journal.pone.0227915

Neilson S, Lusthaus C (2007) IDRC-supported capacity building: developing a framework for capturing capacity changes. Universalia, Montreal. http://hdl.handle.net/10625/29146. Accessed 10 Dec 2020

Newell P, Srivastava S, Naess LO et al (2020) Towards transformative climate justice. Institute of Development Studies, Brighton, UK. http://hdl.handle.net/10625/59197. Accessed 10 Dec 2020

Nightingale AJ, Eriksen S, Taylor M et al (2019) Beyond technical fixes: climate solutions and the great derangement. Climate Dev 12(4):343–352. https://doi.org/10.1080/17565529.2019.1624495

Norström AV, Cvitanovic C, Löf MF et al (2020) Principles for knowledge co-production in sustainability research. Nat Sustain 3(3):182–190. https://doi.org/10.1038/s41893-019-0448-2

O'Neill M (2020) Collaborating for adaptation: Findings and outcomes of a research initiative across Africa and Asia. IDRC, Ottawa. http://hdl.handle.net/10625/58971. Accessed 10 Dec 2020

Pelling M, Garschagen M (2019) Put equity first in climate adaptation. Nature 569:327–329. https://doi.org/10.1038/d41586-019-01497-9

Peters K, Dupar M, Opitz-Stapleton S et al (2020) Climate change, conflict and fragility: an evidence review. Overseas Development Institute, London. https://www.odi.org/publications/17015-climate-change-conflict-and-fragility-evidence-review-and-recommendations-research-and-action. Accessed 12 Dec 2020

Prakash A, Cundill G, Scodanibbio L et al (2019) Climate change adaptation research for impact. CARIAA working paper 23. http://hdl.handle.net/10625/57489. Accessed 12 Dec 2020

Rao N, Mishra A, Prakash A et al (2019) A qualitative comparative analysis of women's agency and adaptive capacity in climate change hotspots in Asia and Africa. Nat Clim Chang 9(12):964–971. https://doi.org/10.1038/s41558-019-0638-y

Rosenzweig C, Horton RM (2013) Research priorities on vulnerability, impacts and adaptation: responding to the climate change challenge. UNEP, Nairobi

Siders AR (2019) Adaptive capacity to climate change: a synthesis of concepts, methods, and findings in a fragmented field. Wires Clim Change 10:e573. https://doi.org/10.1002/wcc.573

Singh C, Tebboth M, Spear D et al (2019) Exploring methodological approaches to assess climate change vulnerability and adaptation. Reg Environ Change 19:2667–2682. https://doi.org/10.1007/s10113-019-01562-z

Smith DM, Matthews JH, Bharati L et al (2019) Adaptation's thirst: accelerating the convergence of water and climate action. Background paper for the global commission on adaptation. Global Commission on Adaptation, Rotterdam and Washington DC. www.gca.org. Accessed 12 Dec 2020

UNFCCC (United Nations' Framework Convention on Climate Change) (2019) 25 years of adaptation. Report by the Adaptation Committee

Vincent K, Cundill G (2021) The evolution of empirical adaptation research in the global South from 2010 to 2020. Climate Dev. https://doi.org/10.1080/17565529.2021.1877104

Vincent K, Steynor A, McClure A et al (2021) Co-production: learning from contexts. In: Conway D, Vincent K (eds) Climate Risk Africa. Palgrave Macmillan, Cham, pp 37–56. https://doi.org/10.1007/978-3-030-61160-6_3

Virji H, Padgham J, Seipt C (2012) Capacity building to support knowledge systems for resilient development. Curr Opin Environ Sustain 4(1):115–121. https://doi.org/10.1016/j.cosust.2012.01.005

Werners SE, Wise RM, Butler JRA et al (2021) Adaptation pathways: a review of approaches and a learning framework. Environ Sci Policy 116:266–275. https://doi.org/10.1016/j.envsci.2020.11.003

Wu L, Wang D, Evans JA (2019) Large teams develop and small teams disrupt science and technology. Nature 566(7744):378–382. https://doi.org/10.1038/s41586-019-0941-9

Webpages

AGNES, African Group of Negotiators Experts Support. https://training.agnes-africa.org/programs/climate-governance-diplomacy-and-negotiations-leadership-program/

CVF, Climate Vulnerable Forum. https://thecvf.org/

LDC Climate Change. http://www.ldc-climate.org/

Part II
Case Studies

Chapter 8
Water-Resilient Places—Developing a Policy Framework for Surface Water Management and Blue-Green Infrastructure

Barry Greig and David Faichney

Abstract This chapter broadly describes Scotland's overarching approach to the management of its water resources with a specific focus on the development of measures that respond to the multiple and complex challenges of surface water flooding in Scotland. It considers the interface with the emerging "net zero emissions" agenda, to offer an example of how Scotland aims to grow the sector sustainably and responsibly while tackling major policy challenges, like the effective integration of mechanisms, structures and interventions to tackle surface water flooding. The chapter also offers an explanation of the Scottish Government's Hydro Nation agenda which seeks to maximise the value of water resources in a sustainable and responsible way and the development of a stakeholder-led vision for the water sector. Finally, given the overarching theme of the main publication in relation to water security, the chapter outlines how Scottish water-related knowledge is shared with developing world nations, with a specific focus on Malawi.

Keywords Surface water flooding · Net zero emissions · Policy challenges · Value of water · Scotland

B. Greig (✉)
Scottish Government, Water Industry Division, Edinburgh, Scotland
e-mail: barry.greig@gov.scot

D. Faichney
Scottish Environment Protection Agency (Seconded To Scottish Government To Develop Water Resilient Places Policy 2019–2021), Stirling, Scotland
e-mail: david.faichney@sepa.org.uk

© The Author(s), under exclusive license to Springer Nature Singapore Pte Ltd. 2022
A. K. Biswas and C. Tortajada (eds.), *Water Security Under Climate Change*,
Water Resources Development and Management,
https://doi.org/10.1007/978-981-16-5493-0_8

8.1 Introduction

Water is the crucial underpinning of Scottish life, with huge social, environmental and economic significance. It is a critical resource in most key sectors of the Scottish economy, particularly manufacturing, agriculture, food and drink, tourism and energy. But it is also a high-performing sector in its own right with a diverse supply chain, an established innovation support ecosystem, world-leading research base and a highly regarded governance and regulatory framework.

This chapter broadly describes Scotland's overarching approach to the management of its water resources with a specific focus on the development of measures that respond to the multiple and complex challenges of surface water flooding in Scotland in the context of climate adaptation and Scotland's legally binding climate change target to cut greenhouse gas emissions to net zero by 2045, five years ahead of the date set for the UK as a whole. It outlines how the sector is supported and developed to add more value to society and the economy, and deliver against the Scottish Government's vision of "Scotland: The Hydro Nation". It considers the interface with the emerging "net zero emissions" agenda, to offer an example of how Scotland aims to grow the sector sustainably and responsibly while tackling major policy challenges, like the effective integration of mechanisms, structures and interventions to tackle surface water flooding. One of the biggest challenges we face in Scotland is how we adapt to our increased exposure to flooding through climate change[1] and the continued densification of our towns and cities. Our "total asset" that needs to be flood resilient continues to increase.

In response to the issues outlined above, *Water-Resilient Places—A Policy Framework for Surface Water Management and Blue-Green Infrastructure* was published in January 2021 to meet two commitments set out in the Scottish Government's 2019–2020 Programme for Scotland, namely:

- to review Scotland's approach to blue-green cities and bring forward proposals by the end of 2020 and
- to support and promote Scottish Water's increased use of natural, blue-green infrastructure to manage surface water away from homes and businesses, and help to create great places to live.

This important policy document has been developed by the Scottish Government in the context of the climate emergency with input from key stakeholders with interests in flood risk management, drainage, blue-green infrastructure, land-use planning and placemaking. It aims to enhance how Scotland can collectively improve the management of surface water flooding by complementing and supporting existing policy and organisational responsibilities as set out in the Flood Risk Management (Scotland) Act 2009. The policy objectives articulated in the document aim to explain the relevance of surface water management to all sectors

---

[1] Climate change-related flooding challenges for Scotland include: increased rainfall (winter rainfall totals and summer rainfall intensities), sea-level rise and more frequent river flooding.

recognising the interface of interests that interact in this area and help make it a core consideration for the range of parties involved in designing for climate adaptation, sustainable placemaking and delivering great blue-green places to live.

8.2 Scotland's Water Environment

Scotland covers an area of 78,000 km^2, including 787 islands, and has a population just under 5.5 million. Generally considered a wet country, Scotland has more than 125,000 km of rivers and streams and over 25,500 lochs (Eng. lakes), and a 220 km canal network. Loch Ness alone, arguably Scotland's most well-known loch, has a volume of 7.4 million m^3, which is greater than all the surface water in England and Wales combined.

Water policy in Scotland is set by the Scottish Government in conformity with European Union water and environmental law, and agricultural law and policy. The government also provides direction to Scottish Water, the country's single, national, publicly owned water utility. Formed in 2002, Scottish Water supplies 2.5 million households and 150,000 business with nearly 1.5 billion litres/day of drinking water through more than 30,000 miles of water pipes and over 200 water treatment works. Wastewater is collected through another 30,000 miles of sewers and treated at nearly 2000 wastewater treatment works before it is returned to the environment (Scottish Water Annual Report 2019).

8.2.1 Context—Scotland, The Hydro Nation: Strategy and Structure

As a small, but responsible nation, Scotland has long recognised the principles of sustainable development, and since 1999, the Scottish Government has worked continuously to mainstream those principles across all policy areas. In respect of water issues, this approach was underpinned in 2013 by a statutory duty placed on the Scottish Ministers within the Water Resources (Scotland) Act to "*take such reasonable steps as they consider appropriate for the purpose of ensuring the development of the value of Scotland's water resources.*" (Scottish Government 2013).

The Scottish Government vision of Scotland as a Hydro Nation is that of a country that views and manages its water resources responsibly, regardless of their relative abundance, and sees our relationship and the ways we work with the water environment and industry as inextricably linked to our national identity. The Hydro Nation *approach* aims to maximise the value of Scotland's water resources by increasing the entire sector's contribution to the national economy.

From this perspective, Scotland's water resources are not limited to the resource itself but also include the contribution of water experts who advise on government policy and create practical solutions for industry or regulators and indeed Scottish Water itself, one of the best-performing water utility companies in the UK.

The values of these water-related assets and resources are themselves viewed broadly, in terms of both economic and non-economic impact and worth. It is acknowledged that while some aspects, such as exports of water technology, can be measured directly in terms of the money they add to the national economy, others—like clean, good-tasting drinking water and how Scotland's water bodies contribute to quality of life and wider national identity through, say the provision of ecosystem services or even the aesthetic pull that helps bring in tourism revenue to the country —must be measured indirectly. These so-called non-economic values are often gauged in terms of the price consumers would be willing to pay for these benefits, so an important element of Hydro Nation is concerned, for example, with increasing public understanding of the value of water bringing solutions to customers in remote areas, generating renewable energy, reducing, or even recovering, priority pollutants from wastewater or indeed, as per the focus of this piece, how we manage and harness water through integrated approaches to reduce the negative impacts of surface water flooding. The Hydro Nation agenda and wider Sector Vision developed by Scottish Water with the input of sector stakeholders help facilitate bringing forward ideas, approaches and infrastructure that improve the quality of the places we live and work and the lives we lead.

8.2.2 Hydro Nation: Strategy and Structure

The Hydro Nation Strategy is currently articulated across four thematic areas: national; international; knowledge; and innovation. The national theme is aimed at preparing, developing and supporting the Scottish Water industry to tackle a wide range of policy areas, including emission reduction epitomised by Scottish Water's net zero road map approach, and the delivery of the wider vision for the water sector, rural supply issues and investing in blue-green infrastructure and energy. The international theme supports the export and exchange of Scottish water-related knowledge, working with other public sector actors to share their expertise with developing world nations, numbering Malawi in particular, as well as Zambia, Tanzania, New Zealand and Australia among its partners. These two themes are described in greater detail below.

The knowledge theme sponsors research across all areas of interest and includes at its core the Centre for Expertise in Water (CREW) which administers the Hydro Nation Scholars Programme on behalf of government. The innovation theme includes the publicly funded Hydro Nation Water Innovation Service (HNWIS) which has supported over 100 Scottish companies developing water and wastewater technologies by providing technical, trial, product development support, networking, market insight and information and connecting them with potential customers. It also includes the Scottish Enterprise Low Carbon Team, which provides business development advice and assesses opportunities of interest to the sector.

All Hydro Nation strategic themes and activities reflect the established principles of sustainability, though this is perhaps especially evident in the national and international themes.

Nationally, the water sector in Scotland is preparing for climate change by reducing its carbon footprint and protecting adaptive resources. While the industry is undoubtedly a significant energy user, it is worth noting that since monitoring of the greenhouse gas emissions associated with Scottish Water's water and wastewater services to customers began in 2007, operational emissions have reduced by 45% to 254,000 tCO_2e in 2019/20 (Scottish Water Annual Report 2020). The business's long-term strategy is to mitigate its contribution to the causes of global heating by reducing its emissions while maximising its contribution to renewable energy generation and the ability to capture and store emissions in our land. This relies on Scotland's natural capital—the stocks of natural resources in Scotland's environment, to deliver services. Considering natural capital can also help adapt to climate change and help address emissions by locking up carbon. Working with nature and adopting nature-based solutions offer the opportunity to deliver multiple benefits, improving the water environment, providing additional benefits for biodiversity, creating better places to live and doing so in a way that supports low emissions. An example of this aspect is preservation of peatlands, which cover over a fifth of Scotland's land area. By collaborating with local landowners and the national Peatland Action Group, led by NatureScot, Scottish Water is helping to restore nearly 500 ha of peatland, with a further estimated 1500 ha identified for restoration in multiple locations, thereby contributing to carbon sequestration as well as to improvements in downstream water quality. In partnership with Forest and Land Scotland and non-governmental organisations (NGOs), Scottish Water plans to implement woodland creation on its land and at operational sites. Taken together, these actions will help put the business on track to meeting its goal of capturing and storing more CO_2 than it produces and supporting this effort. Scottish Water is working with the world-renowned James Hutton Institute to understand how much carbon dioxide is stored on its land and how it can best use these valuable natural resources in meeting its climate change ambitions.

Scottish Water's 25-year strategic plan—Our Future Together—was published in February 2020 and outlines the impact of the changing climate and how the organisation will reduce emissions to become net zero by 2040. The plan highlights how future investment in vital infrastructure and assets, which were not designed to cope with our changing climate, must combine with innovative and sustainable ways of dealing with climate change and supporting economic growth. Scottish

Water's Net Zero Emissions Routemap sets out the steps that can be taken to decarbonise the business across every aspect of its operations by 2040 (five years ahead of Scotland's national targets) as well as committing the intention to go beyond net zero emissions.

The aim for Scottish Water is to eliminate as far as possible all direct and indirect emissions, and invest in Scotland's natural capital (and other technologies as they arise) to balance any emissions that cannot be reduced or eliminated. An Expert Advisory Panel to help Scottish Water deliver on its net zero ambitions has been established, made up of people from the private and public sectors and academia, including the UK Committee for Climate Change.

Its input has already helped to shape the Net Zero Emissions Routemap and will continue to review progress, including suggesting what adjustments may be needed to deliver against targets. Finally, Scottish Water's 2019 Sustainability Report provides further information on actions undertaken by the business to support and promote sustainable development across its operations.

8.3   Scotland's Water Sector Vision

Recognising the interlinked interests of key sector stakeholders and the need for a concerted effort to tackle climate change and develop a sector fit to meet future challenges, Scottish Water, the Water Industry Commission in Scotland (WICS), the Customer Forum, Citizens Advice Scotland, the Scottish Environment Protection Agency (SEPA) and the Drinking Water Quality Regulator (DWQR) are working together in a shared vision for the first time. The vision aims to develop a sector that will be "admired for excellence, secure a sustainable future and inspire a Hydro Nation". These key principles will be at the heart of Scottish Water's future strategy. It was formed by Scottish Water, the Water Industry Commission for Scotland (WICS), the Customer Forum, Citizen's Advice Scotland (CAS), the Scottish Environment Protection Agency (SEPA) and the Drinking Water Quality Regulator (DWQR).

Recognising the enormity of the challenge and the significant transformation required to achieve, the vision follows a call in 2019 from Roseanna Cunningham MSP, Cabinet Secretary for Environment, Climate Change and Land Reform for the water industry to create a common understanding of what the sector should seek to achieve over the next three decades. This development marks the first time a common Vision has been owned by all sector stakeholders and marks a new collaborative way of working between Scottish Water, its regulator and others.

> **Box 1: Scotland's Water Industry Vision**
>
> Scotland's water sector will be admired for excellence, secure a sustainable future and inspire a Hydro Nation.
>
> Together, we will support the health and well-being of the nation. We will ensure that all of Scotland gets excellent quality drinking water that people can enjoy all of the time. Scotland's wastewater will be collected, treated and recycled in ways that generate value and protect the environment. We will enable the economy to prosper.
>
> We will transform how we work to live within the means of our planet's resources, enhance the natural environment and maximise our positive contribution to Scotland achieving net zero emissions.
>
> We will involve and inspire Scotland's people to love their water and only use what they need. We will promote access to the natural environment and encourage communities to enjoy and protect it.
>
> We will be agile and collaborate within the sector and with others to be resilient to the challenges which will face us.
>
> We will keep services affordable by innovating and delivering the greatest possible value from our resources, helping those who need it most. We will serve all customers and communities in a way that is fair and equitable to present and future generations.
>
> We are a vital part of a flourishing Scotland.

8.4   The Hydro Nation Chair

Flowing from the vision, in December 2020 Scottish Water and the Scottish Funding Council (SFC) announced a unique partnership with the University of Stirling to support Scotland's ambitions to become one of the world's leading hydro nations.

The University of Stirling has been appointed to host and lead a £3.5 m initiative for six years from April 2021 to increase Scotland's profile as a global leader in water research and innovation led by the newly created post of the Scotland Hydro Nation Chair funded by Scottish Water. The chair will provide the leadership to forge collaborative partnerships across the sector to deliver solutions for sustainable water management in Scotland supporting the transition to a net zero economy and a green recovery following the COVID-19 pandemic. This important new initiative recognises that research and innovation will be a key enabler in transforming the industry to achieve objectives that will require transformative change to the way the water sector operates to make a positive contribution to Scotland's net zero ambition, deliver service excellence and live within the means of the planet's resources.

The aim is that Scotland's Hydro Nation Chair will complement the existing relevant research and innovation initiatives and infrastructure in Scotland and will play a leading role in maximising Scotland's academic water impact, international reputation and engagement with wider sectors to deliver the knowledge and capability required to enable transformative change and to attain net zero carbon emissions by 2040. New research and innovation stimulated by the creation of a Scotland Hydro Nation Chair will include the recycling of wastewater and ways of enhancing the natural environment. Importantly, it will also work towards ensuring that people across the whole of Scotland will continue to enjoy access to high-quality drinking water.

8.5 Infrastructure Investment and Climate Change Action

Scottish Water invests £700 m each year to maintain and improve services. Investment to maintain and enhance assets and infrastructure which accounts for the bulk of that investment has the potential to drive significant emissions, both directly and from supply chains. This covers the emissions embodied in the concrete, steel and materials used, as well as site activities. Scottish Water is working with design teams, delivery partners and supply chains to find ways to reduce the emission impacts of investment with the goal of reducing the carbon intensity of investments by 75% by 2040.

A Construction Expert Panel has been established to bring external knowledge and innovation support. Chaired by Scottish Water's Director of Capital Investment, the panel includes senior leaders from construction and supply chain partners, Construction Scotland Innovation Centre, and from academia. It provides a forum to set strategic leadership across investment delivery and to identify and promote delivery of low-emission products and practices.

8.6 Surface Water Management and Climate Change

Evidence tells us we have a growing surface water management challenge in Scotland that is set to get worse if we do not take urgent action.

Scotland's second National Flood Risk Assessment estimates that the number of properties exposed to impacts from surface water flooding is set to rise from the current figure of 210,000 to 270,000 by 2080 (SEPA 2018) due to climate change.

Our current approach to surface water management is not delivering improvements at a fast enough rate to reduce our exposure to future floods from this source.

The Water-Resilient Places policy framework was drafted to take forward the imperative to transform our surface water management approach and combine it with opportunities to work with others to help us meet the challenge: including climate adaptation, planning reform, the developing visions for our future towns and cities and the green recovery.

The policy objectives aim to make surface water management relevant to all sectors and make it a core consideration in designing for climate adaptation, sustainable placemaking and delivering great blue-green places to live.

8.7 Water-Resilient Places—Developing a Policy Framework for Surface Water Management and Blue-Green Infrastructure

8.7.1 Background

Since the introduction of the Flood Risk Management (Scotland) Act 2009, Scotland has made strong progress in understanding the impact of flooding in Scotland, where priorities lie and how stakeholders can better work together to manage the impacts of floods on our communities.

Nationally, Scotland is well into the second flood risk management planning cycle and stakeholders are collectively learning more about the size of the challenge ahead through better flood mapping, modelling and analysis. Organisations are gaining valuable experience in bringing forward and implementing actions to deliver real impacts, and as the scale of the challenge becomes more apparent, we are discovering that the rate at which we can reasonably expect to implement actions is likely to be outstripped by the increase in exposure to flooding through climate change.

We recognise that our urban areas in particular face mounting challenges with surface water drainage and related flooding. Despite considerable capital investment, the continued densification of our towns and cities is adding to the pressure on drainage systems that are already at capacity and the "total asset" that needs to be flood resilient continues to increase.

To make certain Scotland's places continue to thrive and remain attractive to people, businesses and investors, we must ensure they are water resilient and set up for the climate challenges ahead. This will require a concerted effort across all sectors to ensure that new development is appropriately sited and designed and that existing buildings can transition to managing rainwater through blue-green infrastructure instead of sewers reducing the pressure on our drainage systems.[2]

Scotland identifies three key challenges in achieving water resilience that must be addressed together and through a cross-sector effort:

---

[2] This is often referred to as "disconnection" or "retrofitting". Scottish Water's Surface Water policy states: *"For sustainability and to protect our customers from potential future sewer flooding, we will not normally accept any surface water connections into our combined sewer system".*

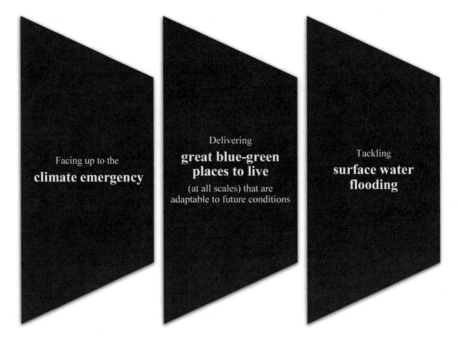

Fig. 8.1 Three challenges

1. Facing up to the climate emergency—both mitigation *and* adaptation;
2. Delivering great blue-green places to live (at all scales) that are adaptable to future conditions;
3. Tackling surface water flooding.

The reward for tackling these challenges as one will be the low-carbon, water-resilient places Scotland needs to flourish under future conditions. Creating water-resilient places where surface water flooding impacts are minimised will not be achieved by making minor adjustments to current processes; it is acknowledged that significant changes to our approach are required (Fig. 8.1).

We believe that success will be achieved by bringing players together behind the common aim of creating great places and by following the Place Principle[3] as adopted by the Scottish Government and the Convention of Scottish Local Authorities (COSLA—the national association of Scottish councils and employers' association for its member authorities) "...*to help overcome organisational and*

---

[3] The Place Principle was developed by partners in the public and private sectors, the third sector and communities, to help them develop a clear vision for their place. It promotes a shared understanding of place and the need to take a more collaborative approach to a place's services and assets to achieve better outcomes for people and communities. The principle encourages and enables local flexibility to respond to issues and circumstances in different places.

sectoral boundaries, to encourage better collaboration and community involvement, and improve the impact of combined energy, resources and investment".

By focusing on creating blue-green places that are sustainably drained and have low exposure to flooding impacts, multiple benefits can be delivered for our communities, contributing to the Scottish Government's National Performance Framework (Scottish Government 2021), and help to deliver Scotland's ambition for a robust, resilient well-being economy post-COVID-19 (Scottish Government 2020).

Attitudes in this topic area are changing with more sectors recognising that blue-green infrastructure can bring multiple benefits and help meet a wider range of policy objectives including flood risk management, city regeneration, environmental improvement and enhanced well-being and health. Scottish Water's "*no more in; what's in, out.*" surface water policy stating that they will not normally accept any surface water connections into their combined sewer system also requires a shift in favour of blue-green interventions.

Public sector, private sector, third sector, community and individuals' activities all have the potential to help ensure that our places are low carbon and resilient to the impacts of climate change. By applying our knowledge of floods and drainage systems to the activities of all sectors, the aim is to design a future where increased rainfall, sea-level rise and more frequent river flooding demand less of our attention, resources and time. This approach recognises that a water-resilient society benefits everyone and therefore we all have a role to play in making sure it happens.

We are therefore responding to the challenging impacts of climate change on surface water flooding by seeking to improve how we manage surface water flooding by complementing and supporting the existing policy and organisational responsibilities as set out in the Flood Risk Management (Scotland) Act 2009. It seeks to make surface water management relevant to all sectors and make it a core consideration in designing for climate adaptation, sustainable placemaking and delivering great blue-green places to live.

The five primary recommendations set out in the framework document listed below are supported by a further 16 subsidiary recommendations. All 21 recommendations are set out later in this chapter.

- *A vision for blue-green cities for Scotland should be established.*
- *Scotland should channel support towards actions that contribute to creating great places that are resilient to future flooding and drainage challenges, and away from activities that add to our future flooding and drainage burden.*
- *We should take a placemaking approach to achieving blue-green cities and water resilience, involving partners in the public and private sectors, the third sector, individuals and communities.*
- *Surface water flooding issues should be solution-focused and addressed by coordinating across organisations and implementing the best integrated sustainable solution (overcoming current legislative responsibilities and debates about ownership and ongoing maintenance).*

- *The drainage of surface water from all new sites, wherever practicable, should be by blue-green infrastructure. Land for blue-green infrastructure should be a site prerequisite, and all designs should presume no rainwater connection to sewer.*

8.7.2 Background to the Water-Resilient Places Policy

Scottish Government has identified the need to improve how we manage surface water in Scotland. The Programme for Government: *Protecting Scotland's Future —The Government's Programme for Scotland 2019–2020*[4] contains commitments to work together to increase Scotland's use of blue-green infrastructure for drainage and flood management and to review our approach to blue-green cities and bring forward proposals by the end of 2020. The framework published in February 2020 contributes to that review by considering what improvements can be made to surface water management in communities across Scotland by building on existing policy and by improving how we work together. The focus on placemaking[5] aims to increase the efficiency and effectiveness of funding and widen support for the statutory stakeholders. It outlines how surface water is currently managed in Scotland, sets out a vision for the future and describes the components that should be brought together to form a coherent framework that will support delivery. It concludes with recommendations for action to improve the delivery of surface water management and flood resilience in Scotland, to support the commitments in the Programme for Government and to help address the relevant recommendations in the Infrastructure Commission for Scotland's Key Findings Report[6]—specifically those focused on climate adaptation, "infrastructure first" and improving regulatory coherence across water provision, flood management and resilience.

---

[4] Protecting Scotland's Future – The Government's Programme for Scotland 2019–2020.

- p51: "*Scottish Water... ...will also take action in climate adaptation and pursue further partnerships with local authorities and others to adapt to increased intensity rainfall events by creating natural, blue/green infrastructure to manage surface water away from homes and businesses and help create great places to live*".

- p91: "*We are also reviewing our approach to Blue-Green cities and will bring forward proposals by the end of this year*".

[5] The placemaking approach as promoted by Scottish Government and supported across all sectors requires an integrated, collaborative and participative approach to decisions about services, land and buildings and is applicable to a place whether it is existing, changing or in the planning.

[6] On 20 January 2020, the Infrastructure Commission for Scotland published its first report Phase 1 Key Findings: A Blueprint for Scotland. The report sets out eight overarching themes and 23 specific recommendations for Scottish Government to consider. Recommendations are presented in Annex 3 with relevant ones highlighted.

Taken together, the recommendations aim to support the transition to water-resilient places where communities can continue to thrive as climate change impacts play out over the coming decades.

The recommendations focus on what is required to improve surface water management in Scotland. Scottish Government is now working with key stakeholders to prioritise them and describe how they will be taken forward.

8.7.3 Managing Surface Water in Scotland in 2020

In Scotland, the management of surface water, including flooding, is a significant and well-known challenge[7] for responsible authorities.[8] Surface water flooding by its nature is complex as it is often caused by a combination of factors. Resolving surface water flooding issues requires a coordinated effort across organisations, and this can be difficult to achieve given the current policy and legislative framework. Activities and actions in this space are predominantly "issue-driven" with responsibility for resolving particular issues sitting with different organisations.

Optimal delivery is rarely achieved due to the small number of organisations that we depend on to carry out the specific (issue-driven) actions and the range of legislation, policy, practice, deadlines, competing priorities and resources at play in this space. This has been well recognised by responsible authorities for some time and is reflected in their enthusiasm for a reform to how we manage our surface water issues.[9]

Current responsibilities for managing surface water in Scotland are outlined in Annex 1 and a list of the main relevant legislation and associated documents in Annex 2.

Responsible authorities generally understand and agree what solutions are required to address specific identified issues, but a nationally consistent approach is lacking and organisations can struggle to achieve multiple benefits or align priorities, resources and finances into truly joined-up services without taking a more outcome-based approach.

There are exceptions to this with some excellent examples of where organisations have come together to deliver joint outcomes including the Metropolitan Glasgow Strategic Drainage Partnership (n.d.), the Edinburgh and Lothians Strategic Drainage Partnership and the Sustainable Growth Agreement between

---

[7] According to the second National Flood Risk Assessment carried out by SEPA in 2018, the number of properties exposed to surface water flooding will increase from 210 000 to 270 000 by 2080. (The 2080 figure is the current estimate of the number of properties at risk from the 1:200-year flood plus climate change.)

[8] Authorities responsible under relevant legislation are principally but not limited to: Scottish Government, SEPA, Scottish Water, local authorities and the National Parks.

[9] The Scottish Advisory and Implementation Forum for Flooding (SAIFF) in 2018 called for "…a transformation in the way we handle surface water…".

SEPA and Scottish Water. There are also integrated catchment studies and surface water management plans that are being taken forward jointly by local authorities and Scottish Water as prioritised in our Flood Risk Management Plans.[10]

Notably, in 2020, the City of Edinburgh completed their Vision for Water (2020) which focuses on integrating design for water and flooding with the urban landscape (blue-green infrastructure). This has been developed in direct response to the climate emergency, and its aims include providing greener and more attractive places for people, improving biodiversity, reducing exposure to floods and improving environmental water quality. Edinburgh sees this strategy as being at the heart of Edinburgh's future success.

Despite these positive examples, a fully unified approach to the management of surface water in Scotland encompassing existing-retrofit and new-build challenges is yet to be achieved. This is perhaps unsurprising considering:

- The number and dispersed nature of surface water management issues[11];
- The range of factors that contribute to surface water flooding;
- The distributed responsibilities for surface water management;
- The diversity of actions that can contribute to surface water management; and
- The fact that many of the current problems and potential solutions are within areas that are already highly developed, making retrofit a complex and challenging issue.

8.7.4   Vision for the Future

Improving how surface water is managed in Scotland requires a bold vision that engages the widest possible range of players and a framework to support delivery. The draft vision set out below Box 2 to stimulate discussion takes as its starting point that Scotland will thrive *because* it is water resilient. It aims to present a powerful ambition that *everyone* can get behind.

> **Box 2: Draft Vision for Water-Resilient Places**
> *Scotland's blue-green towns and cities are thriving water-resilient places designed to adapt to increased rainfall, river flooding and sea-level rise. They attract people, businesses and investors because they are great places to be and because they are resilient to climate change.*

---

[10] Scotland has 14 Flood Risk Management Plans outlining a set of prioritised actions to reduce the impact of floods. They provide detail on the costs, benefits and delivery timetable for actions.

[11] Including flooding, drainage, environmental water quality and the performance of combined sewer overflows (CSO) and their impact on receptors including bathing waters.

> *They provide wide-ranging economic, social, environmental and well-being benefits to individuals, communities and the nation.*

The vision aims to:

- Present a positive image of our future towns/cities/places;
- Make the link between water resilience and thriving successful places;
- Identify that planning and designing for drainage and flood risk management *(through blue-green infrastructure)* drive multiple benefits for our communities;
- Engage a broad range of stakeholders to adapt their activities to contribute to our future water resilience.

Realising this vision will require a fundamental change to how we in Scotland regard our water environment. By considering water first, we can move from battling to overcome its negative impacts, to capitalising on the positive contributions it can make to the realisation of our overarching Hydro Nation ambition (Scottish Government n.d.).

Scotland's approach recognises that the journey towards blue-green places and water resilience will require a shift from the current position where a few organisations are tasked with "fixing" all our water issues to enable others to carry out their activities, to the position where the effective management of water is known to contribute to the success of all our activities and is supported by a broader range of players. Understanding the direct link between an organisation's activities and water resilience will lead to more informed choices that have the potential to benefit those directly and indirectly affected by their decisions and actions.

This draft vision is supported by a framework that describes what needs to come together to make it happen. This includes five elements well known to flood risk management and drainage practitioners and a sixth that came through very strongly in the course of research for the policy paper and discussions with stakeholders and is clearly a very important factor to the success of surface water management in future, i.e. that all decision-makers contribute to water resilience.

8.7.5   A Framework for the Delivery of Water-Resilient Places

The framework in Fig. 8.2 supports the vision and outlines that we need to bring together to ensure a coordinated, cross-sector and sustainable approach to managing surface water and drainage in our cities, towns and smaller settlements.

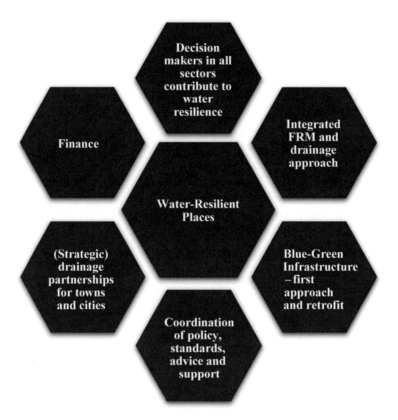

Fig. 8.2 Key elements: what we need to bring together to deliver water-resilient places

Combined with the vision, this framework will provide the policy focus to bring organisations together to increase the delivery of blue-green infrastructure and create more and better water-resilient places in Scotland.

8.8 Recommendations

The following are recommendations of what Scotland should do to improve its surface water management. They are structured around the six key elements from Fig. 8.2 and are presented here in the context of:

- Facing up to the climate emergency;
- Delivering great blue-green places to live (at all scales) that are adaptable to future conditions;
- Tackling surface water flooding (Fig. 8.3).

Our water resilience in Scotland will be improved by making it a core consideration for a broader range of decision makers. Many decisions are currently made without reference to flood risk management or drainage often resulting in an increase in the "total asset" required to be water resilient, more problems that need to be overcome or missed opportunities to make improvements in this space. Few decision-makers understand the impact they can have on their water resilience or on the water resilience of others. The transition to water resilient places would be helped if decision makers accounted for their activity in terms of how it contributed to tackling the climate emergency and how it impacted on flooding and drainage.

Fig. 8.3 Widening the range of decision-makers contributing to water resilience

Recommendation 1
A vision for blue-green cities for Scotland should be established.

Recommendation 2
A strategy and routemap should be set out supported by the key policy changes that are required to drive the transition to blue-green cities and water resilience.

Recommendation 3
Scotland should channel support *towards* actions that contribute to creating great places that are resilient to future flooding and drainage challenges, and *away from* activities that add to our future flooding and drainage burden.

Recommendation 4
We should take a placemaking approach to achieving blue-green cities and water resilience involving partners in the public and private sectors, the third sector, individuals and communities.[12]

Recommendation 5
Relevant decision-makers, including public bodies as part of their climate adaptation duties, should take account of flooding and drainage within their climate planning. (*Public sector bodies are legally required to reduce greenhouse gas emissions and support Scotland's adaptation to a changing climate.*)

---

[12] The transition to blue-green places will require interventions at all scales.

Recommendation 6

Climate impact assessments applying to public policies/activities should include assessing the impact of the proposed policy/activity on water resilience. That is, considering if the activity adds to flooding and drainage issues helps manage flooding and drainage or has no effect on flooding and drainage.

Recommendation 7

A guidance and support package should be made available to policy-makers and investment decision-makers to give them the tools to maximise water resilience and success for their activities. This should include a tool to assess whether their activity has a negative, positive or neutral effect on our water resilience.

Recommendation 8

The land-use planning process (development planning and development management) should, where appropriate, include a requirement for all sites/development proposals to be assessed and report on how they will contribute positively to the climate emergency[13] and water resilience (Fig. 8.4).

Recommendation 9

Surface water flooding issues should be solution-focused and addressed by coordinating across organisations and implementing the best integrated sustainable solution (*overcoming current legislative responsibilities and debates about ownership and ongoing maintenance*).

Recommendation 10

Working links between the flooding, water industry and climate policy teams in Scottish Government should be strengthened to improve coordination and encourage delivery of more and better blue-green actions.

Recommendation 11

Guidance and support should be produced to allow flood risk management prioritisation to factor in the wider benefits of blue-green actions such that progress can be made across all sources of flooding. Current benefit/cost analysis techniques do not adequately account for "other" benefits and favour fluvial and coastal actions.

Recommendation 12

How we measure our success in terms of reducing the impacts of flooding should be reviewed to encourage a wider range of actions. The current approach (*counting properties at risk and damages avoided*) often favours fluvial and coastal protection schemes over surface water flooding management actions. This should include introducing new ways of accounting for the wider benefits that blue-green actions bring to health, well-being, economic prosperity and our natural environment (Fig. 8.5).

---

[13] This should include climate mitigation *and* adaptation.

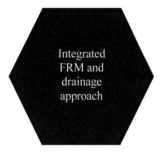

Integrated FRM and drainage approach
The Flood Risk Management (Scotland) Act 2009 established the framework for the integrated delivery of flood risk management. Progressing the joint delivery approach required by the Act has presented some strategic, tactical and operational challenges particularly around surface water management planning. Integrating drainage system requirements with surface water flood risk reduction across organisations and Scottish Government policy areas continues to challenge and requires improvement. An outcome-based approach where "who decides", "who pays" and "who delivers" is established up-front is at the heart of this.

Fig. 8.4 Integrated FRM and drainage approach

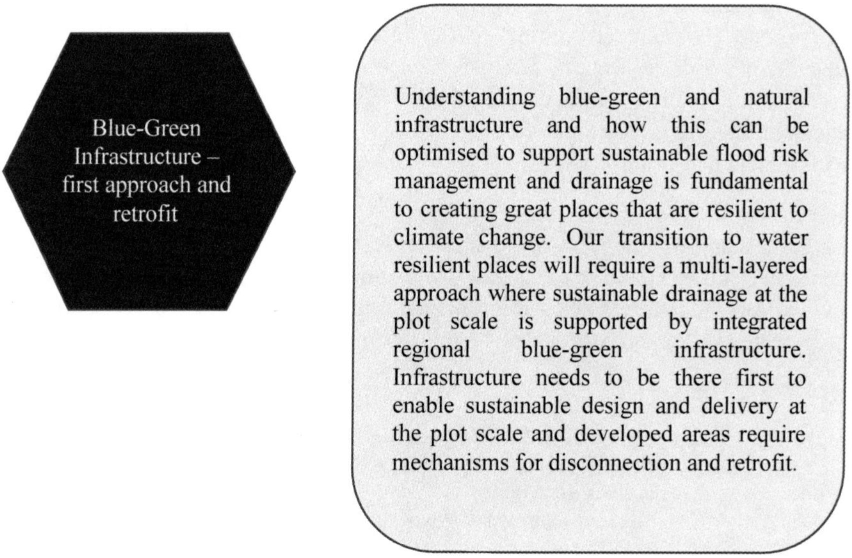
Blue-Green Infrastructure – first approach and retrofit
Understanding blue-green and natural infrastructure and how this can be optimised to support sustainable flood risk management and drainage is fundamental to creating great places that are resilient to climate change. Our transition to water resilient places will require a multi-layered approach where sustainable drainage at the plot scale is supported by integrated regional blue-green infrastructure. Infrastructure needs to be there first to enable sustainable design and delivery at the plot scale and developed areas require mechanisms for disconnection and retrofit.

Fig. 8.5 Blue-green infrastructure: first approach and retrofit

Recommendation 13

Placemaking (*and masterplanning*) should establish blue-green infrastructure needs from the outset where planning authorities' decisions are informed by a comprehensive water strategy[14] where:

- The natural infrastructure is defined;
- Strategic flood risk and drainage assessments are carried out;
- A blue-green infrastructure structure plan is defined.

Recommendation 14

The drainage of surface water from all new sites wherever practicable should be by blue-green infrastructure. Land for blue-green infrastructure should be a site prerequisite, and all designs should presume no rainwater connection to sewer.[15]

Recommendation 15

It should be a priority for existing developed areas to remove as much surface water from sewers as possible through disconnection, retrofitting and diversion to blue-green infrastructure (*incentives and guidance should be put in place to support this*) (Fig. 8.6).

Recommendation 16

Scottish Government should establish a strategic stakeholder group dedicated to promoting and supporting the transition towards blue-green places and water resilience.

Recommendation 17

To support Recommendation 18 of the Infrastructure Commission for Scotland (n.d.) Key Findings Report, Scottish Government should consider how to bring together the quality, standards and value for money elements of flood risk management, coastal erosion and drainage actions, including how they are determined and regulated (Fig. 8.7).

Recommendation 18

Larger towns and cities should be encouraged to establish drainage partnerships to lead a coordinated drive towards blue-green cities and water resilience. Membership of the drainage partnerships should include senior leaders of relevant organisations empowered to make cross-sector strategic commitments (Fig. 8.8).[16]

---

[14] The City of Edinburgh Council Water Management Strategy focuses on integrating design for water and flooding with the urban landscape (blue-green infrastructure). This has been developed in direct response to the climate emergency and aims to inform planning decisions, provide greener and more attractive places for people, improve biodiversity, reduce exposure to floods and improve environmental water quality.

[15] Scottish Water's Surface Water policy states: "*For sustainability and to protect our customers from potential future sewer flooding, we will not normally accept any surface water connections into our combined sewer system*".

[16] There are currently two such partnerships in place. The Metropolitan Glasgow Strategic Drainage Partnership (established in 2002) and the Edinburgh and Lothians Strategic Partnership (established in 2019).

Unifying our approach to surface water management and guiding the many organisations involved in this complex space requires coordination of policy, standards, advice and support. A focal point for surface water management is required where surface water flooding, drainage and blue-green infrastructure can be considered in-the-round. Such a hub would ensure that all legislative requirements are met, set the strategic direction to optimise resources and outcomes and promote and safeguard best practice.

Fig. 8.6 Coordination for resilient places

The Metropolitan Glasgow Strategic Partnership has successfully brought partners together behind their vision 'to transform how the city region thinks about and manages rainfall to end uncontrolled flooding and improve (environmental) water quality.' We are still learning from this approach but there are clear benefits that could be realised in other towns and cities across Scotland if this approach was adopted elsewhere. The recent establishment of the Edinburgh and Lothians Strategic Drainage Partnership shows growing interest in this sort of grouping.

Fig. 8.7 Strategic partnerships

Recommendation 19

Scottish Government should consider how our transition to blue-green places will be funded and where new sources of sustainable finance from a wider range of beneficiaries can be accessed to support the vision.

A financial framework is required to support the transition to blue-green cities and water resilience. This will need to identify the sources of funding and establish funding streams (e.g. from a hybrid of flood risk management funds, water charges and private finance).

This will include understanding and managing established funding sources, seeking new sources of finance and establishing mechanisms to coordinate and direct funding to support the delivery of multiple benefits including optimal improvements to flooding, drainage and blue-green infrastructure. This would not only contribute to meeting our surface water management objectives but also green space, well-being and connectivity objectives for our towns, cities and smaller settlements.

Fig. 8.8 Finance

Recommendation 20

Funding of blue-green infrastructure and water resilience should come from a broader base of public and private contributors reflecting the wide-ranging benefits it provides.

Recommendation 21

Public expenditure should always take into account how to make investments climate positive and water resilient positive.[17]

8.9 Hydro Nation International

The goal of the Hydro Nation International strategy is to share Scotland's knowledge and innovation in a global context. To this end, Hydro Nation Research International (HNRI) coordinates a range of international water-related activities that contribute not only to the Hydro Nation agenda, but also to the United Nations'

---

[17] Ill-informed expenditure can inadvertently add to the "total asset" exposed to flooding and drainage issues.

Sustainable Development Goals, in particular Sustainable Development Goal 6 (SDG6), *"Ensure availability and sustainable management of water and sanitation for all by 2030"*. The sustainable management and stewardship of water resources is a key feature of the Scottish Government's approach, and the same principles apply to Scotland's activities in sub-Saharan Africa; Hydro Nation projects in Malawi and Tanzania are prime examples.

8.9.1 Scotland and Malawi

Scotland and Malawi have had long-standing historical ties stretching back to the expeditions of Dr. David Livingstone, the famous Victorian missionary and abolitionist. This relationship was reconfirmed in 2013 by legislation in both countries and the establishment of a government-to-government cooperation agreement supporting joint work at an official level on water resource management, governance and legislation. The Scottish Government has conducted this work in part through its Climate Justice Fund (CJF) delivery partner, the University of Strathclyde which focuses on Malawi's ability to achieve the UN Sustainable Development Goals as they relate to water, wastewater and infrastructure management. This challenge is readily evident in Malawi, where 10 million people do not have access to adequate sanitation, nearly 2 million people do not have access to safe water, and over 300,000 children under the age of five die each year from diarrhoeal diseases. A safe and effective Malawi water infrastructure is the keystone to population health which in turn promotes educational achievement and entrepreneurial opportunity, helps address issues such as gender equality and provides an important bridge from poverty to prosperity.[18]

Through the CJF Water Futures Programme, the Scottish Government has helped evaluate the sustainability of over 120,000 rural Malawian water supplies and supported active communication between the Ministry of Agriculture, Irrigation and Water Development and other ministries and a host of stakeholders, including industry, NGOs, researchers and rural communities. This longitudinal approach allows research to assess sustainable development needs, while government and community stakeholders develop and evaluate appropriate policy. Initial assessment primarily revealed a lack of capacity to understand and manage the complex nature of groundwater resources which is the primary drinking water supply for more than 80% of its 19 million inhabitants. Malawi's population growth rate of 3% only adds to the urgency of this need, where agricultural development, deforestation and climate change vulnerability require a truly integrated approach to water-resource management.

---

[18] For further reading, see in this volume, Alexander and Cordova, "Alleviating Poverty through Sustainable Industrial Water Use: A Watersheds Perspective".

The following initiatives are representative of the work done by CJF and the Scottish and Malawian governments in this arena:

Asset Management Information Systems and Data Collection

The Water Futures Programme (Box 3) has helped the Malawi government collect and access data to improve their decisions as they develop their water policy. New digital tools developed by CJF for water now provide real-time access to management information for water resources and supplies across all of Malawi. Malawi staff have collected water infrastructure data at surface water, groundwater, gravity-fed rural and peri-urban water points and targeted waste and sanitation infrastructure. CJF is near completion of the first National Dataset for Rural Water Supplies in Malawi which will include information about which sanitation facilities and solid waste sites are co-located with water supply points, posing a potential risk of contamination (Kalin et al. 2019; Kelly et al. 2019; Rivett et al. 2019; Truslove et al. 2020; Addison et al. 2020).

Capacity Building and Training

The programme has delivered training in groundwater resources and rural infrastructure evaluation and management to over 400 government staff across all 28 districts, the 3 regional offices and at the national Ministry of Agriculture, Irrigation and Water Development. Training has covered such technical skills as drilling oversight, hydrogeology, and data collection and analysis. While most training has taken place in Malawi, 38 staff members travelled to Scotland for additional training in team-building and Integrated Water Resource Management.

Research

Science-based policy must underpin the attainment of SDG6 targets in Malawi. Collaboration between Scotland and Malawi has to date produced 70 co-authored research reports and an array of peer-reviewed publications in support of sustainable water management decisions. Topics include borehole forensics, drilling and infrastructure training, water supply contract management, and automated indication and evaluation of facilities in conformance with the UN Joint Monitoring Programme for Water Supply, Sanitation and Hygiene (JMP). In addition, Scotland has provided support for the Government of Malawi Isotope Hydrology facility and Scottish Ph.D. students have worked closely with Malawian partner organisations on fundamental research supporting policy reform.

Policy Exchange and Support

Since the inception of the programme, Scottish and Malawian professionals have shared "best practices" for sustainable long-term management of water resources. The first policy exchange visit was in 2012, and subsequent annual exchanges have included representatives of district (local) and regional senior staff, who share experiences with a range of Scottish agencies and organisations. In addition to meetings with the Scottish Environmental Protection Agency and Parliament, Malawian representatives have worked closely with various Scottish Water agencies and have been embedded within the international Organisation for Economic

Co-operation and Development (OECD) review of the Scottish Water Industry. These activities supported the establishment in 2018 of the Malawi National Water Resources Authority.

Regulatory Engagement
Another aspect of this Hydro Nation International initiative is the collaboration between the Scottish Environment Protection Agency (SEPA) and Malawi's new National Water Resources Authority (NWRA), whose mission is to help ensure adequate and sustainable water supplies, prevent pollution of the water environment, manage water catchments and manage flood risk. Working "regulator-to-regulator", the Malawi Scotland Regulatory Partnership has provided regulatory knowledge, advice and guidance to help establish a framework for environmental regulation in Malawi. The Phase 1 goal of this collaboration was to produce a roadmap for operationalisation of the new authority. Now in Phase 2, this partnership looks to build upon that roadmap in collaboration with the newly formed Malawi Environmental Protection Agency (MEPA) and other stakeholders.

While the scope of these partnerships has related to a range of water uses, all contribute to building Malawi's capacity to encourage and support more sustainable use of water by industry. Strong governance, including the ability to collect and analyse data and create and enforce appropriate policies, is essential for progress in this area.

> **Box 3: Case Study: Climate Justice Fund—Water Futures**
> Internationally, as a Hydro Nation with a global conscience, since 2012 Scotland has been engaged through its Climate Justice Fund (CJF) in water-related projects in Malawi, Zambia, Tanzania and Rwanda, as well as international projects in India and Pakistan. In September 2015, First Minister Nicola Sturgeon confirmed Scotland's commitment to the newly announced UN Sustainable Development Goals, subsequently supporting international trade opportunities in water technology and exporting Scottish expertise in water governance and management. The drive to achieve the United Nations Sustainable Development Goals sets the global context for much of the work undertaken by the Hydro Nation Strategy internationally, especially in Malawi, a country with which Scotland has enjoyed a long historical connection.

8.10 Looking to the Future

Scotland is making great strides working with industry to understand its needs and capabilities. Opening the industrial, commercial and institutional water market to retail competition was an important first step, but much remains to be done.

The state-of-the-art development centres on Scottish Water drinking water and wastewater asset sites are helping to drive forward the introduction of new water and energy reduction technologies to be used at scale by utilities and large industrial water users. As Scotland moves ever closer to its goal of sustainable water use, it will continue to share its expertise with countries throughout the world.

Annex 1: Current Responsibilities for Surface Water Management in Scotland

Scottish Government

The Scottish Government is responsible for making national policy on planning and flood risk management including flood protection, natural flood management and flood warning. It is also responsible for drainage of motorway and major trunk roads, through its agency Transport Scotland.

The following organisations have duties and responsibilities to manage surface water and reduce the impacts of flooding.

Scottish Environment Protection Agency (SEPA)

SEPA is Scotland's national flood forecasting, flood warning authority and strategic flood risk management authority. SEPA produces Scotland's Flood Risk Management Strategies and works closely with other responsible organisations to ensure that a nationally consistent approach to flood risk management is adopted.

Scottish Water

It has the public drainage duty and is responsible for the drainage of rainwater run-off (surface water) from roofs and any paved ground surface within the property boundary. Scottish Water can help protect homes from flooding caused by overflowing or blocked sewers.

Local Authorities

Local authorities are responsible for the drainage of local roads and public highways and for providing flood protection and maintaining watercourses. This includes inspection, clearing and repair of watercourses to reduce flood risk and routine maintenance of road gullies on public roads and highways.

Local authorities are responsible for producing Scotland's Local Flood Risk Management Plans and work in partnership with SEPA, Scottish Water and other responsible authorities to develop these.

Landowners

Landowners are responsible for the management of surface water on their land and must ensure that run-off from their curtilage does not cause flooding problems to their neighbours.

Individuals

Individuals are responsible for managing their own flood risk and protecting themselves, their family, property or business.

Annex 2: Legislation, Regulations and Guidance

There are multiple pieces of legislation, regulation and guidance relevant to managing our water environment including:

- Sewerage (Scotland) Act 1968 (as amended);
- Local Government Scotland Acts 1973 and 1994;
- Control of Pollution Act 1974 (as amended);
- Roads (Scotland) Act 1984;
- Environment Act 1995;
- Town and Country Planning (Scotland) Act 1997;
- Building (Scotland) Acts 2003, together with the relevant technical standards;
- The Water Environment & Water Services (Scotland) Act 2003;
- Planning (Scotland) Act 2019;
- Sewers for Scotland (4th Edition) 2018;
- Flood Risk Management (Scotland) Act 2009;
- Water Environment (Controlled Activities) (Scotland) Regulations 2011;
- The development planning and development management process;
- The National Planning Framework (NPF) (currently NPF3 soon to be NPF4);
- Scottish Planning Policy (SPP);
- Planning Advice Notes on Flooding, Sustainable Urban Drainage Systems, Water and Drainage and Designing Streets;
- SuDS for Roads 2010;
- The SuDS Manual (C753) 2015;
- Water Assessment and Drainage Assessment Guide 2016;
- SuDS regulatory method—WAT-RM-08 (SUDSWP);
- https://www.susdrain.org/

References

Addison MJ, Rivett MO, Robinson H, Fraser A, Miller AM, Phiri P, Mleta P, Kalin RM (2020) Fluoride occurrence in the lower East African Rift System, Southern Malawi. Sci Total Environ 712:136260. https://doi.org/10.1016/j.scitotenv.2019.136260

Infrastructure Commission for Scotland (n.d.) Key findings report. Recommendation 18. https://infrastructurecommission.scot/storage/280/Phase1_PartC.pdf. Accessed 15 Feb 2021

Kalin RM, Mwanamveka J, Coulson AB, Robertson DJC, Clark H, Rathjen J, Rivett MO (2019) Stranded assets as a key concept to guide investment strategies for sustainable development goal 6. Water 11:702. https://doi.org/10.3390/w11040702

Kelly L, Kalin RM, Bertram D, Kanjaye M, Nkhata M, Sibande H (2019) Quantification of temporal variations in base flow index using sporadic river data: application to the Bua Catchment, Malawi. Water 11(5):901. https://www.mdpi.com/2073-4441/11/5/901

Metropolitan Glasgow Strategic Drainage Partnership (n.d.) https://www.mgsdp.org. Accessed 7 April 2021

Rivett MO, Budimir L, Mannix N, Miller AVM, Addison MJ, Moyo P, Wanangwa GJ, Phiri OL, Songola CE, Nhlema M, Thomas MAS, Polmanteer RT, Borge A, Kalin RM (2019) Responding to salinity in a rural African alluvial valley aquifer system: to boldly go beyond the world of hand-pumped groundwater supply? Sci Total Environ 653:1005–1024. https://doi.org/10.1016/j.scitotenv.2018.10.337

Scottish Government (n.d.) Policy water. Hydro nation strategy. https://www.gov.scot/policies/water/hydro-nation/. Accessed 6 April 2021

Scottish Government (2013) Water resources (Scotland) Act, 2013 asp (5). http://www.legislation.gov.uk/asp/2013/5/enacted. Accessed 6 April 2021

Scottish Government (2020) Towards a Robust, Resilient Wellbeing Economy for Scotland: report of the advisory group on economic recovery. https://www.gov.scot/publications/towards-robust-resilient-wellbeing-economy-scotland-report-advisory-group-economic-recovery/. Accessed 6 April 2021

Scottish Government (2021) National performance framework. https://nationalperformance.gov.scot/. Accessed 6 April 2021

Scottish Water Annual Report (2019) https://www.scottishwater.co.uk/Help-and-Resources/Document-Hub/Key-Publications/Annual-Reports. Accessed 6 April 2020

Scottish Water Annual Report (2020) https://www.scottishwater.co.uk/Help-and-Resources/Document-Hub/Key-Publications/Annual-Reports. Accessed 6 April 2020

SEPA (Scottish Environment Protection Agency) (2018) National flood risk assessment. https://www.sepa.org.uk/data-visualisation/nfra2018/. Accessed 6 April 2021

Truslove JP, Coulson AB, Nhlema M, Mbalame E, Kalin RM (2020) Reflecting SDG 6.1 in rural water supply tariffs: considering 'affordability' versus 'operations and maintenance costs' in Malawi. Sustainability 12:744. https://www.mdpi.com/2071-1050/12/2/744

Vision for water management in the City of Edinburgh (2020) https://www.edinburgh.gov.uk/downloads/file/28665/water-vision. Accessed 7 April 2021

Chapter 9
Supporting Evidence-Based Water and Climate Change Policy in Scotland Through Innovation and Expert Knowledge: The Centre of Expertise for Waters (CREW)

Robert C. Ferrier, Rachel C. Helliwell, Helen M. Jones, Nikki H. Dodd, M. Sophie Beier, and Ioanna Akoumianaki

Abstract The Centre of Expertise for Waters (CREW), established in 2011, develops and commissions research projects, analysis and synthesis that directly informs ongoing water policy and regulatory processes in Scotland. The Centre is an established and trusted knowledge broker, informing policy, agency and other relevant stakeholders of current understanding and future expected changes across the water sector. All activities are supported by academic excellence and expertise from Scotland's universities, research institutes and other UK centres. The "here-and-now" delivery within CREW is built around keeping pace with developing and emerging policy challenges, drawing on experts within the water community to deliver timely outcomes. Projects are co-constructed with relevant policy stakeholders to ensure cross-organisational priorities are met.

R. C. Ferrier (✉) · R. C. Helliwell · N. H. Dodd · M. S. Beier · I. Akoumianaki
Centre of Expertise for Waters (CREW), Hydro Nation International Centre,
James Hutton Institute, Aberdeen AB15 8QH, UK
e-mail: bob.ferrier@hutton.ac.uk

R. C. Helliwell
e-mail: rachel.helliwell@hutton.ac.uk

N. H. Dodd
e-mail: nikki.dodd@hutton.ac.uk

M. S. Beier
e-mail: sophie.beier@hutton.ac.uk

I. Akoumianaki
e-mail: joanna.akoumianaki@hutton.ac.uk

H. M. Jones
Scottish Government Rural and Environment Science and Analytical Services Division,
Victoria Quay, Edinburgh EH6 6QQ, UK
e-mail: helen.m.jones@gov.scot

© The Author(s), under exclusive license to Springer Nature Singapore Pte Ltd. 2022
A. K. Biswas and C. Tortajada (eds.), *Water Security Under Climate Change*,
Water Resources Development and Management,
https://doi.org/10.1007/978-981-16-5493-0_9

CREW is governed according to key policy areas in Scotland's water sector including flooding, coastal erosion, catchment and natural resource management, rural sustainability and water quality management. Cross-cutting activities focus on managing impact and adapting to climate change, land use and urbanisation, promoting the circular economy and resource efficiency, a post-COVID green recovery, the move to net zero and a just transition for communities.

The Centre has supported EU and national policy development and implementation strategies and Scotland's ambition to deliver the United Nations' Sustainable Development Goals both nationally and internationally. This chapter uses examples to highlight the central principles of the Centre in that the knowledge generated should improve understanding and communication between science and policy, promote improved networking and win-win solutions, and deliver tangible impact and longer-term outcomes for the water environment and society.

Keywords Water · Policy research · Expertise · Scotland · Knowledge broker

9.1 Introduction

Global water resources are under extreme and increasing pressure from human climate change, urbanisation, over-abstraction, pollution, wider degradation, biodiversity depletion and loss of natural capital. Under a business-as-usual strategy, there will be a potential 40% gap between the required global demand and the safe available supply of freshwater by 2030 (World Economic Forum 2012; UN Water 2020). Global drivers are fairly ubiquitous, but the way in which the regional expression of those drivers differs depends upon the availability of the natural resource, its use and overuse.

Sustainable resource management benefits from the development of robust policy and societal governance of water resources and their use. There is an urgent need to accelerate the application and translation of scientific evidence into policy. This can be a complex endeavour, as the uncertainties and complexity of scientific findings do not always lend themselves to concluding clear policy decisions. For this to be done effectively, the Centre of Expertise for Waters (CREW) acts as a knowledge broker, ensuring that the right information gets to the right people in the right format and at the right time. Ensuring confidence in the evidence used to shape policy given its broad societal reach is a critical function of the Centre.

CREW was established in 2011 by the Rural and Environment Science and Analytical Services (RESAS) Division of the Scottish Government to provide capacity and independent scientific advice at the research: policy interface and scientific and technical solutions to underpin water and wider environmental policy development and implementation in Scotland. CREW is one of several Centres of Expertise established by the Scottish Government, which cover additional domains such as Animal Disease Outbreaks (Bowden et al. 2020), Climate Change (Wreford et al. 2019) and Plant Disease (Plant Health Centre of Expertise 2020). These

Centres were established in response to Scottish Government's need for a mechanism which offered rapid and easy access to high calibre scientific evidence and advice. The aim was to create virtual centres which would bring together partnerships or networks of established research experts from a range of research organisations and which would be able to provide coordinated evidence and advice on demand.

The primary objectives of CREW are to:

- Deliver timely and high-quality advice, synthesis and analysis to the Scottish Government and its policy divisions (through "Capacity Building" projects).
- Maintain a network of expertise across all of Scotland's water research community (including all universities and research institutes) to provide support on policy issues.
- Ensure that knowledge generated from the activities of the Centre reaches the widest possible relevant policy, stakeholder audience and communities of interest through a programme of knowledge exchange and engagement.
- Provide a rapid response function responding to pressing policy issues and unforeseen events, linking specialist and policy teams (through "Call Down" activities).

The Centre operates across all water policy domains, providing advice and synthesis in relation to policies focused on catchment management (land and water), flooding and drought, coastal erosion, urban challenges and green–blue infrastructure, the challenge of provision and treatment of drinking water and waste water for rural communities, water quality and pollution, circular economy and waste, the move to a zero carbon, and societal understanding and awareness of water issues and challenges. All these topics have active and ongoing policy deliberations, establishing new principles and revision through updating existing frameworks, revising guidance and establishing new legislation. A consistent challenge for CREW is to ensure that utility of information produced is accessible and has tangible resonance with the broad range of information "users" and communities.

9.2 The Water Resources of Scotland

Scotland is a water-rich country, with over 90% of the UK's standing water. The Scottish Environment Protection Agency (SEPA 2019) details that Scotland has over 125,000 km of rivers and streams and a canal network of approximately 220 km. There are over 25,000 "lochs" (lakes) mainly in the North and West of the country. Most are small, but notable larger lochs include Loch Lomond (71 km^2), Loch Ness (with a volume of over 7 Mm^3) and Loch Morar (UK's deepest at 310 m). Scotland has significant groundwater reserves which support 75% of private drinking water in predominantly rural areas. More than 80% of Scotland's groundwater is considered to be in good condition, though there are some areas of concern.

Although considerable progress has been made over the last forty years in relation to gross pollution of watercourses and the impact of point source pollution, there are still ongoing challenges affecting surface and groundwater resources in Scotland including; hydro-morphological barriers to fish migration, physical alteration of stream networks (including land drainage and loss of wetlands), point and diffuse pollution from both rural and urban environments, abstraction and industrial discharges. The programme of activity outlined for the next 6-year cycle of the River Basin Management Plans for the Scottish River Basin Districts (2021–2027) aims to increase the number of water bodies (over 3000 in Scotland) in good or better condition (or recovering to good or better status) to nearly 95% and is presently undergoing consultation (SEPA 2021a).

Significant water management issues in Scotland's river basin districts are generated from several sectors (Table 9.1), though their spatial and temporal impacts are different across the county and are closely linked with major land use types and the distribution of industry and urbanisation.

Climate change is having a significant impact on Scotland's environment. The climate is changing with the ten warmest years being recorded since 1997, and the annual temperature in the last decade (2009–2018) was 0.67°C warmer than the long-term (1961–1990) average. In addition, annual rainfall was 15% higher than the long-term record and average winter rainfall was 25% higher. Scotland has also

Table 9.1 Significant water management issues in Scotland

Pressure	Sector
Diffuse pollution	Agriculture
	Forestry
	Urbanisation
	Transport
	Private rural wastewater
Point source pollution	Wastewater management
	Aquaculture
	Industrial production
	Disposal of waste
	Mining and quarrying
Abstraction and flow regulation	Power generation
	Public water Supply
	Agriculture (mainly seasonal)
	Industry
Changes to morphology	Historical engineering
	Agriculture
	Power generation
	Urbanisation
	Land reclaim
Invasive alien species	All sectors

Source SEPA (2013)

experienced some notable extremes, with significant flooding in the winters of 2015–2016 and 2018–2019. The summer of 2018 was unusually dry and warm, and many private water supplies ran dry leaving people needing assistance from their local public authority. Projections highlight that such phenomenon will become increasingly frequent, with a trend towards wetter weather in the West and drier in the East (Rivington et al. 2020).

9.3 The Water Policy Landscape in Scotland

Scotland's journey to becoming a "Hydro Nation" began with the publication of the Hydro Nation Prospectus document by the Scottish Government in 2012, which described its vision of Scotland: The Hydro Nation, as a country that:

> … recognise(s) water as part of our national and international identity. We understand the sustainable management of our water resource is crucial to our future success and a key component of the flourishing low-carbon economy and the basis of growing international trade opportunities.

The subsequent Water Resources (Scotland) Act (Scottish Government 2013) made it a statutory duty for Scottish Ministers to "take such reasonable steps as they consider appropriate for the purpose of ensuring the development of Scotland's water resources". This included a reference to the "value" of water resources:

(A) Means the value of the resources on any basis (including their monetary or non-monetary worth).
(B) Extends to the economic, social, environmental or other benefit deriving from the use of the resources (or any activities in relation to them).

Scotland's unique Hydro Nation agenda comprises a broad suite of activities supporting the Scottish Government's vision of Scotland as a world leader in the sustainable use and responsible management of water, and the important incorporation of both monetary and non-monetary values was an international exemplar (Martin-Ortega et al. 2013).

This globally significant legislation and statutory duty underpinned the philosophy and implementation of an Hydro Nation *duty* covering all aspects of the water sector, water environment and public ensuring that as a nation that recognises the importance and value of water within our national and international identity, it manages the water environment to the best advantage, developing and promoting more efficient resource use and employing its knowledge and expertise at home and internationally in ways which contribute to a flourishing low-carbon economy. In addition, Scotland has set a world-leading target of *"net zero carbon emissions by 2045"*, stimulating a drive within Scotland's water industry to achieve these targets by 2040.

CREW sits centrally within the framework of the Hydro Nation agenda providing the boundary organisation supporting the flow of knowledge and understanding from academic and stakeholder partners to policy. The path length between these communities in Scotland is short, and this engenders a broad community and collective approach to delivering knowledge to policy.

The Scottish Government passed the ambitious Climate Change (Scotland) Act in 2009 establishing a commitment to reducing greenhouse gas emissions by 75% (compared to 1991 baseline values) and a move to net zero by 2045. Recently, the Scottish Climate Change Adaptation Programme (2019–2024) has set out policies to prepare the country for the challenges faced in the delivery of the act, and the risks set out in the UK Climate Change Risk Assessment (UK CCRA 2017). This Adaptation Programme takes an outcome-based approach focusing on what the policy should achieve rather than solely inputs and outputs (Scottish Government 2019). Such an approach is highly complementary to the impact strategy of CREW and one in which the focus is on increasing transparency and accountability in the delivery of knowledge supporting policy.

In December 2020, the Scottish Government updated the 2018–32 Climate Change Plan to embrace and secure a "just" and "green" recovery in a post-COVID world (Scottish Government 2020). Clearly in early 2021, the world is still facing a significant challenge from the ongoing pandemic, but the commitment prioritises economic, social and environmental well-being and responds to the twin challenges of climate change and biodiversity loss. Increasingly, water and wastewater play a central role in this challenge and have become a significant feature of the circular economy and renewable energy options, including manufacturing and the developing hydrogen economy.

Circular economy principles—improving water efficiency, reuse of clean water discharge, increasing resource recovery and reducing energy used to clean, pump, heat and cool water—have stimulated innovation in new water processing and treatment technology.

Impacts of water stress are already being felt by wider UK business, acknowledged by the Food & Drink Federation's "Courtauld Commitment" to reduce water use by 20% (2015–2025) and a continued drive for greater efficiency in use and reuse. The Scottish Whisky Association's Environmental Strategy has a target to improve distilling water efficiency by 10%, and there are over 300 manufacturers and enterprises located in vulnerable water-stressed (DEFRA 2007). More stringent environmental legislation and regulation in markets regarding supply and use is a significant market opportunity for new technologies and processes.

Future proofing Scotland's water-intensive industries (Food and Drink being the highest Scottish and UK industrial water consumptive sector) provides a blueprint for other industries in water challenged regions where water demand is expected to rise with water stress becoming a significant development constraint. Currently, the Scottish Government and SEPA are developing policy on water stress, such as Scotland's National Water Scarcity Plan (SEPA 2020) and Water Resources Management Plan which is currently under consultation (SEPA 2021b).

9.4 CREW's Operational Model

Central to the ambition of research, policy interaction is the development of a community of practice around water science and knowledge. The importance of linking the "here-and-now" need of policy through the provision of information synthesis and analysis is very different from the discovery science ambition of many academic organisations; however, it is those strong academic credentials that also make the information and guidance delivered by researchers to be of such value (Fig. 9.1).

CREW provides actionable recommendations gathered through research, synthesis, analysis and expert knowledge. CREW supports key stakeholders in water policy in Scotland (Scottish Government (SG), Scottish Environment Protection Agency (SEPA), Scottish Water (SW), NatureScot (NS, formerly Scottish Natural Heritage), Water Industry Commission for Scotland (WICS), the Drinking Water Quality Regulator for Scotland (DWQR), Zero Waste Scotland (ZWS), Scottish Canals (SC), Food Standards Scotland (FSS) and contribution from Consumer Advice Scotland (CAS) as well as local authorities, regional stakeholders and communities). The research partnership involves over 300 academics and

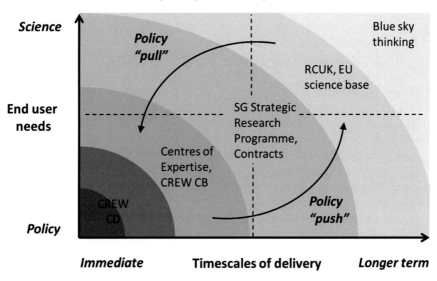

Fig. 9.1 Relationship between timescales of delivery and end-user needs, within the policy push–pull landscape (SG: Scottish Government; CB: Capacity Building projects; CD: Call Down activities; RCUK: UK Research Councils; EU: European Union)

researchers from all of the relevant Scottish Higher Education Institutes (universities), Scottish Research Institutes (James Hutton Institute, Moredun Research Institute, Scotland's Rural College (SRUC), Rowett Research Institute, Biomathematics and Statistics Scotland, and Royal Botanic Garden Edinburgh) and includes UK centres such as the UK Centre for Ecology and Hydrology (UKCEH), British Geological Survey (BGS) and the Met Office. Individual businesses and consultancies also contribute as subcontractors to the university sector.

CREW delivers research policy advice underpinned by some core principles, namely:

- **Co-construction**: The development of a joint understanding of the challenge. Defining the detail of the request to ensure that academic researchers have a clear pathway to delivery. One based on a shared understanding of the reach and utility of the expected outcome, a project that is co-constructed is likely to be targeted to specific needs or policy outcome and therefore make an impact.
- **Communicability**: To achieve the maximum impact of CREW projects, outputs must be available in a format which aligns with the requestors' needs and expectations. Looking for cross-agency benefit is part of the co-construction phase where perspectives from different communities ensure that the output has maximum reach.
- **Trustworthiness**: Evidence is rarely uncontentious and has clarity on conclusions and recommendations derived and against what empirical data and current thinking is critical to the adoption of knowledge into policy. Reed and Meagher (2019) highlight that moral and ideological arguments alongside practicalities can impact such transfer. The importance of a respected community of knowledge generators, as well as a collegiate Centre project facilitation team, integrated with the user community engenders confidence and clarity around societal benefits and potentially dis-benefits of different strategies.
- **Timeliness**: Policy support tends to require the synthesis and assessment of contemporary issues and consideration of future trends and ambitions rather than the development of new primary information and novel data. Whilst not ubiquitous across all the Centre activities, most Centre projects follow this model. It is especially pronounced in Call Down activities which tend to focus on critical issues needing support with an immediacy of a "days to weeks" timescale.
- **Inter-disciplinarity**: Policy problems often require a combination of economics, social science and natural science to address what can be complex, real-world challenges. The Centre needs to ensure there is a holistic framing of the policy challenge and in many instances ensure that an inter- or trans-disciplinary approach is used to effect a positive outcome.
- **Impact, outcomes and "value"**: Although there is an immediacy for many requests from policy, this has to be contextualised against the potentially considerable time lags involved in the production of knowledge and its final use in policy. Legacy impacts are difficult to monitor when the demand for analysis is constantly refreshing, but the impact of capacity building projects can

sometimes only be realised after several years. The Centre's impact model is built around four pillars:

- Developing an inclusive contemporary water community;

  Bridging the supply and demand communities.

- Ensuring collective understanding of the challenge and of the approach needed;

  Maximising benefit to the broadest community of practice through ensuring the policy need and ask is clearly defined.

- An agreed consensus on the action(s) delivered;

  Evidenced action by partners in response to outputs.

- Longer-term impacts evaluated;

  Evidenced by externalities such as improved social cohesion, changed behaviours, enhanced employment and training opportunity, or financial return based on the Centre activities.

- **Cost-effectiveness**: Capacity Building project requests are prioritised by policy teams on an annual basis and tailored to meet both specific demands and the wider interests of the cross-agency and stakeholder communities through co-construction as outlined previously. There is only a finite resource available however to support projects, and therefore, a simple cost–benefit is considered at the start of the annual commissioning cycle to ensure the most pressing challenges are agreed and addressed and that a rolling programme of projects is undertaken. Resources also must be retained to support Call Down and highly responsive requests for analysis and support that emerge at short-term notice.

9.5 Creating Impact: Supporting Policy Outcomes

A Scottish Government review in 2018 found that CREW had delivered significant benefits to its users, was highly valued by them and had effectively strengthened the interaction between science and policy.

(A) Climate change and diffuse pollution impacts on catchment resources

Historical and ongoing issues with diffuse pollution management in Scotland stimulated several projects through which to evaluate the risk to water bodies and to establish principles for management in a changing world. Lilly and Baggaley (2014) developed soil risk maps to assist SEPA Catchment Managers in identifying spatial locations which were prone to erosion, the consequence of impact run-off leaching and compaction. A tablet-based version allowed "on-the-ground" identification of hot-spots of risk by SEPA staff and other stakeholders which were used to inform the community about the relationship between soil types and water quality and implement locally based mitigation measures.

This work contributed to a wider effort to establish a practical design guide for rural sustainable drainage systems (Rural SuDS) (Duffy et al. 2015) developed to reduce the impact from diffuse pollution. In addition, the guidance provided a blueprint for farmers and landowners to apply for grants to the then Agri-Environmental Climate Scheme (AECS) and to assist in achieving Good Agricultural and Environmental Conditions (GAEC), in accordance with guidance and legislation such as the Prevention of Pollution from Agricultural Activities (PEPFAA) code (Scottish Government 2005) and the subsequent Diffuse Pollution General Binding Rules in the Water Environment (Diffuse Pollution) (Scotland) Regulations (Scottish Government 2017) which provide a general level of environmental protection and which contribute significantly to water quality improvements. Collectively, this guidance assisted farmers and landowners in understanding how they could use rural SuDS to protect both the environment and their assets, ensure compliance and reduce operational costs.

The wider contribution to pollutant loading from Onsite Wastewater Treatment Systems (OWTS) which provide treatment for non-networked rural households and businesses (c. 16,000 in Scotland) was examined by O'Keeffe et al. (2016) and gave an insight into the potential for phosphorus and faecal microbial loads which can potentially impact on the status of Water Framework Directive water bodies and protected areas. Mitigation measures were proposed through which to reduce this environmental loading which concurrently identified the opportunity within rural provision for increased energy generation, reduced greenhouse gas emission and/or nutrient recovery at both household and local scales. The governance and management of rural supplies were further explored by Hendry and Akoumianaki (2016), who highlighted that rural provision issues were a global challenge, but that better guidance and support for individuals and communities, the role of "trusted intermediaries", improved data collection and risk assessment, potential certification, and clarity about institutional engagement and the policy landscape [e.g. EU Drinking Water Quality (DWQ) (Directive 1998/83/EC) and (Directive 2015/17 87/EU)] would be beneficial to consumers, stakeholders and policy teams.

This work informed the development of the Sustainable Rural Communities concept promoted by the Scottish Government, which aims to facilitate a paradigm shift in delivering affordable energy, treatment and disposal of waste and the provision of drinking water supplies and aims to deliver a closed-loop approach which is carbon neutral, cost effective and resilient.

A critical question for rural environments and the management of diffuse pollution was would agricultural land affect the flood risk and impact water quality especially under the projected scenarios of increased winter rainfall? Hallett et al. (2016) reviewed data from monitored catchments over winter 2015/16 and found a 30% increase in the occurrence of severely degraded soils. These degraded areas generated a run-off, erosion and nutrient loss of about 10 times greater than from undamaged field areas. This analysis complements the guidance developed by CREW and available for farmers and other stakeholders in managing soil and land

(Cloy et al. 2018), where the promotion of good soil structure and field drainage systems are key to achieving good water quality and minimising flood risk. Additionally, they are also vital for agricultural productivity and for the greenhouse gas balance of many agricultural systems. Such advice is actively encouraged by agency and other staff as good practice across the agricultural sector.

(B) Climate change

Although there is a degree of uncertainty about how global warming trends will be expressed at a local or regional scale, evidence from the 1960s until the present day confirms that Scotland's average climate has become wetter (especially in the West of the country) and warmer. This trend is expected to continue going forward with on average, hotter and drier summers and milder wetter winters (Werritty and Sugden 2012; Scottish Government 2019).

A recent review by Scotland's Centre of Expertise connecting climate change research and policy (CXC 2016) (Rivington et al. 2019) identified the change in winter snow cover in the Cairngorm Mountains, N.E. Scotland, and noted a decline over the period 1969–2005, consistent with other mountainous regions. In the near term, the estimates indicate a potential for a continuation of snow cover at the current range of variation, but with a substantial decline from the 2040s. These findings are in line with results from the UK Meteorological Office and Intergovernmental Panel on Climate Change (IPCC 2019).

In addition to the decline in winter snow coverage and storage, the number and frequency of extreme rainfall events have increased, and recent years have seen numerous, widespread and significant flood events in Scotland. By way of summary:

- High river flow run-off from Scottish rivers has increased by over 20% and winter river run-off by nearly 45% over the last 4 decades (Hannaford 2015).
- Under a high emission scenario, peak river flows for some Scottish river catchments could increase by more than 50% by the 2080s (Kay et al. 2021).
- Projected increases in intense heavy rainfall events in both summer and winter will also increase the risk of extensive and significant river and surface water flooding (Met Office 2019).

Although temperatures are projected to increase in both summer and winter, warming is expected to be greatest in summer. Notably, in 2018 Scotland experienced anomalously high summer temperatures and a widespread drought causing major impacts across the rural sector, impacting local water supplies, crop production and the incidence of wildfire (Undorf et al. 2020). Climate change projections identify an increased chance of experiencing a summer as hot as the summer of 2018 to between 12 and 25%. With future warming, hot summers by mid-century could become even more common, somewhere closer to a 50% increase (Met Office 2019).

CREW and its sister centre CXC (Climate XChange 2016), along with other partners, have undertaken several analyses and syntheses in support of the

Implementation of the Flood Risk Management (Scotland) Act 2009 and the development of the Climate Ready Scotland: Second Scottish Climate Change Adaptation Programme 2019–2024 providing a bridgehead of knowledge through which to implement sustainable practices across planning, construction and development.

(i) Resilience and land use planning

Measures for Natural Flood Management (NFM) have gained considerable momentum as options to lessen the consequences of flooding on downstream landscapes and communities with the dual benefits of storage water to recharge groundwater and enhance the resilience to drought.

The overriding concept of "slowing the flow" provides a useful benchmark for a whole range of different both generic and site-specific interventions. Spray et al. (2015) undertook an extensive survey of farmers' attitudes to NFM and the potential use of different policy instruments to promote uptake. They found that many parties were interested in the potential use of NFM measures but needed robust evidence of efficacy and clarity on incentivisation before implementation. Favoured measures included woodland planting along riparian corridors and in less productive areas to minimise the impact on productive areas, agricultural flexibility and income. Further research (Marshall et al. 2019) explored factors that affect community support for NFM, especially for those in flood-prone areas. The authors recommended that to encourage wider community support for NFM the following principles should be supported, namely.

- **Provide information about NFM measures**. Effectively communicate the multiple benefits of NFM (e.g. improved water quality, reducing soil erosion, biodiversity gains, amenity value), even if these might not be included in a standard cost–benefit analysis (CBA). Identify and communicate how different NFM measures may be used as part of local FRM schemes to mitigate flood risk.
- **Promote community engagement around NFM**. Find opportunities for continuous and constructive engagement with communities. Provide relatable evidence of the efficacy of NFM schemes, e.g. establishing local pilot studies, describing existing NFM projects or facilitating fact-finding visits to these. Consider involving communities in river monitoring or other aspects of scheme planning and implementation.
- **Build and maintain trust around Flood Risk Management processes**. Agencies and authorities need to engage effectively and consistently to build and maintain trust with communities. Provide clearer cross-links between parts of the process, i.e. so those seeking information due to specific flood events can also see information on Local Flood Risk Planning processes, Potentially Vulnerable Areas (PVAs), etc. (Marshall et al. 2019).

Following the establishment of the NFM Network Scotland (NFM 2018) membership amongst practitioners, researchers and communities has grown, increasing the knowledge and best practice on NFM. The provision of case studies, practical operational guidance along with agency support and incentivisation is increasing the uptake of NFM in Scotland.

Understanding the mechanisms and timing of how urban floods are generated in response to both climate change and rapidly changing urban environments has been a focus of considerable research in recent years. One important factor affecting surface water flood risk is the conversion of porous surfaces (e.g. gardens, green spaces and open land) to impervious surfaces (e.g. buildings, paving and car parking). Rowland et al. (2019) identified that within the City of Edinburgh "urban creep" (the loss of porous surfaces) averaged nearly 6.5 ha per year (from 1990 to 2015), and the rate of urban expansion (mainly in the peri-urban fringe) over the same period was nearly 5 ha per annum. Although these values seem modest for a significant urban conurbation such as Edinburgh with a population of c.0.5 M (City of Edinburgh Council area), the history and legacy of city development and the change in land cover and porosity of surfaces could be a significant generator of surface run-off into existing urban infrastructure.

Surface water flooding in general is difficult to predict as events are often very localised, develop quickly and can be extremely intense. The risk in Scotland is particularly high with over 100,000 properties identified as being vulnerable in the SEPA National Flood Risk Assessment. This is a high-profile risk with Scottish cities experiencing surface water flooding in recent years (e.g. Glasgow in 2002 and with subsequent smaller events experienced in 2007, 2011, 2012 and 2013, Aberdeen in 2001 and 2015, and Edinburgh and Stirling in 2019) (Speight et al. 2019).

Moore et al. (2015) produced the UK's first operational surface water flood risk forecast with a 24-h lead time, which was successfully piloted in Glasgow at the Commonwealth Games. It delivered a novel method for forecasting the impacts of flooding in real-time and increased knowledge on communicating uncertainties in flood risk. The 2014 pilot paved the way to several significant developments in forecasting and modelling real-time surface water flooding. These have built on the upgrade of the UK radar network enabling improved observations of heavy rainfall, improvements to numerical weather prediction, growth in the use of citizen observations for flood risk modelling and monitoring, cloud computing and the development of faster inundation models, and the strengthening of the urban flood forecasting community of practice. Although there are still limited examples of operational surface water flood forecasting models at urban scales in the UK, the challenge is less driven by numerical and computational constraints but the ongoing difficulty of communicating uncertainties around the location and timing of the heavy rainfall forecasts and the resulting impacts at an urban scale (Speight et al. 2019).

At a different spatial scale, the Metropolitan Glasgow Strategic Drainage Partnership (MGSDP) is a cross-institutional partnership that aims to make North Glasgow more resilient to the impacts of climate change by working together, breaking silos and developing innovative approaches to urban water management which focus on the delivery of both technical and socio-economic outcomes for the development and regeneration of that part of the city (Allan et al. 2015). Urban water planning utilised Sustainable Urban Drainage Systems (SuDS) principles and infrastructure such as the Forth and Clyde canal as a conduit for receiving storm water, as well as the development of the canal itself to promote green–blue infrastructure delivering multiple benefits to the surrounding community. The MGSDP is part of Scotland's National Planning Framework 3 (NPF3), involving an investment of around £38 M, and is a nationally significant exemplar of catchment-scale water and drainage infrastructure planning, aiming to better service existing communities, unlock potential development and build greater resilience to long-term climate change (MGSDP 2021).

Scotland's National Coast Change Assessment (Fitton et al. 2017) provided a contemporary understanding of the resilience and vulnerability of Scotland's coastal assets. It identified the important role of natural features (beaches, dunes and coastal habitats) in protecting approximately £13B of assets, many of which were under significant pressure from future climate change and prone to increased erosion. The project also highlighted the need for better integration of coastal vulnerability within existing policy to ensure complementarity of planning and future management, and the outputs of the next phase of this project will be launched in summer 2021.

A decade on from the Flood Risk Management (Scotland) Act (Scottish Government 2019), and through the new evidence provided, the Scottish Government has committed over £42 M each year to increase flood resilience for communities across the country. The ongoing risk-based plan is focused on:

- Reducing the number of people and properties at risk and continued investment focused on protecting the most vulnerable communities.
- Attenuating flow in rural and urban landscapes ("slowing the flow").
- Implementing sustainable surface water management and mitigating pluvial flooding.
- Dynamic place-based coastal management.
- Ensuring that public and communities understand risk and protection measures and promote behaviours to safeguard their properties and businesses.
- Ensuring that flood management is climate-proofed for future long-term change and impacts.

The act has created a framework for the coordination and cooperation and identified the key working relationships and responsibilities of the agencies involved in flood risk management and continues to build on scientific (both biophysical and social) knowledge supporting policy development generated by the activities within CREW, CXC and other partner organisations (Scottish Government 2021a).

Most recently, the Scottish Government (2021b) has developed a policy framework entitled Water-resilient places—surface water management and blue-green infrastructure. This framework outlines how surface water is currently managed in Scotland, sets out a vision for the future and describes the components that should be brought together to form a coherent framework that will support delivery of water-resilient places.

(ii) Managing the impacts of climate extremes

Many areas of Great Britain were badly affected by flooding in the winter of 2015/2016. Over a fourteen-week period commencing in early November 2015, a "persistent and exceptionally mild cyclonic episode" brought "severe, extensive and protracted flooding which impacted most damagingly on northern Britain, Northern Ireland and parts of Wales." The flooding had considerable impacts on numerous communities, including private homes, business premises, transport infrastructure and agricultural land with damage estimated to be more than £1.3 billion (Marsh et al. 2016). In Scotland, severe flooding also affected the North-East of Scotland in late December 2015 and early January 2016. Some flooding was experienced in Aberdeen city, but most flooding and associated disruption was experienced across Aberdeenshire, in remote towns, villages and the open countryside. These flooding events in Scotland were some of the worst in living memory, and a CREW project was undertaken to establish the long-term impacts of this flooding and to better understand what types of support and advice people and communities needed and at what times as part of a recovery strategy (Philip et al. 2020). This study was one of the first of its kind, with a particular focus on how resilience changes over time given that significant impacts were still felt three years after the floods.

The analysis highlighted that many people were unable to return to their homes for more than six months after the flood, and that many moved more than once when using temporary accommodation, with some people being moved up to five times. This placed a significant burden on people's well-being post-flood, and more than 60% of the respondents in the survey indicated a deterioration in personal health and well-being. In addition, flooding affected the whole community, and those not directly impacted also faced disruption to utilities, transport and local service. Detailed recommendations from the study highlighted the need for flood preparation in potentially vulnerable communities and resilience planning, improved communication and liaison between residents, statutory authorities and emergency services during the flood progression and post-flood implementation of support targeted at both immediate practical assistance and longer-term well-being. The impact of such catastrophic flooding events leaves a significant legacy across the whole community (Currie et al. 2020), and outputs directly inform the development of the "Living with Flooding" Campaign.

Scotland is generally thought of as a water-rich country and as one of the wetter areas in the UK and Europe. However, dry summers have occurred several times in recent years and the droughts of 2003, 2012, 2018 and 2020 being significant enough to result in water scarcity. Gosling (2014) identified that future drier summers are expected in Scotland, leading to reduced supply, drier soils and lower river flows during the summer (Brown et al. 2012; Rivington et al. 2020), especially in vulnerable areas. Such vulnerability is defined as a combination of the adaptability, sensitivity, and exposure of areas. Adaptability is the capacity for an area to minimise negative effects; sensitivity being the amount of change for a given amount of exposure; and exposure being the amount of land use or climate change expected (Sample et al. 2016). The East of Scotland is drier, more densely populated, has more agricultural land than the west and is considered more vulnerable to droughts than the west (Waajen 2019).

A most recent example of this vulnerability was the 2018 drought, where northern and eastern regions of Scotland received less than 75% of the long-term average rainfall (lowest in circa 40 years), and this followed the winter of 2017–2018 which was the third consecutive dry winter period (October–March). The drought event of summer 2018 was marked by severe water shortages in private water supplies (PWS) and treatment, raising awareness about their vulnerability to water scarcity. The drought also resulted in significant impacts on agriculture and businesses, ecological pressures caused by record low flows in several Scottish rivers and wildfires in many locations (Hare et al. 2020).

To further assess this, CREW commissioned a project to review the likely impacts of climate change in terms of the amount, frequency and spatial distribution of precipitation (rainfall and snow). The study analysed the most recent UKCIP18 climate change projections to assess future spatial and seasonal changes in precipitation and mapped drought risk (i.e. meteorological drought) against the known density of PWS. This produced a bespoke Meteorological Drought Risk Indicator, which identified areas at risk from drought. Further review highlighted the importance of hydrological droughts (low flow/water levels in rivers, lochs and groundwater) which can last for months after a meteorological drought (the prolonged absence of precipitation) has subsided. PWS depend heavily on groundwater and are particularly affected once a hydrological drought occurs. The study highlighted that too little is currently known about hydrological droughts in Scotland to effectively predict subsequent impacts on PWS. Therefore, developing adaptive measures to increase the resilience of PWS has been identified as a key future priority (Rivington et al. 2020).

Understanding the nature and extent of water scarcity and the potential for future droughts has stimulated cross-policy considerations. Water scarcity was previously not considered within Scotland's Land Use Strategy Phase 1 (2011–2015), nor in Phase 2 (2012–2021). However, the current drafting of Phase 3 will consider water resources management in a holistic way including wider consideration of the impact of extremes on landscapes, communities and businesses. Scotland's National Water Scarcity Plan (Scottish Environment Protection Agency 2020) sets out how water resources will be managed in advance of and during periods of prolonged dry weather. The inter-agency plan aims to balance environmental protection with human resourcing and economic activity. This is based on the development of action plans to be implemented at local level which involve several specific mitigation measures and actions, and details expected inter-agency and wider stakeholder roles, responsibilities and behaviours. This recognises the increasing importance of water scarcity in the delivery of Scotland's River Basin Management Plans and the need to respond to the growing climate challenge.

9.6 Future Direction and Closing Remarks

A key requirement for knowledge brokerage within a Centre of Expertise such as CREW is the need for adaptive management and for the articulation of the knowledge transfer process within a policy landscape that is dynamic, seeking to integrate across sectors and delivering against pressing schedules and deadlines. Self-evaluation and critical appraisal are important elements of the Centres learning, and this has been an active process since CREW's establishment. The relationship with knowledge providers through Scotland's wider academic community is also dynamic, and constantly evolving there is a concomitant requirement to embrace new expertise and skills, insights and concepts, technologies and know-how, to address the ever-changing policy landscape and need for evidence-based decision supporting social, environmental and economic outcomes.

Knight and Lyall (2013) identified that interactions between researchers and policy can take a number of forms ranging from simple one-to-one engagement to more formalised interaction exchange of information and through shared events, to the co-construction of outputs and outcomes. CREW's operational structures aim to promote that interaction from rapid dialogue with experts, provision of analysis and synthesis of knowledge, to outcome-focused delivery.

Critical to delivery is maintaining and promoting relationships between both the research and policy communities. Key to those relationships is the Centre's Facilitation Team (CFT). Their role is to ensure awareness of both evolving policy and advances in research pertinent to meeting the contemporary challenges faced by decision-makers. Within individual projects, there is a need for flexibility and responsiveness to address changing circumstances in both policy and scientific knowledge, but also an important requirement to avoid "mission drift" as knowledge and understanding evolves. A CFT lead may spend up to 30% of their time on project managing/enabling the co-construction of projects, ensuring all stakeholders are consulted and that the final project specification provides clear information on: the policy need/project background; anticipated impact of the project; aims and objectives; outputs; key dates; and that the project is delivered on budget. The Centre's overall ambition managed through the efforts of the CFT is to "*ensure that the right knowledge is shared with the users of that knowledge, in the right way and at the right time*".

Evaluating the impact of projects and wider Centre activities is sometimes difficult to access as deliberative conflict resolution and problem framing which underpin the work of the Centre are not necessarily amenable to simple metrics. Regular self-evaluation is used to appraise the engagement and delivery model. This review process highlighted the (positive) impact of CREW balanced by the need for reflexive/critical evaluation of CREW's practices and delivery. CREW maintains a quality manual covering all governance and operational processes which is adaptively managed and amended along with an annual review of risk. Such shared learning has informed improvements and enhanced the impact of the Centre over time.

CREW was recently involved in an independent comparative analysis of international water research organisations involved in agenda setting practices, conducted by KWR, the Dutch Watercycle Research Institute. The report highlighted that CREW's *"interactive agenda setting"* driven by *"demand articulation"* increased *"the cognitive and social proximity"* between the end-user and provider communities. In addition, the review considered that a successful knowledge provider needs *"to maintain sufficient flexibility to act upon emerging issues"* supporting CREW's two streams of engagement (Capacity Building and Call Down) and engagement approach to project prioritisation.

CREW tracks impact through post-project review and records end-user narratives captured by questionnaires sent to customers and research providers on the utility of the outputs produced in the annual reporting to RESAS. CREW's internal evaluation has highlighted the need for qualitative and quantitative data and assessment over time rather than simple snapshots, particularly in relation to longer-term *outcomes* at both project and thematic level. Evaluating the final impact of a particular activity is fraught with difficulties yet is the most sought after for delivering a true "impact agenda".

The ultimate goal of *changed behaviours* as a result of CREW activity is most difficult to evaluate and takes longer to achieve, however as previously highlighted in the CREW submission on the Value of Scotland's Water Resources (Martin-Ortega et al. 2013) which informed the Parliamentary debate preceding the establishment of the Water Resources (Scotland) Act 2013, and which subsequently placed a duty on Ministers to "ensure the development and value of water resources," and "their sustainable use" (and forms the basis of the ongoing Hydro Nation strategy), provides an tangible legacy example of how knowledge delivered at the science-policy interface and at the most appropriate time can generate significant future outcomes.

Future water policy challenges

There is no doubt that meeting the global challenges of the climate and biodiversity crises will be the most pressing global need. The impact of climate change (even in a water-rich country as Scotland) manifests itself in many ways, affecting environmental functioning and sustainability. People must be central to the development of robust policies, and the Centre of Expertise model established by the Scottish Government aims to ensure that the knowledge generated has the widest benefit to communities and society.

The challenges we face are global in nature but are regionally expressed in different ways. Critical to this challenge is the move towards a low-carbon future. Recently, the Scotland Government updated the 2018 Climate Change Plan to a post-COVID vision of building a wider green recovery whilst delivering on the ambitious target of net zero by 2045. This included a recognition of the impact of COVID on society and the need for a "just" transition.

Future research will be needed to support cross-sectoral policy, and a strategic challenge is the adoption of a holistic vision of the interplay between water, energy, production and the environment, and the development of a truly circular economy.

For example, Troldborg et al. (2017) highlighted the potential for developing a foundation for reclaimed water use in Scotland, and increasingly there has been a consolidated focus on identifying opportunities for resource recovery from the water cycle. This includes energy and/or cost savings through more efficient use of water, using energy produced currently as a waste product as a resource, or the recovery of nutrients and other components of value. Similarly, Scotland is extremely well placed to realise the potential opportunities of a developing green hydrogen economy delivering to both climate change objectives and rural and urban community development.

Wider strategies to reduce greenhouse gas emissions and address climate change must also contribute to halting biodiversity decline. The concept and adoption of 'nature-based solutions and whole ecosystem restoration focusing on soil carbon management, peatland restoration, afforestation and riparian woodland establishment, diffuse pollution management on-the ground mitigation measures, deliver the added benefit of environmental enhancement (improvements in water quality and flood protection, carbon sequestration, soil health etc.) whilst delivering social benefits through job creation, improved wellbeing and social cohesion.

Water is a central component of all our lives, and the water cycle is the most important earth system process sustaining life on this planet. The development of Hydro Nation and the duties placed on Scottish Ministers to ensure sustainability of our resources place the water crisis central to the national ambition both home and globally. The Centre of Expertise for Water delivers knowledge to support the sustainable development of water resources for the benefit of both citizens and communities.

Acknowledgements The Centre of Expertise for Waters (CREW) is supported by Scottish Government's Rural and Environment Science and Analytical Services (RESAS) Division. In addition, CREW acknowledges the considerable commitment of the scientific (university and research institutes), policy and stakeholder communities in their collaborative efforts to deliver the work of the Centre.

References

Allan R, Wilkinson M, Dodd N (2015) The North Glasgow integrated water management system: a review. https://www.crew.ac.uk/publication/integrated-water-management. Accessed 14 April 2021

Bowden LA, Voas S, Mellor D, Auty H (2020) EPIC, Scottish governments centre for expertise in animal disease outbreaks: a model for provision of risk-based evidence to policy. Policy. Front Vet Sci 7:119. https://doi.org/10.3389/fvets.2020.00119

Brown I, Dunn S, Matthews K, Poggio L, Sample J, Miller D (2012) Mapping of water supply-demand deficits with climate change in Scotland: land use implications. CREW report 2011/CRW006. Accessed 14 April 2021

Climate XChange (2016) Centre of expertise on climate change. https://www.climatexchange.org.uk/. Accessed 14 April 2021

Cloy J, Audsley R, Hargreaves B, Ball B, Crooks B, Griffiths B (2018) Valuing your soils: practical guidance for Scottish farmers. https://www.farmingforabetterclimate.org/wp-content/uploads/2018/02/Valuing_Your_Soils_PG.pdf. Accessed 14 April 2021

Currie M, Philip L, Dowds G (2020) Long-term impacts of flooding following the winter 2015/16 flooding in North East Scotland: summary report. CRW2016_02. Scotland's Centre of Expertise for Waters (CREW). https://www.crew.ac.uk/sites/www.crew.ac.uk/files/publication/CRW2016_02_Impact_OF_Flooding_Summary_Report.pdf. Accessed 14 April 2021

DEFRA (2007) Future water: the governments water strategy for England. https://assets.publishing.service.gov.uk/government/uploads/system/uploads/attachment_data/file/69346/pb13562-future-water-080204.pdf. Accessed 14 April 2021

Duffy A, Berwick N, Dello-Sterpaio P (2015) How do we increase public understanding of the benefits provided by SUDS? CRW2014/14. https://www.crew.ac.uk/publication/public-understanding-suds. Accessed 14 April 2021

Gosling R (2014) Assessing the impact of projected climate change on drought vulnerability in Scotland. Hydrol Res 45(6):806–816

Hallett P, Hall R, Lilly A, Baggaley B, Crooks B, Ball B, Raffan A, Braun H, Russell T, Aitkenhead M, Riach D, Rowan J, Long A (2016) Effect of soil structure and field drainage on water quality and flood risk. CRW2014_03. https://www.crew.ac.uk/publication/soil-structure-field-drainage. Accessed 14 April 2021

Hannaford J (2015) Climate-driven changes in UK river flows: a review of the evidence. Prog Phys Geogr Earth Environ 39:29–48

Hansom JD, Fitton JM, Rennie AF (2017) Dynamic coast—national coastal change assessment: methodology. CRW2014/2. http://www.dynamiccoast.co.uk/files/reports/NCCA%20-%20Recommendations.pdf. Accessed 14 April 2021

Hare M, Helliwell RC, Ferrier RC (eds) (2020) Exploring Scotland's resilience to drought and low flow conditions: world water day 22nd March 2019—full report. Produced on behalf of CREW and Hydro Nation International Centre. ISBN: 978-0-902701-72-4

Hendry S, Akoumianaki I (2016) Governance and management of small rural water supplies: a comparative study. CRW2015/05. https://www.crew.ac.uk/sites/www.crew.ac.uk/files/sites/default/files/publication/CRW2015_05%20Final%20report.pdf. Accessed 14 April 2021

IPCC (Intergovernmental Panel on Climate Change) (2019) 6th assessment report. https://www.ipcc.ch/. Accessed 14 April 2021

Kay AL, Rudd AC, Fry M, Nash G, Allen S (2021) Climate change impacts on peak river flows: combining national-scale hydrological modelling and probabilistic predictions. Clim Risk Manag 31:100263. https://doi.org/10.1016/j.crm.2020.100263

Knight C, Lyall C (2013) Knowledge brokers and the role of intermediaries in producing research impact. Evid Policy 9(3):309–316

Lilly A, Baggaley NJ (2014) Developing simple indicators to assess the role of soils in determining risks to water quality. CREW project number CD2012_42. https://www.crew.ac.uk/publication/indicators-soils-water-quality. Accessed 14 April 2021

Marsh TJ, Kirby C, Muchan K, Barker L, Henderson E, Hannaford J (2016) The winter floods of 2015/2016 in the UK—a review. Centre for Ecology & Hydrology, Wallingford, UK. 37 pages. ISBN: 978-1-906698-61-4

Marshall K, Waylen K, Wilkinson M (2019) Communities at risk of flooding and their attitudes towards natural flood management. CRW2018_03. Scotland's Centre of Expertise for Waters (CREW)

Martin-Ortega J, Holstead K, Kenyon W (2013) The value of Scotland's water resources. https://www.hutton.ac.uk/sites/default/files/files/publications/water-resources-bill-leaflet-feb2013.pdf. Accessed 14 April 2021

Met Office (2019) UK climate projections: headline findings September 2019 Version 2. https://www.metoffice.gov.uk/binaries/content/assets/metofficegovuk/pdf/research/ukcp/ukcp-headline-findings-v2.pdf. Accessed 15 April 2021

MGSDP (2021) Metropolitan Glasgow strategic drainage partnership. https://www.mgsdp.org/. Accessed 15 April 2021

Moore RJ, Cole SJ, Dunn S, Ghimire S, Golding BW, Pierce CE, Roberts NM, Speight L (2015) Surface water flood forecasting for urban communities. CREW report CRW2012/03. The James Hutton Institute, Aberdeen, UK

NFM (2018) Natural Flood Management Network Scotland. https://www.nfm.scot/. Accessed 15 April 2021

O'Keeffe J, Akunna J, Olszewska J, Bruce A, May L, Allan R (2016) Practical measures for reducing phosphorus and faecal microbial loads from onsite wastewater treatment system discharges to the environment: a review. https://www.crew.ac.uk/sites/www.crew.ac.uk/files/sites/default/files/publication/CREW_septic_tanks.pdf. Accessed 15 April 2021

Philip L, Dowds G, Currie M (2020) Long-term impacts of flooding following the winter 2015/16 flooding in North East Scotland: comprehensive report. CRW2016_02. Scotland's Centre of Expertise for Waters (CREW)

Plant Health Centre of Expertise (2020) https://www.planthealthcentre.scot/. Accessed 15 April 2021

Reed M, Meagher L (2019) Using evidence in environment and sustainability issues. In: Boaz A, Davies H, Fraser A, Nutley S (eds) What works now? Evidence-based policy and practice. The Policy Press, Bristol, pp 151–223

Rivington M, Akoumianaki I, Coull M (2020) Private water supplies and climate change: the likely impacts of climate change (amount, frequency and distribution of precipitation), and the resilience of private water supplies. CRW2018_05. Scotland's Centre of Expertise for Waters (CREW). https://www.crew.ac.uk/publication/PWS-water-scarcity. Accessed 15 April 2021

Rivington M, Spencer M, Gimona A, Artz R, Wardell-Johnson D, Ball J (2019) Snow cover and climate change in the Cairngorms National Park: summary assessment. Centre of Expertise for Climate Change (ClimateXChange). https://www.climatexchange.org.uk/media/3900/cxc-snow-cover-and-climate-change-in-the-cairngorms-national-park_1.pdf. Accessed 15 April 2021

Rowland C, Scholefield P, O'Neil A, Miller J (2019) Quantifying rates of urban creep in Scotland: results for Edinburgh between 1990, 2005 and 2015. CRW2016_16. https://www.crew.ac.uk/publication/urban-creep. Accessed 16 April 2021

Sample JE, Baber I, Badger R (2016) A spatially distributed risk screening tool to assess climate and land use change impacts on water-related ecosystem services. Environ Model Softw 83:12–26

Scottish Government (2005) Prevention of environmental pollution from agricultural activities: a code of good practice. https://www.gov.scot/publications/prevention-environmental-pollution-agricultural-activity-guidance/. Accessed 15 April 2021

Scottish Government (2013) Water Resources (Scotland) Act 2013. https://www.legislation.gov.uk/asp/2013/5/enacted. Accessed 14 April 2021

Scottish Government (2017) The Water Environment (Miscellaneous) (Scotland) Regulations. https://www.legislation.gov.uk/ssi/2017/389/introduction/made. Accessed 14 April 2021

Scottish Government (2019) Climate ready Scotland: Second Scottish Climate Change Adaptation Programme 2019–2024. Laid before the Scottish Parliament by the Scottish Ministers under Section 53 of the Climate Change (Scotland) Act 2009, September 2019. SG/2019/150. https://www.gov.scot/binaries/content/documents/govscot/publications/strategy-plan/2019/09/climate-ready-scotland-second-scottish-climate-change-adaptation-programme-2019-2024/documents/climate-ready-scotland-second-scottish-climate-change-adaptation-programme-2019-2024/climate-ready-scotland-second-scottish-climate-change-adaptation-programme-2019-2024/govscot%3Adocument/climate-ready-scotland-second-scottish-climate-change-adaptation-programme-2019-2024.pdf?forceDownload=true

Scottish Government (2020) Securing a green recovery on a path to net zero: climate change plan 2018–2032—update. https://www.gov.scot/publications/securing-green-recovery-path-net-zero-update-climate-change-plan-20182032/pages/2/. Accessed 15 April 2021

Scottish Government (2021a) Implementation of the Flood Risk Management (Scotland) Act 2009. Report to the Scottish Parliament. Laid before the Scottish Parliament by the Scottish Ministers under Section 52 of the Flood Risk Management Scotland Act, January 2021.

Report SG/2021/11. https://www.gov.scot/publications/implementation-flood-risk-management-scotland-act-2009-report-scottish-parliament-2019/. Accessed 15 April 2021

Scottish Government (2021b) Water-resilient places—surface water management and blue-green infrastructure: policy framework. https://www.gov.scot/publications/water-resilient-places-policy-framework-surface-water-management-blue-green-infrastructure/pages/1/. Accessed 16 April 2021

SEPA (2013) Scottish Environment Protection Agency: an introduction to the significant water management issues in the Scotland River Basin District. Scottish Environment Protection Agency. https://www.sepa.org.uk/media/38319/an-introduction-to-the-significant-water-management-issues-in-the-scotland-river-basin-district.pdf. Accessed 16 April 2021

SEPA (2019) Scottish Environment Protection Agency. Scotland's Environment. https://www.environment.gov.scot/our-environment/water/scotland-s-freshwater/. Accessed 15 April 2021

SEPA (2020) Scottish Environment Protection Agency. Scotland's National Water Scarcity Plan. https://www.sepa.org.uk/media/510820/scotlands-national-water-scarcity-plan-july-2020.pdf. Accessed 15 April 2021

SEPA (2021a) Scottish Environment Protection Agency. The draft river basin management plan for Scotland 2021–2027. https://consultation.sepa.org.uk/rbmp/draft-river-basin-management-plan-for-scotland/. Accessed 15 April 2021

SEPA (2021b) Scottish Environment Protection Agency. Water resources management plan (currently under consultation) https://consultation.sepa.org.uk/circular-economy/water-resources-management-plan-consultation/consult_view/. Accessed 15 April 2021.

Speight L, Cranston M, Kelly L, White CJ (2019) Towards improved surface water flood forecasts for Scotland: A review of UK and international operational and emerging capabilities for the Scottish Environment Protection Agency. University of Strathclyde, Glasgow

Spray CJ, Arthur S, Bergmann A, Bell J, Beevers L, Blanc J (2015) Land management for increased flood resilience. CREW CRW2012/6. https://www.crew.ac.uk/publication/land-management-increased-flood-resilience. Accessed 16 April 2021

Troldborg M, Duckett D, Hough RL, Kyle C (2017) Developing a foundation for reclaimed water use in Scotland. CREW Report: CRW2013_16. https://www.crew.ac.uk/sites/www.crew.ac.uk/files/publication/CRW2013_16_Foundation_For_Reclaimed_Water_Use_Main_Report.pdf. Accessed 15 April 2021

UK CCRA (2017) UK Climate Change Risk Assessment 2017 Evidence Report https://www.theccc.org.uk/uk-climate-change-risk-assessment-2017/synthesis-report/. Accessed 16 April 2021

Undorf S, Allen K, Hagg J, Li S, Lott FC, Metzger MJ, Sparrow SN, Tett SFB (2020) Learning from the 2018 heatwave in the context of climate change: are high-temperature extremes important for adaptation in Scotland? Environ Res Lett 15(3):034051. https://iopscience.iop.org/article/10.1088/1748-9326/ab6999. Accessed 15 April 2021

UN Water (2020) UN World Water Development Report 2020. https://www.unwater.org/publications/world-water-development-report-2020/. Accessed 16 April 2021

Waajen AC (2019) The increased risk of water scarcity in Scotland due to climate change and the influence of land use on water scarcity: issues and solutions. CXC Report. https://www.climatexchange.org.uk/media/3680/cxc-water-scarcity-climate-change-and-land-use-options.pdf. Accessed 15 April 2021

World Economic Forum (2012) The Water Resources Group 2030: background, impact and the way froward. Briefing report prepared for the World Economic Forum Annual Meeting 2012 in Davos-Klosters, Switzerland 26th January 2012. http://www3.weforum.org/docs/WEF/WRG_Background_Impact_and_Way_Forward.pdf. Accessed 15 April 2021

Werritty A, Sugden D (2012) Climate change and Scotland: recent trends and impacts. Earth Environ Sci Trans R Soc Edinb 103(2):133–147

Wreford A, Peace S, Reed M, Bandola-Gill J, Low R, Cross A (2019) Evidence-informed climate policy: mobilising strategic research and pooling experience for rapid evidence generation. Clim Change 156:171–190. https://doi.org/10.1007/s10584-019-02483-w

Chapter 10
Building A Resilient and Sustainable Water and Wastewater Service for Scotland

Mark E. Williams, Gordon Reid, and Simon A. Parsons

Abstract Climate change affects the managed water cycle in terms of drought, water quality and the need to manage flood risk in the urban environment. In Scotland, we already experience environmental events consistent with climate projections. Seasonal temperature and rainfall anomalies have seen droughts and floods in parts of the country, and Scottish Water is focussed on delivering water and wastewater services resilient to more extreme and variable conditions. Working with others and managing our landscapes are vital to climate change adaptation in water catchments and to sustainably drain our cities. Carbon emissions from services are high—water collection, treatment and supply, and the collection, treatment and recycling of wastewater are asset and energy-intensive processes. Scottish Water has committed to deliver net zero emissions across operational and investment activities by 2040 through reducing power consumption, generating renewable electricity, reducing the emission intensity of investment and increasing the carbon stored in its landholdings.

Keywords Climate change resilience · Net zero emissions · Adaptation · Mitigation · Water services · Wastewater · Sustainable drainage

10.1 Introduction

Scottish Water is the national public water and wastewater service provider for Scotland, serving over five million people. Our services are delivered through an extensive asset base of over 230 water treatment works, 1800 wastewater works and 100,0000 kms of water and wastewater pipes. Owned by the Scottish Government and accountable to ministers, Scottish Water operates within a regulatory frame-

M. E. Williams (✉) · G. Reid · S. A. Parsons
Scottish Water, Dunfermline, UK
e-mail: Mark.Williams@scottishwater.co.uk

S. A. Parsons
e-mail: Simon.Parsons@scottishwater.co.uk

work that requires close collaboration working with drinking water, environmental and economic regulators, as well as with customer interests and government.

Scotland's environment is critical to the delivery of resilient water and wastewater services. We depend on sufficient good quality water resources for treating and serving to customers and healthy water catchments for recycling treated wastewaters. In considering climate change, it is therefore vital that we understand and manage the implications for the natural environment on which we depend, as well as for the integrity and performance of our assets.

In this chapter we share some of our experience of adapting to the challenges of climate change and our role in mitigating the impact providing these services has on the environment.

10.2 Climate Change Projections in Scotland

Global heating is already causing an increasing frequency of extreme weather events. The summer of 2018 was the driest and warmest in Scotland for 25 years, following just four months after a prolonged spell of freezing, wintery weather characterised by a storm system nicknamed the 'beast from the east'. Projections suggest that by 2050, a summer as warm as, or warmer than, 2018 would be expected roughly every two years (Fig. 10.1).

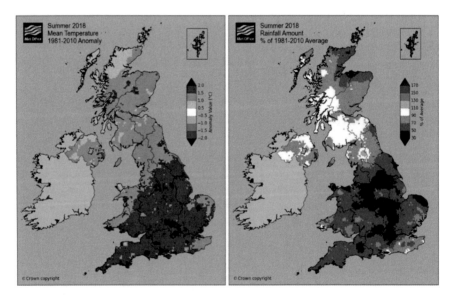

Fig. 10.1 2018 summer temperature and rainfall anomalies showing elevated temperatures across Scotland and regions of Scotland where there was a significant decrease in rainfall compared with the 1981–2010 average (Met Office 2018)

The effects of climate change are seen most clearly in the water cycle and, therefore, the water environment on which we depend. In recent years, Scotland has experienced an increase in average annual rainfall, with the last decade (2009–2018) around 15% wetter than the 1961–1990 average and winters 25% wetter.

However, changes are not uniform, and examination of the seasonal winter rainfall during 2019 and 2020 shows stark differences (Fig. 10.2). 2019 was characterised by significantly less rainfall than might be expected, whilst 2020 saw large areas of Scotland experience significantly more rainfall than the average and some regions saw a decrease.

Climate change projections suggest we will see an increase in the frequency and intensity of extreme weather events in Scotland, from extended periods of dry weather presenting challenges to water resources, and significant increases in the intensity of rainfall events that would cause flood risk to assets and to our ability to drain urban areas, presenting flood risk to customers (Climate Change Committee 2019).

Current projections indicate we may experience an increase in the rainfall intensity of some events of approximately 45% over the next 30 years. This could see a 90–135% increase in water volume in our sewers, and as our sewers were built in the nineteenth and twentieth centuries, they will not always be able to cope (Scottish Water 2018; UKWIR 2017).

The variability of the weather will also increase, with our assets having to manage across a wider range of extremes. This variability in conditions has a further impact on the environment, with periods of wetting and drying increasing

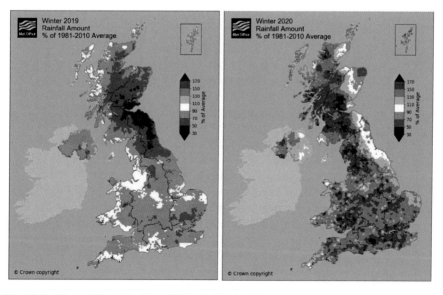

Fig. 10.2 Winter 2019 and winter 2020 rainfall showing highly variable patterns between the years and for 2020 significant differences across small geographic ranges (Met Office 2019)

the risk of soil erosion which can impact on the quality of surface waters we abstract from. This in turn can present challenges to the treatment of water to a suitable standard.

10.3   Adaptation—Serving Customers in a Changing Climate

Over recent years, Scottish Water has taken a number of steps to ensure that our assets and services are able to serve customers in a changing climate.

We have taken a 'high-emission pathway' as a planning assumption, and using adaptation risk assessment and planning tools developed within the UK water industry (UKWIR 2013), we have reviewed the key implications for our assets and concluded that:

- Climate change presents risks to service including water quality, quantity (supply), flooding of assets and customers, and low flows/high temperatures leading to treatment and odour risk.
- Impacts may be greater than previously understood, particularly for drainage and flooding.
- Scenarios do not differ markedly over the next 20 years irrespective of carbon emission trajectory.
- There will be increased incidence of extreme events, and the range of conditions will broaden.
- There are climate change interdependencies with, e.g., population growth, land use changes (water demand, etc.) and urban creep (run-off of surface water), and these will need to be considered in adaptation planning.

Service resilience is a key focus for our water and wastewater teams to put in place appropriate measures that will address service challenges from a number of change factors. These include growth, urbanisation, land use changes, asset deterioration and new technologies/demands. Climate change is a further element that will exacerbate some of these effects.

Our strategy for climate change adaptation is to secure a service resilient to climate change by using the latest scenarios, projections, tools and data and applying them to asset and service risk and resilience planning.

We aim to ensure that planned improvements to water supply resilience and sewer system capacity and operation will increase our ability to deal with a changing climate. To do this, we are continuing to develop our understanding of how climate change impacts water availability and our sewer network, and to develop solutions to improve longer-term resilience to such change.

Key adaptation responses we have undertaken so far include:

- **Monitoring**—models and projections need to be augmented by observed trends and impacts, and improving the monitoring of our raw water sources for both flow and quality will help us to understand how our catchments may be changing and to identify trigger points for us to respond.

- **Modelling**—using the latest tools to integrate UK climate projections into water resource models, and to develop tools to integrate climate projections into drainage models to manage storm events.
- **Investment**—we have improved treatment in some water treatment works in response to deteriorating water quality, increased the resilience of our water resources through greater connectivity and invested in significant schemes to manage and reduce flood risk to our wastewater treatment works and from our sewer network.
- **Catchment** —increased understanding of the quality of water catchments and their vulnerability to climate change is informing how we are working with farmers and landowners to improve riparian habitats and restore peatland to support water quality.
- **Integrated Drainage Management**—recognising the challenges presented by increased storm intensity, we have adopted a strategy to manage future capacity risks by preventing further surface flows into sewers and where we can remove surface water drainage. We are working in partnership with Scottish Environmental Protection Agency (SEPA), local authorities and others across Scotland to address the twin challenges of growth and climate change to ensure our towns and cities are resilient to climate change. A good example of this is the Metropolitan Glasgow Strategic Drainage Partnership (2020) which is transforming how the Glasgow city region thinks about and manages rainfall to end uncontrolled flooding and improve water quality.

A summary of the climate change hazards across our service and the responses we are taking is shown in below (Table 10.1).

10.4   Planning and Adapting to Extreme Rainfall

Significant increases in the frequency and intensity of extreme rainfall events are potentially one of the most impactful consequences of climate change. In the urban environment, increased rainfall intensity has the potential to overwhelm drainage systems and cause flooding to properties and urban spaces.

In rural water catchments, increased rainfall intensity can lead to greater run-off from the land to watercourses. This presents a risk to the quality of water we abstract for drinking in terms of both natural contaminants such as organic matter from soil erosion and increased transport of contaminants such as nutrients and pesticides from land management.

To help us to better understand the way in which rainfall patterns and intensity are changing, and the impact it is having on the flow and quality of our water sources, we have significantly increased the scale of our monitoring of rainfall, water quality and water flows within our catchments.

Table 10.1 Summary of climate change hazards and our response

Climate change hazard/risk	Resilience response
Water resources	
Periods of lower rainfall or reduced snow cover increase risks of water supply–demand deficits Smaller surface water sources become more vulnerable	Apply climate change projections to water resource models for 52 priority zones Increase rainfall and river flow monitoring Use outcomes to inform 25-year water resource and resilience plans Continue work to increase connectivity
Water quality	
Raw water quality deteriorates through climate change—soil and bank erosion, organic matter, bacteria, nutrients, eutrophication risk	Extend raw water monitoring programmes to improve trend analysis Data inform future interventions in terms of treatment and catchment management Research in organics, bacteria and water chemistry supports resilience decisions
Water networks	
Climate change leads to more variable ground conditions with risk of landslides and soil movements	Risk assessment work on climate scenarios includes landslip appraisals System-by-system resilience assessments to identify and manage critical sections
Wastewater networks	
Changes in frequency and intensity of rainfall events lead to an increase in sewer flood risk	Develop and deploy new tools to integrate climate change into future rainfall projections Use tools in network resilience planning for future investment Apply storm water strategy to prevent new surface water connections Develop drainage and surface water management masterplans informed by climate models
Wastewater treatment	
Variable WW flow conditions lead to risks of washouts (high flow) and septicity (low flow) Higher risk of low river flows leads to tighter or variable standards	Treatment capability assessments inform risks and future responses
Asset/service flood risk	
Climate change leads to increased asset flood risk through rainfall/rivers, or sea level rise, which impacts service levels	Service risk and resilience planning (water and waste) informed by climate projections for flood/storm surge return periods
Climate change leads to indirect service impacts through loss of power/comms/travel to sites, etc.	Service risk and resilience planning (water and wastewater) consider secondary vulnerability, service impact and measures to reduce risk

To mitigate the risk of deteriorating water quality, over the past 10 years we have created a catchment management programme to work with landowners and farmers to address the risk of run-off impacting water bodies. This has seen extensive investment in measures such as buffer strips along water courses and farm management plans to improve riparian habitats and reduce erosion risk.

Working with NatureScot and the Peatland Action Group, we have also invested extensively in the restoration of this vital habitat. As well as loss of an important habitat, peatland erosion is a significant risk to water quality through the release of organic matter and the greenhouse gases driving climate change. By working to reprofile land and ensure continued wetting, we protect our source waters and support biodiversity and climate change mitigation (Scottish Water 2019).

Future projections of flow in large catchments are relatively straightforward to create from climate change models, but this is much more difficult for the urban environment, where short duration storms are more significant. Working with the water industry's partners, we have developed tools to integrate climate change projections into future rainfall time series (UKWIR 2017). The goal was to develop tools to understand future rainfall events, particularly in urban drainage planning, and we have now developed projections for key parts of Scotland across three future epochs—2030s, 2050s and 2080s.

In the urban environment, a large proportion of our sewers are 'combined'—they carry both foul sewage and surface water drainage from properties and roads. This is the proportion of drainage that is most affected by climate change, and the scale of storm event uplift projected by climate change will increasingly lead to the risk of future flooding from surface water flows.

Traditional responses to flood risk have seen investment in increasing the capacity of our sewers and drainage networks. Over the past 10 years, we have invested significantly in sewer infrastructure, for storage tanks both to help to deal with storm events and to upsize sewers.

The most significant project was the creation of the Shieldhall Tunnel, a new 4.7 km sewer on the south side of Glasgow to alleviate flood risk across large parts of the Glasgow network. At almost 5 m in diameter, this was opened in 2018 and was the largest construction project in Scotland during its creation (BBC 2018).

Whilst traditional civil engineering is vital to address acute drainage and flooding problems, a key problem is that it delivers a system that is larger but is still finite. Future climate change risk can be incorporated up to a point, but the scale of potential change is impossible to fully accommodate. Instead, we must find different ways to manage the flow of water in the urban environment.

Achieving this is not straightforward. We cannot simply disconnect roofs and roads from the sewer system without providing alternative routes to carry water. Water management in the urban environment is fundamentally about land management and the way we use landscapes. Future extreme events risk overwhelming both combined sewers and the surface water systems leading to problems across the urban environment (Fig. 10.3).

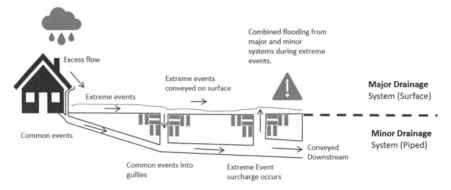

Fig. 10.3 Water management in the urban environment

10.5   Building Sponge Cities

We have seen both across Scotland and globally that managing surface water flows demands a partnership approach between the water industry, local authorities, road authorities and developers to help us to shape the urban landscape to deal with future flood risk. A key goal is the creation of 'blue-green' approaches to urban water management. This means seeking to manage water flows on the surface—through the creation of detention basins and flow channels, and through soakaways, permeable surfaces and natural systems that direct water flows and absorb or hold the excesses of storm events without overwhelming drainage capacity.

This presents an opportunity to reprofile the landscape and to create attractive places with green space, sustainable and green ponds to hold storm water. This approach can provide a number of benefits in terms of supporting future climate change adaptation, improving connectivity to nature and increasing the attractiveness of the urban environment.

We see this with new developments across Scotland through the creation of Sustainable Urban Drainage detention basins that take surface flows from the development and create an attractive landscape (Fig. 10.4).

The big challenge is in retrofitting such approaches into the existing urban drainage system. This demands partnership working with local authorities, developers and road authorities to understand the challenge and to look holistically at how water might be managed in the urban catchment. There are a range of approaches that can be taken to mitigate the impact of storm events on our infrastructure:

Fig. 10.4 Examples of sustainable urban drainage

Scottish Water is actively pursuing such partnerships across Scotland to understand and to develop such approaches to create more resilient urban environments. Below we highlight a truly innovative project—Glasgow's Smart Canal.

10.6 Smart Canals—A Low-Carbon Surface Water Management Adaptation Measure

Effective collaboration between Glasgow City Council, Scottish Canals and Scottish Water developed an innovative solution to move surface water safely from north Glasgow in a low-emission way, reducing flood risk impact now and into the future, and enabling massive regeneration. In the first scheme of its kind in Europe, the North Glasgow Integrated Water Management System (or Smart Canal) uses sensor and predictive weather technology to provide early warning of wet weather. Advanced warning of heavy rainfall will automatically trigger a lowering of the canal water level to create capacity for surface water run-off (Scottish Canals 2021).

Using forecasts, ahead of periods of heavy rain canal water will be moved safely, using gravity, through a network of newly created urban spaces such as sustainable urban drainage ponds and channels. These absorb and manage the water in a controlled way. This approach provides 55,000 m^3 (equivalent to 22 Olympic swimming pools) of extra capacity in the canal for floodwater.

A conventional approach to removing this volume of surface water would have been to build a tunnel.

The estimated emissions of the tunnelling approach would have been over 3500 tonnes of carbon dioxide equivalent, coupled with the further operational emissions associated with pumping the storm water through the conventional drainage system. The innovative smart canal approach has therefore avoided substantial emissions.

110 ha across the north of the city is now suitable for investment, regeneration and development, allowing more than 3000 new homes to be built. Whilst this approach benefits from existing canal and drainage infrastructure, partnership working between Scottish Water, local authorities and others is key to helping us to sustainably manage the impact of climate change on the sewer network, and we are working with local authorities.

10.7 Mitigating Our Impact on the Environment

Whilst we rely on a healthy environment to deliver our services, we also require significant infrastructure and energy. Consequently, we are one of Scotland's larger users of electricity, with a demand of some 570 GWh of electricity per annum. This means we are a significant contributor to Scotland's carbon footprint.

Whilst we have been taking action to mitigate our greenhouse gas emissions for well over a decade, we are focussed on doing all we can to reduce the impact we have on the environment. In 2020, we committed to getting to net zero emissions by 2040 and to go beyond—to become a carbon sink for Scotland.

10.8   Operational Emissions

Emissions from our water and wastewater services have fallen by almost 50% since we began measuring and managing them in 2006–2007, to just over 250,000 tonnes carbon dioxide equivalents (tCO$_2$e) in 2019–2020. Large reductions were due to reduction on leakage, energy efficiency and renewable programmes and external factors such as the greening of the electricity grid.

Around two-thirds of our emissions are associated with electricity use. Reducing electricity consumption is a significant element of our emission reduction strategy, but water is heavy. Where we cannot use gravity, it requires energy for it to be pumped across large parts of Scotland; we continue to rely on energy intense processes to produce high-quality drinking water and to protect our environment. Our operational footprint also includes process emissions from wastewater treatment and waste management, gas and fuel oil use, and emissions associated with our fleet, which covered 19 million miles in 2019–2020 (Fig. 10.5).

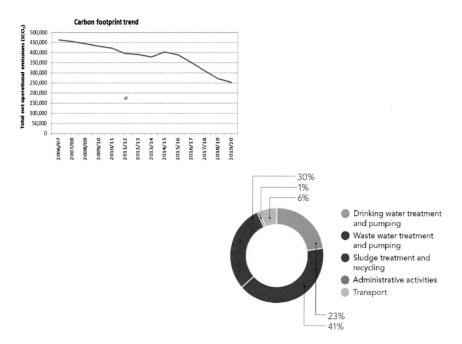

Fig. 10.5 Scottish Water operational carbon footprint

10.9   Investment Emissions

Each year, we invest significantly in our asset base, and our capital programme is a major source of emissions. We have a big focus on working with our supply chain and delivery partners to transform greenhouse gas emissions in our capital investment. Investment emissions comprise the carbon embodied within the extraction materials, processing into products such as cement, steel, pipes and equipment, and construction of assets (Fig. 10.6).

We have created tools to assess emissions during investment planning and delivery, and have determined that each £1 m invested yields 200–300 tCO_2e depending on the investment type. Whilst this is less mature than for operational emissions, we are now able to use this approach in strategic and detailed investment planning and are working with supply chains to target reduced emission intensity from investment.

Collectively, operational and investment emissions amount to around 450,000 tCO_2e per annum.

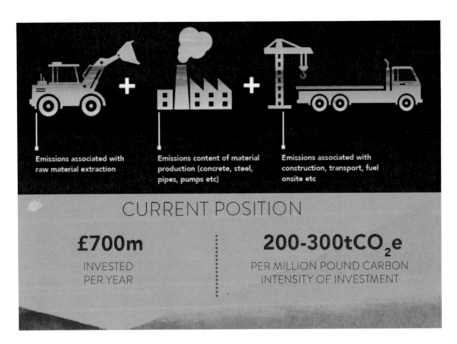

Fig. 10.6 Scottish Water investment carbon footprint

10.10   Net Zero Emission Challenge

In September 2020, we published our Net Zero Emissions Routemap (Scottish Water n.d.) setting out the strategic steps we would take to deliver net zero emissions by 2040. The routemap seeks to maximise emission reduction for those emissions we control—day-to-day operational emissions from our services, and those we influence—emissions in the supply chain supporting our capital investment programme.

Net zero emission is a major challenge that Scottish Water will embrace, both in our day-to-day service provision and in our capital investment. It demands close working, across the business and with our wider supply chain, to bring innovation for emission reductions into capital investment and to identify opportunities to avoid the emissions associated with the products we use.

We also need to find ways to 'lock up' greenhouse gases in our landholdings to help balance those emissions that cannot today be eliminated. We own over 22,000 ha of land with significant opportunity to grow carbon sequestration through afforestation, peatland restoration and improved land management.

Our net zero goal is therefore to balance the residual emissions we cannot eliminate with emissions stored on our land, and our strategy is focussed across five key areas:

Electricity

- Reduce consumption.
- Generate renewable power: self-generate more electricity than we consume.

Process emissions

- Optimise the performance of wastewater processes.
- Invest in energy generation at sludge assets.

Transport

- Reduce the distance we travel.
- Transition to a zero-emission fleet.

Investment

- Across design, procurement of materials and onsite construction, reduce intensity of investment by 75%.

Carbon storage

- Invest in peatland restoration, land management and afforestation.

Below we set out some examples of ongoing work in reducing our energy demand, generating renewable energy and restoring peatland.

10.11 Reducing Our Demand

Wastewater treatment is the most carbon-intense part of our service.

Since 2010, we have delivered a programme of energy saving initiatives at our wastewater treatment works. The cumulative savings to 2018–2019 were 26 GWh of electricity and over 10,000 tonnes of carbon dioxide equivalent (tCO2e). We aim to deliver a further 8 GWh efficiencies by 2021 and to further expand this programme in our next investment programme (2021–2027).

Initiatives include:

- Process improvements through real-time control system.
- Replacement of pumps and aeration equipment with more energy efficient models.
- Heating and lighting improvements.

Real-time control (RTC) is an operating philosophy that can be used to improve the energy efficiency and process control of the wastewater aeration processes. Wastewater that is treated with biological processes that rely on us adding air, up to 60% of the total energy consumption, occurs during the aeration phase.

Reducing energy consumption during aeration can therefore yield large energy savings. With RTC, the amount of air delivered to treatment is optimised by using a network of measurement probes and modelling, linked to variable speed drives on the aeration blowers. This delivers only what is required for treating the wastewater at that time, avoiding excessive aeration and saving energy.

We are already seeing a number of benefits from this, including:

- Reduced greenhouse gas emissions—RTC typically saves 20–30% of the electricity used in the aeration process. Where there is electricity use, there are greenhouse gas emissions, so there are obvious savings here.
- Financial savings—reduction in our electricity use means we save on operational costs.
- Improved treatment stability and compliance—because the system is taking measurements constantly, it can react rapidly to any peak loading and maintain compliance.
- Asset life and maintenance costs—because the aeration pumps are not operating constantly at high speed, wear and tear on the motors is reduced, along with maintenance costs, downtime and repair costs.

The recently completed RTC installation at one of Scotland's largest wastewater treatment works at Shieldhall in Glasgow has delivered great results. The first year of RTC operation at the site yielded 2.2 GWh of annual efficiencies, with £240,000 savings on energy costs and almost 850 tCO_2e reduction. RTC systems have also been installed at wastewater treatment works at Laighpark, Philipshill, Erskine, Perth and Hamilton; we plan to invest significantly in rolling this out at other sites.

10.12 Generating Renewable Energy

Where we have both the opportunity to generate energy and a demand from our assets and services, we invest to install and operate renewable generation ourselves. This displaces grid electricity used by Scottish Water and has a dual benefit: it directly reduces our carbon footprint and provides benefit to our customers in reducing the cost of service. Since 2013, Scottish Water and Scottish Water Horizons have trebled the amount of renewable capacity on our sites to more than 75 GWh/year.

Historically, we focussed mainly on electricity from hydro schemes and combined heat and power (CHP) utilising the biogas produced at some wastewater assets. We have diversified this portfolio with the advent of new technologies, and we now employ hydropower, small-scale wind, solar PV, solar thermal, solar + battery, combined heat and power and biomass technologies.

More than 100 of our water and wastewater treatment works are either fully or partly self-sufficient in their electricity requirements. This is helping make our service more sustainable by reducing our use of grid electricity and lowering operating costs.

For example, at Stornoway Wastewater Treatment Works we installed 15 small wind turbines on the land around the works. They now generate 35% of the site's electricity consumption (equivalent to 83 households) and export electricity to the grid when the demand from the treatment works drops.

10.13 Reducing the Greenhouse Gas Intensity of the Electricity Grid

The UK's electricity is generated from multiple sites across the country and distributed via the national grid to wherever it is needed. Greenhouse gas emissions associated with grid electricity vary depending on which fuels are used to generate electricity at any given time. Fossil fuels (gas, oil and coal) have high emissions, whilst renewables (wind, hydro and solar) have low or no emissions.

The more renewables that 'feed' the grid, the lower the greenhouse gas intensity of grid electricity. Where we have no electricity demand from our own assets, we can partner with energy companies to generate renewable electricity on our land to supply the national grid.

Through our renewables programme, we now host three large-scale wind farms and a hydro generation plant. Together, they have a design capacity of 832 GWh/year. We continue to work with generating companies to progress new renewable opportunities: at Lower Glendevon Reservoir, we have a lease agreed for a hydro plant, and a wind farm in the Daer water resource catchment is now in development.

10.14   Heat from Sewers

In Scotland, the way we heat our buildings and homes comprises 50% of the total energy consumed across society. The Scottish Government has made a commitment to decarbonise heat sources and is supporting projects that demonstrate innovative, low-carbon ways of generating heat.

Thanks to things like showers and washing machines, the average temperature of wastewater that runs through the sewers beneath our feet is 15°C. This energy can be captured to create sustainable, low carbon heat. Scottish Water Horizons has been leading the roll out of groundbreaking 'heat from wastewater' technology. The first project of its kind in the UK was commissioned at Borders College, Galashiels, in 2015, and to date it has saved the college 223 tCO_2e by exploiting the renewable heat value of wastewater.

One of our most recent projects is the Stirling District Heat Network. Working in collaboration with Stirling Council, and with support from the Scottish Government's Low Carbon Infrastructure Transition Programme, the project is providing low-carbon heat to a number of the council's customers, including a High School, offices and conference centre and the head offices of the Water Industry Commission for Scotland and Zero Waste Scotland. The project uses heat from wastewater technology alongside a combined heat and power (CHP) engine to generate low-emission energy—the first time these two technologies have been used together in this way in the UK. The CHP engine is powered by mains gas and provides the majority of the electricity demand from the wastewater treatment works (WWTW), as well as providing the electricity required to power the heat pumps (Scottish Water 2021).

10.15   Peatland Restoration

Peatland covers more than 20% of Scotland's land area, which equates to around 1.8 million hectares and is 60% of the UK's total peatland. In terms of global habitat, Scotland boasts 15% of the world's blanket bog (peatland can be blanket bog, raised bog or lowland bog).

Peatland delivers three important ecosystem services:

- Water—the majority of Scotland's public drinking water supplies start in peatland, and globally peatland holds around 10% of the world's freshwater.
- Carbon sequestration—Scotland's peatland holds an estimated 1620 million tonnes of carbon dioxide equivalent (tCO_2e).
- Biodiversity—peatland's share of the estimated social and economic value of biodiversity in Scotland is £23 billion per year.

Studies suggest that 80% of Scotland's peatland is degraded, either through natural means or through modification such as artificial drainage. Where drinking water sources are fed from degraded peatland, this can lead to declining water quality and more treatment required to comply with drinking water quality standards.

Peatland restoration can improve the quality of the water used for drinking water, as well as providing a wide range of other ecosystem services. We have been collaborating with the Peatland Action Group, Scottish Natural Heritage and land managers/owners to restore peatland in our drinking water catchments. To date, we have funded or co-funded the restoration of 190 ha across two locations. In addition, we have supported the restoration of a further 250 ha with in-kind contributions (such as water quality monitoring and site inspections).

A recent example is Loch Orasaigh for North Lochs WTW on the Isle of Lewis, covering 11 ha. Once peatland has been restored to a healthy growing status, it can sequester between 0.7 and 2.8 additional tCO$_2$e per hectare per year, but the principal benefit is that it avoids releasing up to 23 tonnes of carbon per year.

The emissions sequestered through this work are not formally accounted for in our carbon footprint at this stage; however, it has been restored to Peatland Code standards and therefore will contribute to Scottish national targets.

10.16   Investment Emissions

We invest around £700 million per year in our water and wastewater assets to maintain and improve our services. Today, we estimate that each £1 m invested generates 200 to 300 tCO$_2$e. With investment expected to increase in the coming years, emissions resulting from capital investment will also increase unless we act.

From the extraction of raw materials all the way to the delivery of a project on site, there are emissions released at every step. Regardless of where these come from in the world and who produces them, we will account for them and work hard to drive them down. Our approach is set out below (Fig. 10.7). Examples to date have seen up look at low carbon pipe materials, cement-free alternatives to conventional concrete, significantly reduce the steel used in our construction and embrace no-dig technologies.

Although these emissions are generally considered the responsibility of our supply chain, we have a big role to play in designing and procuring low-emission materials and working with our delivery and supply chain partners to find more sustainable ways to deliver projects.

10.17   Summary

Scotland's environment is critical to the delivery of resilient water and wastewater services. We depend on sufficient good quality water resources for treating and serving to customers, and healthy water catchments for recycling treated wastewaters.

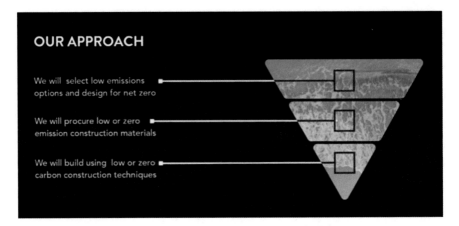

Fig. 10.7 Our approach to reducing our investment emissions

We have made considerable progress in understanding the risks associated with a changing climate and implementing mitigations across much of our asset base. Intense rainfall and long dry periods though will continue to put our services under pressure. Collaborative working with partners across Scotland is key to managing the risks over the long term.

We have a significant carbon footprint and are committed to getting to net zero emissions across all our activities by 2040 and going beyond thereafter. Reducing our consumption, generating renewable energy, restoring peatland and reducing the carbon intensity of our investment are key steps to achieving this ambition.

References

BBC (2018) Shieldhall Tunnel now operational as Scotland's biggest sewer. https://www.bbc.com/news/uk-scotland-glasgow-west-44998611. Accessed 20 Feb 2021

Climate Change Committee (2019) Final assessment: the first Scottish climate change adaptation programme. https://www.theccc.org.uk/publication/final-assessment-of-scotlands-first-climate-change-adaptation-programme/. Accessed 20 Feb 2021

Met Office (2018) An assessment of the weather experienced across the UK during Summer 2018 (June, July, August) and how it compares with the 1981 to 2010 average. https://www.metoffice.gov.uk/binaries/content/assets/metofficegovuk/pdf/weather/learn-about/uk-past-events/summaries/uk_monthly_climate_summary_summer_2018.pdf. Accessed 20 Feb 2021

Met Office (2019) An assessment of the weather experienced across the UK during Winter 2018/19 (December, January and February) and how it compares with the 1981 to 2010 average. https://www.metoffice.gov.uk/binaries/content/assets/metofficegovuk/pdf/weather/learn-about/uk-past-events/summaries/uk_monthly_climate_summary_winter_2019.pdf. Accessed 21 Feb 2021

Scottish Canals (2021) Glasgow's smart canal. https://www.scottishcanals.co.uk/placemaking/north-glasgow/glasgows-smart-canal/. Accessed 20 Feb 2021

Scottish Water (nd) Net zero emissions routemap. https://scottishwaternetzero.co.uk/. Accessed 20 Feb 2021

Scottish Water (2018) Shaping the future of your water and waste water services. https://www.scottishwater.co.uk/en/help%20and%20resources/document%20hub/key%20publications/reports. Accessed 25 Feb 2021

Scottish Water (2019) Funding announced for first outer hebrides peatland restoration project. https://www.scottishwater.co.uk/about-us/news-and-views/260719-lewis-peatlands-project. Accessed 25 Feb 2021

Scottish Water (2021) Heat from waste water. https://www.scottishwater.co.uk/About-Us/Energy-and-Sustainability/Renewable-Energy-Technologies/Heat-from-Waste-Water. Accessed 15 April 2021

The Metropolitan Glasgow Strategic Drainage Partnership (2020). https://www.mgsdp.org

UKWIR (UK Water Industry Research) (2013) Updating the UK Water Industry Climate Change Adaptation Framework. www.ukwir.org/eng/water-research-reports-publications. Accessed 25 Feb 2021

UKWIR (UK Water Industry Research) (2017) Rainfall intensity for sewer design—stage 2, 17/CL/10/17. https://ukwir.org/rainfall-intensity-for-sewer-design-stage-2-0. Accessed 26 Feb 2021

Chapter 11
What Are the Key Enablers in Pursuing Both Disaster Risk Reduction and Climate Change Adaptation? Practical Lessons from Asian River Basins

Megumi Muto

Abstract The key to integrating DRR and climate adaptation is ex-ante investment to reduce risk, the core message of the 2015 Sendai Framework for DRR, along with the ability to adapt to changing climate situations. Past experiences in Indonesia in implementing DRR suggest that institution building at the river basin level is important. Secondly, even with the long-established implementation of river basin management for DRR, many challenges remain in adapting to extreme climate events. The case of Japan, when hit by Typhoon Hagibis, revealed that coordination with local governments needs further improvement. Lastly, the Philippines is a case of an unfinished institutional and financial framework for river basin management. The weakness in the system in terms of institutions and financing makes adapting to extreme climate events a challenge. Finally, issues in forward planning and financing for highly uncertain climate events, both in terms of structural and non-structural measures, are discussed.

Keywords Disaster risk reduction · Climate adaptation · River basin management · Flood risk · Resilience finance

11.1 Introduction

As we approach the 26th UN Climate Change Conference in Glasgow in November 2021 (COP26), countries are raising their aspirations to cut carbon emissions towards net zero. A wide array of government-led and market-based mechanisms is in place, thanks to the fact that carbon emissions from projects, policies and even lifestyles have become tangible and measurable. Government policies, market

M. Muto (✉)
Japan International Cooperation Agency (JICA), Tokyo, Japan
e-mail: muto.megumi@jica.go.jp

© The Author(s), under exclusive license to Springer Nature Singapore Pte Ltd. 2022   207
A. K. Biswas and C. Tortajada (eds.), *Water Security Under Climate Change*,
Water Resources Development and Management,
https://doi.org/10.1007/978-981-16-5493-0_11

design and innovative technologies are cross-collaborating towards the common goal of a sustainable future. Furthermore, such collaborations have led to a significant increase in the realisation of green finance because the impacts and returns on investments are clearer than ever (OECD 2020).

11.1.1 What About Water?

Water is at the core of sustainability. How societies use, share and therefore govern water have been keys to the survival of human beings and ecosystems for millenia. Water also manifests in the form of natural hazards, most notably in the form of floods. With the intensification of flood events in recent years, such as the 2011 flood in Thailand, increased attention is being paid to addressing both disaster risk reduction (DRR) and climate adaptation to protect people's lives and economic assets. The key to integrating DRR and climate adaptation is ex-ante investment to reduce risk, the core message of the 2015 Sendai Framework for DRR, along with the ability to adapt to changing situations. It is not only about preparation to manage in the aftermath of a hazard. Disaster risk "reduction" should also include a combination of ex-ante measures, such as non-structural and structural components to reduce the risk attributed to hazards, exposure and vulnerabilities. In addition, under a changing climate, risks have become "moving targets", with high levels of uncertainty, including extreme events. Although conceptually, DRR and climate adaptation have different origins, at the implementation level, they need to be better integrated (Fig. 11.1).

Because of the large externalities involved, promoting ex-ante investment to reduce risk has been a typical public goods problem that cannot be solved with market mechanisms or driven by optimisation at personal level. Even more difficult is finding ways to adapt to changing climate (both on average and at the extremes)

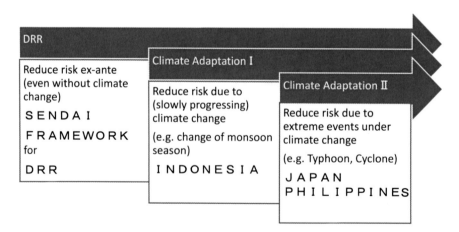

Fig. 11.1 Disaster risk reduction and climate adaptation

in ways that reduce the risks, a process that is itself a public good. Yet, the literature is thin on the ways that governance frameworks, planning methodology and finance mechanisms may work altogether to provide the public good: ex-ante investment to reduce flood risk under a changing climate. I intend to contribute to this literature by identifying the baseline enablers that can generate an implementable set of actions for DRR and climate adaptation for water.

The chapter is structured as follows. First, I review past experiences in Indonesia in implementing DRR for river basins. A river basin is typically an ensemble of multiple watersheds with one main river. The central issue is institution building at the river basin level, because floods—the most frequently discussed water-related hazard—are best understood through such river systems. Secondly, I discuss how, even with the long-established implementation of river basin management for DRR, many challenges remain in adapting to extreme climate events. The case of Japan, when hit by Typhoon Hagibis, revealed that coordination with local governments remains a challenge. The country is trying to overcome this gap through the massive utilisation of its national budget. Lastly, the Philippines is a case of an unfinished institutional and financial framework for river basin management. The weakness in the system in terms of institutions and financing makes adapting to extreme climate events a genuine challenge. Finally, I conclude by suggesting some baseline enablers to make DRR and adaptation happen at the same time and discuss ways forward on financing.

11.2 Review of Past Experiences to Manage River Basins for DRR: Indonesia

Indonesia has an exemplary history in terms of designing governance for water. As an archipelago of islands with multiple watersheds in varying geographies, with contrasting water conditions between wet and dry seasons, and an increasing population pressure, the country has been adopting progressive water management policies that provide valuable lessons for other countries. It is a case of the central government implementing a river basin DRR, backed by law and institutions.

The centrepiece of the institutional design in Indonesia is the Water Law of 2004. River basin organisations (RBOs) have been institutionalised, despite the general policy of massive decentralisation since the Local Government Law of 1999. This suggests that there was an elaborate political decision in Indonesia to work on water issues based on the country's geography and natural environment not according to human-made political jurisdictions. In 2006, the Ministry of Public Works (PU) issued a regulation that the 5590 watersheds nationwide would be grouped into 133 river basins, and that the most important 30 river basins would be directly managed by the PU. In addition, the newly established river basin offices (RBOs) under PU would be in charge of water resource development and management, maintenance and operation of facilities, as well as the coordination of various water user interests (JICA 2010).

Accumulated experiences of JICA-assisted flood control projects in Indonesia (planning, technical assistance and finance for infrastructure) in several major river basins since the 1970s may have led to ground-breaking experiments related to the above set of decisions and conditions. The two illustrative cases below are based on findings from JICA's ex-post project evaluation reports.

11.2.1 DRR in Indonesia: The Case of the Downstream Solo Flood Control Project

Solo is the largest river basin in the most heavily populated island of Java. In the initial flood control planning, the JICA Master Plan of 1974, the overall target hazard level was set at the 10-year return period.[1] The implementing agency was the Directorate General of Water Resources (DGWR) under the Ministry of Public Works and Housing. Upstream infrastructure works were covered by a JICA loan signed in 1985. The hard component for the downstream area, including more than 130 km of river dikes, was financed by a JICA loan signed in 1995, with the weir and retarding ponds to be covered from the Indonesian budget. A state-owned enterprise specialising in water operations undertook the operation and maintenance of the weir (Jasa Tirta I), while the maintenance of river dikes was assigned to an RBO (Figs. 11.2 and 11.3).

At the time of the ex-post evaluation in 2017, efficiency was evaluated as low, due to the project delays arising from difficulties in land acquisition. However, effectiveness was considered high, as it improved to the range of 10–50 year return periods, thanks to additional structural components. The case of downstream Solo is an example of central government-led DRR investment with no explicit consideration of climate change at the time of master planning in the 1970s. Looking retrospectively through the ex-post-evaluation reports over the past 50 years of master planning, from project implementation to completion, we can see that government actors and other stakeholders may have utilised this project as a driving force to adapt to gradually changing climate conditions.

11.2.2 DRR in Indonesia, the Case of Semarang Flood Control Project

This project, commenced in 2006 in Semarang, Java, encompasses the period from the introduction of an improved concept of river basin management explicitly in the Water Law of 2004. At the same time, this approach increased the target return

---

[1] Return period of a flood is the interval between flood events, where an event is any streamflow discharge exceeding a known threshold.

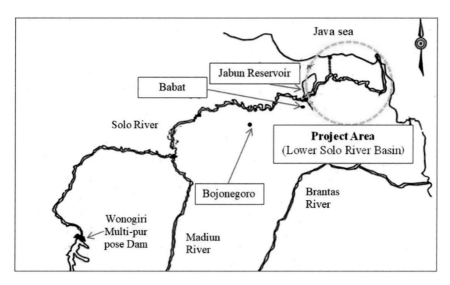

Fig. 11.2 Map of Solo River Basin (JICA 2017)

Fig. 11.3 Babat weir (JICA 2017)

period considerably, to 50 years for the main river. For inner city areas, the return period was set at five years. As one of the projects embraced at the beginning of the 2004 Water Law, the planning and construction of the multipurpose dam (designed for electricity generation, water supply, and flood control) and urban drainage facilities were conducted centrally by the two Directorate Generals under the Ministry of Public Works, DGWR and DGHS (Directorate General of Human Settlement). After completion, an RBO was assigned to oversee the operation and maintenance of the dam. Financing of the dam was provided by both a JICA loan and the central government budget. The maintenance facilities for the inner city drainage were constructed and financed by the corresponding local governments. It turned out that, at the time of ex-post evaluation, only the flood control functions of the dam had been constructed due to delays in the completion of the demand side structures of the other two functions (electricity and water supply). Nevertheless, according to the ex-post evaluation report, the division of roles and coordination mechanisms among different operations and maintenance organisations for flood control has worked well (JICA 2018a) (Figs. 11.4 and 11.5).

The two Indonesian examples highlight key aspects of one project that started before the decentralisation of 1999, and the other after, establishing the importance of institutional design. Throughout, the Indonesian authorities were persistent on the recognition of river systems as the unit for implementing the governing framework. The policy was well-founded by the Water Law of 2004. This was because of the strong conviction that river systems cut across political jurisdictions and that, to be successful in river management, a river basin governance framework is imperative. Even so, under the context of decentralisation, some selected components were assigned to the control of local governments. Moreover, successful water DRR requires strongly integrated river basin-level governance and a solid legal underpinning for integrated planning and implementation. The two project cases also suggest that, over time, through various modes of decision-making, consideration for climate has been reflected to scale up the level of return periods.

11.3 Review of Past Experiences to Manage River Basins for DRR and Additionally for Extreme Climate Events: Japan Under Hagibis 2019

Climate change intensifies extreme events that previously occurred once in 100 years. These are typically manifested as tropical storms (typhoons or cyclones) that travel over heated sea surfaces and intensify. These tropical storms, such as the 2019 Typhoon Hagibis that hit Japan, bring torrential peak rain that falls into the watersheds and river basins on a catastrophic scale, causing flash floods midstream and longer-lasting flooding further downstream.

The impact of Typhoon Hagibis on Japan (950 hPa at time of landing) provides an interesting example of how ex-ante DRR activities in river basin systems generally work well. However, the challenges of extreme events remain, even when the

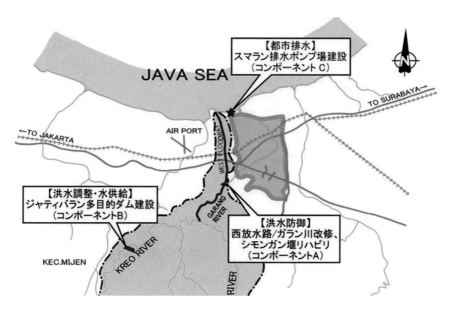

Fig. 11.4 Map of Semarang (JICA 2018a)

Fig. 11.5 Simongan weir (JICA 2018a)

Fig. 11.6 Flooding caused by Typhoon Hagibis (Japan Broadcasting Corporation 2020)

institutional and legal frameworks of river basin management are relatively well established (Fig. 11.6).

The first institutional framework for river management in Japan was established in the late nineteenth century, reflecting a centrally commanded governance structure. The former laws on river management had a strong focus on flood control. However, the institutional framework was based on political and administrative considerations. In 1964, at the beginning of the era of rapid economic growth, the river law was revised to reflect river basin-based governance. Consequently, water resource management was added as an additional purpose. The river systems across the country were categorised into Grade 1, Grade 2 and others. Grade 1 river systems were the responsibility of the national government, Grade 2 were under the prefectures, while the other rivers were under the control of cities and municipalities.

Hagibis travelled across the east part of the archipelago, leaving 25,000 ha inundated (counting Grade 1 rivers only). The death toll was 96, with a further four people unaccounted for. According to the Japan Meteorological Agency (JMA 2019), out of the 18 river systems in the east, in seven cases, the maximum level of 48 h precipitation beforehand was exceeded (by up to 105%). Metropolitan Tokyo was largely saved from major flooding, apart from some inundation occurring due to the overflow from the Tama River in the south. Ishiwatari (forthcoming) succinctly describes "the actual effects of the flood protection systems, consisting of dams constructed upstream, reservoirs midstream, and diversion channels downstream", namely river basin-based management that generally worked during Hagibis. However, nationwide, there were 140 reported cases of (at least partial) collapse of river dikes (Advisory Council on Infrastructure for MLIT 2020).

Subsequent research results (JSCE 2019) suggest that ex-ante DRR in river systems generally worked in the case of Typhoon Hagibis. However, several challenges were also identified: (i) river systems under the jurisdiction of the local governments tended to show weaknesses, (ii) even in river systems under the central government, extraordinary rainfall at the upper watershed translated into overflow into downstream areas with less rainfall and (iii) in some cases the overflow was discharged into flood plains used for agriculture (thus the plains acted practically as a retarding basin). It was also pointed out that, since the main Grade 1 rivers were managed at full capacities aimed at protecting the downstream, Grade 2 ones may have experienced difficulties in discharging into the main river, thus causing inundation in areas close to the confluences.

In response to these experiences under extreme events, the MLIT has issued a new policy framework[2] aimed at an integrated river basin DRR plan that explicitly takes climate change into consideration. The new policy aims to reduce risk by working on all of the three elements of risk: hazard, exposure and vulnerabilities. For hazards, measures will be strengthened by utilising structural measures to better manage floods in anticipation of forthcoming changes in climate. For exposure, emphasis will be on land use, moving people and assets away from high-risk areas. Third, to address vulnerability, proposals involve investing ex-ante in the resilience of assets, business continuity, as well as improving preparedness for people and communities. In addition, in August 2020, Prime Minister Suga announced a major shift in the management of multipurpose dams (such as electricity generation, water resources, flood control) in the same watersheds, in order to maximise the DRR potential of dams. The framework and extent of joint efforts for river basin management are intensifying, mobilising all possible actors under the common objective of DRR ex-ante in anticipation of climate-induced extreme events.

To efficiently implement the new integrated policy framework, the Ministry of Finance is promoting the vertical alignment of project outcomes (such as adapting to expected changes in climate) between the national government, prefectural governments and local governments. The budget transfer from the national government to prefectural and local governments, amounting to 728 billion yen for infrastructure and 785 billion yen for DRR, was available for FY 2020 (MOF 2020).

Nevertheless, since the beginning of rapid urbanisation in the 60s and 70s, fiscal space in local government for DRR has been a critical issue. The risk increasing most rapidly has been exposure due to people and assets proliferating in higher-risk areas. For example, the City of Yokohama, an area that has experienced rapid urbanisation, has mobilised in the following areas using DRR financing. For the rivers under the jurisdiction of the central government and prefectures (Grades 1 and 2), a cost-share system has been utilised, with financing provided through budget transfer from the national government at each layer of governance. For the rivers solely under their jurisdiction (the other river category), they have had to rely on their own budget. For example, they negotiated with private land developers based on "beneficiary/user pays" principles. More generally, land tax for the city

---

[2] "River Basin Disaster Resilience and Sustainability by All".

was also considered part of the "beneficiary/user pays" principle, and measures were taken to capture the land value increases resulting from better flood protection.

Based on these historical experiences nationwide, the national government has been developing public–private schemes for DRR (World Bank 2019). In 2015, there was an improvement in the subsidy scheme for private developers, so that they could construct water-regulating facilities on their premises. These facilities will be managed by the local government in an integrated manner. Recently, prefectures such as Kanagawa have been raising funds through green bonds for the purpose of flood control.

11.4   Review of Past Experiences to Manage River Basins for DRR and Additionally for Extreme Climate Events: The Philippines

The Philippines is located on the typhoon corridor. Every year, more than 20 tropical storms hit the country, including recent super typhoons such as Haiyan in 2013 and Goni in 2020. Since the massive urbanisation that took place after World War II and the deadlier typhoons experienced in the 60s and 70s, the government has been keen to build up the capacities to undertake flood control measures.

Flood control in the major river basins of the Philippines was planned and implemented by the central government—specifically by the Department of Public Works and Highways (DPWH). The largest and earliest flood control infrastructure in the Pasig–Marikina River Basin, which hosts the region of Metro Manila, was the Manggahan Floodway. The Manggahan Floodway, supported by JICA financing through the central government in the 1970s, diverts water from the upper watershed of the Marikina River to Laguna Lake, thereby avoiding the centre of Metro Manila (Fig. 11.7).

After a difficult political transition at the end of the 1980s, the Government of the Philippines restarted the implementation of flood control projects for Metro Manila, making use of the JICA Master Plan of 1990. There were three fronts to work on: capacity building, securing the proper operation of the diversion point of the old Manggahan Floodway and the implementation of dikes starting from downstream areas. At the same time, massive decentralisation took place, similar to that of Indonesia. The critical issues to explore were the ways of implementing and operating flood control and under what framework of governance.

11.4.1   Downstream Pasig–Marikina (KAMANAVA) Flood Control Project

The implementation of flood control projects by DPWH started from the downstream part of the river. The Master Plan of 1990 indicated that the coastal area would

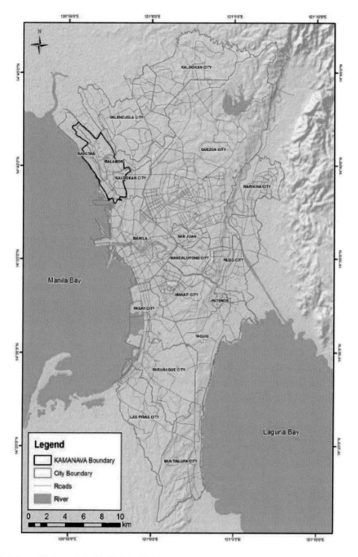

Fig. 11.7 Map of Metro Manila(JICA 2014)

be protected with 10-year return periods, the low and middle stream of the Pasig–
Marikina River with 30-year return periods. Finally, with the completion of the
upper-stream Marikina Dam, the whole system was supposed to provide disaster risk
reduction against once in 100-year events (JICA 1990) (Figs. 11.8 and 11.9).

KAMANAVA is an abbreviation used to designate the area including Kaloocan,
Malabon, Navotas and Valenzuela, with each having its own independent local
government. The area along the coast that suffers from sea level rise receives flood
overflow from the Pasig–Marikina River combined with flow from the Malabon and

Navotas rivers. Historically, a row of fishing villages was converted to commercial and industrial use, and the area hosts clogged canals from the colonial era that add to the complexity. The KAMANAVA flood control project of DPWH, which consisted of flood control dikes and gates, drainage works and risk communication, took place starting from 2000 until completion in 2009 (JICA 2014). Furthermore, the middle stream of the Pasig–Marikina River Basin was covered by a JICA loan (Pasig–Marikina Phase I). It seemed that DRR along the river basin was progressing.

In contrast to Indonesia, the Philippines is prone to extreme climate events. Typhoons Ondoy and Pepeng, which hit the Philippines in 2009 were typical examples. Ondoy and Pepeng were once in 150-year events (PAGASA 2020), the massive overflow from the Pasig–Marikina River Basin, as well as the Malabon and Navotas rivers, inundated the whole KAMANAVA area for weeks. Although the flood control project was implemented by the central government, the response was purely under the jurisdiction of each local government. Therefore, many local governments were overwhelmed, pumping excess water into neighbouring local government areas (Porio 2014). This was not helpful in reducing water levels in the KAMANAVA area as a whole.

River basin management tends to deteriorate when decentralisation takes place and there is no strong institutional design to work on cross-jurisdictional issues. When Ondoy and Pepeng hit Metro Manila, DPWH and the local governments had their own roles, but there was no institutional design or legal authority to effectively coordinate the overall responses. DRR investment to adapt to the changing climate especially was left without anyone to lead responsibly. As a result, LGUs ended up doing what they deemed appropriate from their own perspectives after experiencing extreme events. For example, the local government of Malabon started to invest in additional structures to protect themselves from floods using their own design standards (JICA 2014). However, from a river basin management lens, joint adaptation efforts, design and resource allocations of the central and local governments could have been spent more effectively. The common target could have been in the acceleration of flood control projects towards the upstream area of Pasig–Marikina (under a different local government).

11.5   What Are the Baseline Enablers for Joint Realisation and Spreading Out of DRR and Climate Adaptation?

Synthesising the above case studies from Indonesia, Japan and the Philippines, the following practical lessons can be considered essential in promoting DRR and adaptation:

1. Setting up an institutional framework at the river basin level is important. As water transcends human-made jurisdictions, river basin-based management is a proven practical solution. Integrated planning based on a thorough hydraulic analysis of the river basin and implementation of the structural and

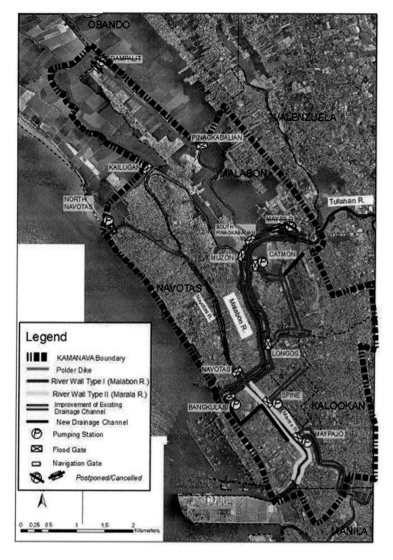

Fig. 11.8 Map of KAMANAVA (JICA 2014)

non-structural measures under a common framework is essential. The governance structure between central and local governments may vary by country, but methods for dealing with water, laws and institutions should be designed and enshrined to ensure the long-term integrity of river management.

2. Even when river basin management is firmly in place, adapting to the changing climate can be challenging. As changes in climate progress incrementally, with some degree of certainty, it may be possible to reflect this in the long-term river

Fig. 11.9 Pumping station (JICA 2014)

basin plan and invest in an adaptive manner, as the Indonesia cases have shown. However, extreme and highly uncertain events, such as intensified typhoons, will require a set of special considerations. First, it is quite difficult to reach a consensus on how to anticipate highly uncertain extreme events. Even when equipped with sophisticated downscaling (physical or statistical) science-based methodologies, engineers who have legal responsibilities hesitate to implement new design parameters unless they are certified by relevant authorities. Different development partners may offer incompatible methodologies to different layers of authorities, adding to the confusion. Capacity building and support for the design standard-setting body are also essential.

3. The third area of challenges concerns the financing for investing ex-ante in DRR for extreme events. The Japanese case demonstrates the goal of intensifying the structural and non-structural DRR of all river basins, mainly through allocation of the national budget at different layers. This leads to the question of what can be done in the case of weaker river basin management and budget allocation. There are also concerns over the justification for ex-ante investment in events that are anticipated to occur only rarely (but when they do hit, can be devastating). Support for the finance/DRR/ climate authorities to advance risk communication and to make the case for DRR investment is also fundamental.

4. Fourth, the importance of coordination between central and local governments should not be underestimated. In the Japanese case, when Hagibis hit, there was a difference in the pace of ex-ante DRR investment and the unfortunate misalignment of the DRR investment design between the central and local governments in charge of respective rivers. This issue was identified as a challenge in the context of extreme climate events.
5. Finally, DRR and adaptation cannot be realised by investing in structures only. Risks consist of hazards, exposure and vulnerabilities. Hazards may be lessened by structural measures, but there are limits both in terms of finance and the environment. Exposure and vulnerabilities are reduced by planning how the society and the people should change investments and behaviours and developing ways to incentivise them through financial measures.

11.6 The Way Forward

As a way forward, there is a need to look more comprehensively into how to better integrate financing DRR and climate adaptation. The current discussion on climate adaptation finance tends to focus on infrastructure projects to be built to promote resilience. The guidelines for climate adaptation finance developed by the Climate Adaptation Commission and the Asian Development Bank (ADB and Global Center on Adaptation 2021) provide the most recent example of this. However, the international discussion on how to plan and implement infrastructure projects to build resilience using a DRR lens, for example, flood control and coastal management, are still in their early stages (JICA 2018b with ASEAN). The G20 principle for quality infrastructure (G20 2019) has a strong policy message on resilience. In addition, apart from infrastructure, there is potential in the area of financial incentives to nudge firms, households and individuals to reduce and risk and improve preparedness. For example, the insurance industry can play an important role in this type of finance that sends price or non-price signals and solicits behavioural change. Using resilience as a core concept in financing DRR and climate adaptation could be the key to forging a powerful synergy between the Paris Agreement and Sendai Framework for DRR.

References

ADB and Global Center on Adaptation (2021) A system-wide approach for infrastructure resilience. Technical note. https://gca.org/wp-content/uploads/2021/01/A-System-wide-Approach-for-Infrastructure-Resilience.pdf. Accessed 10 Feb 2021

Advisory Council on Infrastructure for MLIT (2020) Climate change and water disasters: towards a new integrated policy on river basin banagement. https://www.mlit.go.jp/river/shinngikai_blog/shaseishin/kasenbunkakai/shouiinkai/kikouhendou_suigai/pdf/03_honbun.pdf. Acccesed 10 Feb 2021 (in Japanese)

G20 (2019) G20 principles for quality infrastructure investment. https://www.mof.go.jp/english/international_policy/convention/g20/annex6_1.pdf. Accessed 10 Feb 2021

Ishiwatari M (forthcoming) Effectiveness of investing in flood protection in metropolitan areas: lessons from 2019 Typhoon Hagibis in Japan. Int J Disaster Resilience Built Environ

Japan Broadcasting Corporation (2020) Typhoon hgibis (image). https://www.nhk.or.jp/politics/articles/lastweek/46314.html. Accessed 10 Feb 2021

JICA (Japan International Cooperation Agency) (1990) Report on flood control plan in Manila for the Government of the Philippines. https://openjicareport.jica.go.jp/pdf/10834661_01.pdf. Accessed 10 Feb 2021 (in Japanese)

JICA (Japan International Cooperation Agency) (2010) Mid term review report on capacity building project for practical water management in river basin institutions. https://openjicareport.jica.go.jp/pdf/11999620_01.pdf. Accessed 10 Feb 2021 (in Japanese)

JICA (Japan International Cooperation Agency) (2014) Ex-post evaluation report on KAMANAVA flood control and drainage project in the Philippines. https://www2.jica.go.jp/ja/evaluation/pdf/2014_PH-P212_4_f.pdf. Accessed 10 Feb 2021 (in Japanese)

JICA (Japan International Cooperation Agency) (2017) Ex-post evaluation report on downstream solo flood control project phase 1 in Indonesia. https://www2.jica.go.jp/ja/evaluation/pdf/2017_IP-450_4_f.pdf. Accessed 10 Feb 2021 (in Japanese)

JICA (Japan International Cooperation Agency) (2018a) Ex-post evaluation report on semarang water resource development and flood control project in Indonesia. https://www2.jica.go.jp/ja/evaluation/pdf/2018_IP-534_4_f.pdf. Accessed 12 Feb 2021 (in Japanese)

JICA (Japan International Cooperation Agency) (2018b) Project for stengthening institutional and policy framework on disaster risk reduction and climate change adaptation integration (ASEAN) final report. https://openjicareport.jica.go.jp/pdf/12303509.pdf. Accessed 12 Feb 2021

JMA (Japan Meteorological Agency) (2019) JMA news. https://www.jwa.or.jp/news/2019/11/8535/. Accessed 12 Feb 2021 (in Japanese)

JSCE (Japan Society for Civil Engineers) (2019) Report on typhoon no. 19 extreme rainfall event. https://committees.jsce.or.jp/report/taxonomy/term/63. Accessed 12 Feb 2021 (in Japanese)

MOF (Ministry of Finance) (2020) Highlights of MLIT FY 2020 budget. https://www.mof.go.jp/about_mof/councils/fiscal_system_council/sub-of_fiscal_system/proceedings_sk/material/zaiseier20201019/01.pdf. Accessed 12 Feb 2021 (in Japanese)

OECD (2020) Green infrastructure in the decade for delivery: assessing institutional investment. OECD Publishing, Paris

PAGASA (2020) About tropical cyclone. http://bagong.pagasa.dost.gov.ph/information/about-tropical-cyclone. Accessed 12 Feb 2021

Porio E (2014) Climate change vulnerability and adaptation in Metro Manila: challenging governance and human security needs of urban communities. Asian J Soc Sci 42(1–2):75–102

World Bank (2019) Learning from Japan's experience in integrated urban flood risk management: a series of knowledge notes. World Bank, Washington, DC

Chapter 12
The Great Glacier and Snow-Dependent Rivers of Asia and Climate Change: Heading for Troubled Waters

David J. Molden, Arun B. Shrestha, Walter W. Immerzeel, Amina Maharjan, Golam Rasul, Philippus Wester, Nisha Wagle, Saurav Pradhananga, and Santosh Nepal

Abstract The glacier- and snow-fed river basins of the Hindu Kush Himalaya (HKH) mountains provide water to 1.9 billion people in Asia. The signs of climate change in the HKH mountains are clear, with increased warming and accelerated melting of snow and glaciers. This threatens the water, food, energy and livelihood security for many in Asia. The links between mountains and plains and the differential impacts of climate change on societies upstream and downstream need to be better established to improve adaptation measures. This chapter sheds light on climate change impacts on the cryosphere and mountains, the impact on river systems and the social consequences of such changes in mountains, hills and plains. In high mountains and hills, the impact of climate change is clear, as seen in changes in agropastoral systems and the increasing occurrence of floods and droughts, with losses and damages already high. Moving downstream, the climate change signal is harder to separate from other environmental and management factors. This chapter outlines how climate change in the mountains will impact various sectors in the hills and plains, such as hydropower, irrigation, cities, industries and the environment. It discusses how climate change will potentially lead to increased disasters and out-migration of people. The chapter concludes by highlighting necessary actions, such as the need to reduce emissions globally, build regional cooperation between HKH countries, increase technical and financial support for adaptation, and more robust and interdisciplinary science to address changing policy needs.

Keywords Climate change · HKH mountains · River basins · Environment · Adaptation · Disasters · Agriculture · Hydropower · Cities

D. J. Molden (✉) · A. B. Shrestha · A. Maharjan · G. Rasul · P. Wester · N. Wagle · S. Pradhananga · S. Nepal
International Centre for Integrated Mountain Development (ICIMOD), Kathmandu, Nepal
e-mail: djmolden@gmail.com

W. W. Immerzeel
Utrecht University, Utrecht, The Netherlands

A. K. Biswas and C. Tortajada (eds.), *Water Security Under Climate Change*,
Water Resources Development and Management,
https://doi.org/10.1007/978-981-16-5493-0_12

12.1   Introduction

Images of melting glaciers have become iconic representations of the impacts of climate change, whether in the Arctic or the high mountains of the world. While these images are striking and emotive, they tell only part of the story about the impacts on people, biodiversity and the environment. For countries in Asia that share high mountains, the concern about melting glaciers is largely related to water supply and hazards. We also recognise there are many other factors impacting our water systems, such as snow melt and precipitation as well as rising demand for water in the region. This is a major concern for the 240 million people living in the mountains and hills of the Hindu Kush Himalaya (HKH) as well as people living downstream, as 1.9 billion people live in the ten river basins that have their source in the HKH mountains (Wester et al. 2019).

While the relation between climate change and melting glaciers and snow is clear and supported by outstanding science (IPCC 2019), the impacts of cryospheric change on water, people and the environment are less clear, although the evidence is growing. A few key questions arise: How will the hydrology of water systems change? How will these changes affect water-dependent sectors like hydropower, agriculture, cities and industries? What are their implications for key ecosystems and for people? What kinds of actions and policies will help societies to respond in the short and long run?

This chapter explores the impact of climate change on water resources and subsequent impacts on people and the environment in the Hindu Kush Himalaya region. It starts by identifying climate change impacts on the hydrology of river systems. Since changes in hydrology and precipitation have differing impacts and consequences moving downstream, we use an organising framework that discusses impacts on mountains, hills and plains separately. The chapter concludes with a range of potential responses for adapting to hydrologic changes in these river basins.

12.2   Warming in the HKH Mountains

The HKH, like other mountain systems of the world, is highly sensitive to climate change. Over the past six decades, mean temperature in the HKH has increased at a rate of 0.2°C/decade, while the global temperature increase for the same period was 0.13°C/decade. At the same time, extreme warm events have increased, and extreme cold events have decreased (Krishnan et al. 2019). In mountains, temperatures rise faster than at sea level, a phenomenon known as elevation-dependent warming (e.g. Hock et al. 2019; Pepin et al. 2015). From historic measurements, we see that the temperature in the mountains, including in the HKH, has already risen faster than the global average (DHM 2017; Diaz and Bradley 1997; Krishnan et al. 2019; Liu and Chen 2000; Shrestha et al. 1999).

The warming in the HKH is projected to continue and intensify in the future. With the Paris Agreement (NRDC 2017), the global community set the goal to limit temperature rise by 2100 to well-below 2°C, and ideally to 1.5°C, compared to pre-industrial levels. A special report by the IPCC highlighted the importance of limiting global warming to 1.5°C for adaptation and poverty alleviation (IPCC 2018). However, unless rapid action is taken to lower emissions, or new technologies deployed to remove carbon from the atmosphere, we are not likely to hit this mark. However, even if global warming is contained at 1.5°C, the HKH is likely to warm by 1.8 ± 0.4°C by the end of the century due to elevation dependent warming (Krishnan et al. 2019). In some specific mountain ranges such as the Karakoram, the warming is projected to be 2.2°C ± 0.4°C. Climate projections using Representative Concentration Pathway (RCP) 4.5 yield a warming of 2.2–3.3°C, while for RCP 8.5, which is close to the current emission trends, warming is likely to be 4.2–6.5°C by 2100 (Krishnan et al. 2019).

12.3 Changes in Precipitation

While snow and ice melt are significant factors in the hydrology of HKH rivers, rainfall is in many areas the most significant input to land and water systems. The monsoon and westerly climatic systems and their rhythms are major determinants of river flow hydrology, but also terrestrial ecosystems. Ecological composition and patterns are highly dependent on rain, which influences the density of flora, time of flowering and migration of animals. Agricultural patterns are based largely on these weather phenomena, which determine the timing of planting and harvesting, and what can be cultivated. High variability of rainfall in terms of timing and amount increases uncertainty for water system management. Drastic changes would disrupt the functioning of these ecosystems as we know them.

In spite of the importance of rainfall, we know less about the impact of climate change on rainfall patterns than we do about glacier melt. This is because rainfall patterns in the HKH are innately highly variable, making it hard to differentiate a climate change signal from normal variability of rainfall patterns. The addition of aerosols from air pollution to the atmosphere is also known to impact weather and longer-term climatic patterns. For example, it has been observed that winter fog across the Indo-Gangetic Plains has increased substantially over the last 35 years, with a threefold increase in the number of days with fog, due in part to these aerosols (Saikawa et al. 2019). In addition, there are important feedback loops between land use changes, in particular irrigated agriculture, moisture recycling and precipitation, that further complicate detecting the impact of climate change on rainfall (de Kok et al. 2020; Tuinenburg et al. 2012).

Long-term trends in precipitation have been difficult to discern (Krishnan et al. 2019), but there is some evidence that annual precipitation and annual mean daily precipitation intensity in past decades have increased. Also, annual intense precipitation days (frequency) and annual intense precipitation intensity are experiencing increasing trends (Ren et al. 2017). Looking to the future, there is a divergence among models in projecting changes in precipitation in the HKH. In general, it is projected to increase, but with strong regional differences. Precipitation projection over the HKH region is subject to larger uncertainties both in the global circulation models (GCMs) and RCMs (e.g. Choudhary and Dimri 2017; Hasson et al. 2013, 2015; Mishra 2015; Sanjay et al. 2017). Similar results are shown by Krishnan et al. (2020), although the latter shows that the projected increase is greater in RCMs compared to GCMs. Increasing extremes in precipitation in the HKH was also suggested by Panday et al. (2014), with more frequent extreme rainfall during the monsoon season in the Eastern Himalaya region and a wetter cold season in the Western Himalaya region. The region is naturally prone to water-induced hazards owing to the climate, topography and geological formations. The projected increase in extreme precipitation is likely to exacerbate this situation in the future (Vaidya et al. 2019). In summary, more evidence is needed, and there is still a high degree of uncertainty. Models point to more rainfall in the future, but more importantly, high variability in rainfall patterns with more intense rain and more drought periods.

12.4   Impact on the Cryosphere

While glaciers in High Mountains of Asia (HMA) have been retreating since the end of the Little Ice Age in the late nineteenth century (Bräuning 2006; Kayastha and Harrison 2008; Kick 1989; Mayewski and Jeschke 1979), this retreat has become much faster over the last few decades (Bolch et al. 2019; Maurer et al. 2019; Shean et al. 2020). Despite scattered studies of HKH glaciers, not much was known about the overall glacier status in the HKH 20 years ago. Studies on a handful of glaciers were not enough to draw conclusions about the 54,000 glaciers in the region (Bajracharya and Shrestha 2011).

However, over the last decade there has been an increasing number of studies on glaciers, and the picture is now becoming clearer. Aided by more fieldwork, remote sensing and modelling, we now know that the majority of glaciers in the region are wasting in area and volume at different rates (Zemp et al. 2019). Figure 12.1 shows the rate of loss of glaciers across the HKH between 2000 and 2018. In contrast to this general trend, a more complicated picture emerges in the Karakoram, where some glaciers have shown advances (Bhambri et al. 2013; Hewitt 2011; Paul 2015). This contrast between the Karakoram and the rest of the HKH region has been termed the 'Karakoram anomaly' (Hewitt 2005). While several hypotheses have been put forward (de Kok et al. 2018; Forsythe et al. 2017; Wiltshire 2014), there is not yet a convincing explanation of why this happens. Farinotti et al. (2020) suggest that under the projected climate change it is unlikely that the anomaly will persist in the long term. It is projected that under the 1.5° scenario, which is RCP 2.6 for some climate models, about 30–35% of the volume of HKH glaciers will be lost by the end of the century, while under current emission trends, which is close to

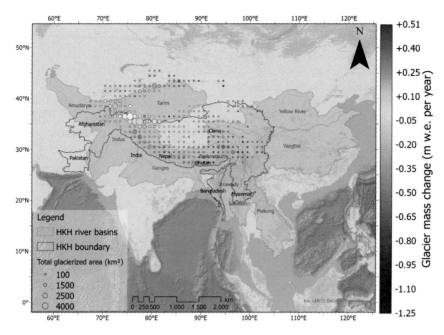

Fig. 12.1 Specific glacier mass balance change in metres of water equivalent per year (m w.e. yr^{-1}) for the period 2000–2018, based on data from Shean et al. (2020). The size of the circle represents the total glacierised area

RCP 8.5, this loss could be as high as 65% (Kraaijenbrink et al. 2017; Rounce et al. 2020).

Although downstream hydrology is influenced by various components such as rainfall, permafrost, snow and glaciers and human interventions, glaciers receive most attention in climate change discussions in the scientific community and the media. For a complete picture, we need to know more about all of these components. Snow is important for water supply in the region (Bookhagen and Burbank 2010; Lutz et al. 2014), as well as for the influence it has on atmospheric circulation and the Asian monsoon (Bansod et al. 2003; Qian et al. 2011; Wu and Zhang 1998; Zhang et al. 2019). Ice cores (Kang et al. 2015; Thompson et al. 2000) and remote sensing-based studies (Gurung et al. 2011) indicate generally decreasing snow cover but with differences between the sub-regions. There is evidence of receding snowlines and of areas receiving increasing amounts of rainfall and less snowfall. This implies that storage from snowpack and retention of water in snow will be reduced, which will impact hydrologic patterns.

Less is known about permafrost, with most permafrost studies concentrated in the Tibetan Plateau (Gruber et al. 2017). The review conducted by Gruber et al. (2017) suggested that the area of permafrost in the HKH largely exceeds the area of glaciers. While detailed studies are lacking, as a general trend, most permafrost in the HKH is likely to have undergone warming and thaw during the past decades

(Zhao et al. 2010). Studies on the Tibetan Plateau suggest multiple impacts of permafrost thaw, such as release of carbon to the atmosphere, desertification and damage to infrastructure (Wang et al. 2008; Yang et al. 2010). As permafrost decreases, there is a major concern that mountain slope stability will be weakened, leading to more landslides (e.g. Gruber and Haeberli 2007).

Air pollution including black carbon can reach high mountain areas. In their assessment of the literature, Saikawa et al. (2019) conclude that air pollutants originating near and within the HKH amplify the effects of global warming and accelerate the melting of the cryosphere through the deposition of black carbon and dust. Sources of black carbon and air pollution emissions include forest fires, coal burning, rubbish burning, diesel fumes, dust and a number of other factors and often come from across borders to reach mountain areas. Clearly, a way to reduce temperature rise and glacier and snow melt is to reduce air pollution and black carbon emissions in the region.

12.5   Cumulative Impact on River Hydrology

The combination of changes in the cryosphere and rainfall, together with changing groundwater patterns, will result in changing river flow patterns over time. In glacier- and snow-fed river basins in many regions, it is projected that melt water yields from glaciers will increase for decades but then decline (IPCC 2019). As glaciers shrink, annual glacier runoff typically first increases till it reaches a turning point, often called 'peak water', after which the runoff declines. The timing of peak water is positively correlated with extent of glaciation in the basin. In most basins in the HKH with major contributions of glacier melt, annual glacier melt runoff is projected to increase until roughly the middle of the century under RCP 4.5 and later in the century under RCP 8.5, followed by steadily declining glacier runoff thereafter (Huss and Hock 2018; Nie et al. 2021). In the Upper Indus Basin (UIB), the peak water is projected to occur around 2045 $\pm$ 17 years under RCP 4.5 and around the middle of the century in most headwaters of the Ganges, while it is suggested that peak water has already occurred, or is close to doing so, in the headwaters of the Brahmaputra (Huss and Hock 2018; Nie et al. 2021). For more extreme scenarios (e.g. RCP 8.5), the peak water is further delayed due to intensified melting.

The change in basin runoff, however, depends on contributions of cryosphere melt and precipitation. Figure 12.2 shows the present contribution of rainfall runoff and melt runoff to total discharge in the upstream basins of HKH rivers (Khanal et al. 2021). The Upper Indus and Amu Darya located in more arid areas have a higher percentage of melt runoff contribution to their flows, while the rivers in the eastern and central region such as the Mekong, Salween and the upper Ganges have a higher contribution from rainfall. In the Upper Indus basin, melt runoff contributes about 45% of the total discharge compared to 13 and 15% in the Ganges and the Brahmaputra at the outlet (Khanal et al. 2021). In some rivers of the Upper Indus, the melt contribution

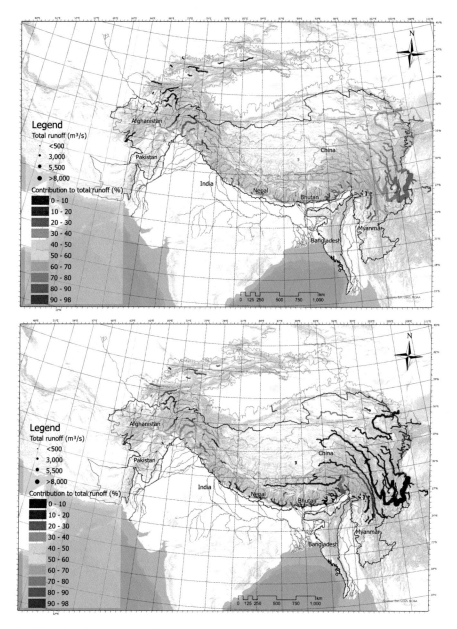

Fig. 12.2 Contributions of rainfall runoff (top) and snow and glacier melt runoff (bottom) to total river flow coming from the mountains from 1985 to 2014 (Khanal et al. 2021)

to total discharge is as high as 90%. In the eastern basin of the HKH like the Yellow and the Yantze, rainfall runoff contributes up to 80% of the total discharge. Table 12.1 shows the rainfall runoff and melt runoff contribution to total discharge in the ten river systems originating from the Hindu Kush Himalayan region.

However, by 2050, the total runoff in the Upper Indus Basin is likely to change by −5 to 12% based on discharge simulations performed using eight GCM runs (four each for RCP 4.5 and RCP 8.5) (Lutz et al. 2014). Most of the changes in the total runoff are directly related to changes in precipitation. In the Upper Ganges, the total runoff will likely increase by ∼1–27% by 2050. The share of melt water is projected to decrease, while the share of rainfall runoff in the total runoff is projected to increase as the rising temperature is most likely to change the rainfall–snowfall dynamics. The case of the Upper Brahmaputra is quite interesting, as it picks up a lot of glacier and snow melt on the Tibetan Plateau, but as it turns to the Bay of Bengal, rain becomes more dominant. In the Upper Brahmaputra, too, the total runoff will likely increase by 2050 (0–13%). The share of melt water is decreasing, while the share of rainfall runoff increases here as well. The trends are in general similar for RCP 4.5 and RCP 8.5 although the spread between the GCM runs is large, especially for the Upper Ganges, mainly due to the larger spread in precipitation projections in the GCM runs used for RCP 8.5 compared to RCP 4.5.

HKH rivers in the west, like the Indus, receive more contribution from snow and glaciers than those in the east, like the Ganges, Brahmaputra, Salween and Mekong, where rainfall runoff contribution is higher. In all basins, the contribution of meltwater decreases, and rainfall becomes more significant as we move downstream. This implies that high mountain ecosystems, agriculture and people will most likely be impacted most by changes in glacier and snow melt. Moving downstream, as contributions from rain become larger, changes in the timing and amount of all flow components will be important when considering impacts of climate change. Contributions to flow vary across the annual cycle, and meltwater contribution during April and May is important when rainfall contribution is low and temperatures are high.

Table 12.1 Rainfall runoff and melt runoff contribution to total discharge of ten major river basins of the HKH region from 1985 to 2014 (Khanal et al. 2021)

Basin	Area (km²)	Glacier area (%)	Rainfall runoff (%)	Melt runoff (%)
Amu Darya	268,280	4.4	5.4	78.8
Brahmaputra	400,182	2.7	62.1	15
Ganges	202,420	4.4	64.7	13.4
Indus Basin	473,494	6.3	43.9	44.8
Irrawaddy	49,029	0.2	78.2	5.1
Mekong	110,678	0.3	55.1	7.7
Salween	119,377	1.5	55.7	16.1
Tarim	1,081,663	3.1	47.3	27.0
Yangtze	687,150	0.4	71	5.7
Yellow	272,857	0.1	63.9	9.7

12.6 Impact of Hydrological Changes on the Uses of Water

Knowing the possible impacts of these hydrologic changes on uses of water is essential for developing responses for different sectors. While we are getting a clearer picture of what will happen to river flows in the future, the implication of these changes is less clear, as only a few scattered studies have explored the downstream impact in detail. Carey et al. (2017, p. 350) argue that people are aware of the potential impact with 1.9 billion people living in HKH river basins, but often conclude, that… 'Unfortunately, research focusing on the human impacts of glacier runoff variability in mountain regions remains limited, and studies often rely on assumptions rather than concrete evidence about the effects of shrinking glaciers on mountain hydrology and societies.' According to a review by Rasul and Molden (2019), a growing number of field-based local studies, both observation and perception based, report that many high-elevation areas of the HKH region are already experiencing water shortages and uncertainties that are considerably impacting agriculture, livelihoods, economy and society.

This knowledge gap is partly attributable to a lack of work across different disciplines. While glaciologists and hydrologists predict changes in river flow patterns, understanding the impacts of these changes on various sectors requires engagement with water management specialists, economists and social scientists. The situation in mountains is very different from that on the plains, so a combined mountain to plain perspective considering multi-scale upstream–downstream linkages is needed.

In a recent important study linking mountain water supply and dependence of societies on mountain water, Immerzeel et al. (2020) developed a water tower index and applied it in 78 watersheds globally to assess the role of mountains in supplying water and the downstream dependence on water for ecosystems and society. Of the river basins they assessed, they found the Indus Basin most vulnerable and most of the ten HKH river basins to be highly vulnerable due to high population densities, high dependence of their populations on water for irrigation, industries and cities and their vulnerability to climate change. Another reason that made the situation in the Indus River Basin most critical was the high dependency of the downstream population on meltwater. Moreover, eight of the ten HKH rivers are transboundary in nature, in a geopolitically sensitive region. The study clearly showed the importance of mountain water for societies around the world and the importance of mountains for HKH rivers.

To discuss impacts, we use a simple framework of water flowing from the mountains to the hills and downstream to the plains (Fig. 12.3). Water use patterns are distinctly different in these three zones. In the high mountains, people are heavily dependent on cryospheric resources. Moving downstream to the hills, river beds cut deep into valleys. Because of the hill topography, this is the location for dams, reservoirs and hydropower projects. The hills receive little snow compared to the high mountains, so people in the hills use water from rivulets and springs. Most

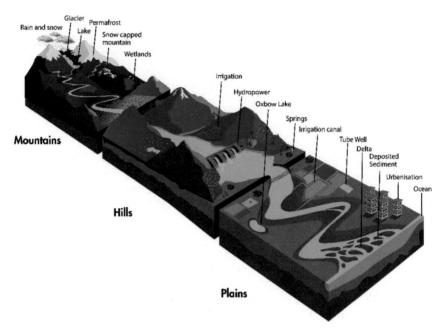

Fig. 12.3 Flow of water from mountains to hills to plains to oceans. Each zone has its unique physiographic and climate change characteristics and is discussed separately in the chapter (adapted from Nepal et al. 2018)

of the HKH population lives in the hills and depends less on glacier and snow melt, rather from rainfed springs. In spite of this, the cryosphere receives much more attention than climate change impacts on water resources in hill region. Population densities increase in the plains, where water is used intensively by cities, the agricultural sector and industries. People depend on water mainly for floodplain agriculture and fishing. Rainfall becomes the primary contributor to river flows, and groundwater use is prevalent. This shows strong upstream–downstream linkages in the HKH river basins (Molden et al. 2016; Nepal et al. 2018), and these linkages may be affected by environmental and socioeconomic changes across different scales.

In high mountains where people rely on glacier and snow melt, changes in the cryosphere will have direct impact on water use, and the climate signal is direct—glaciers and snow melt, and the supply of water changes. Moving downstream, long-term changes in runoff could also be a result of anthropogenic changes such as changes in land use patterns, agricultural practices and upstream water infrastructure like dams. The hills are also intensively used for hydropower development due to year-round availability of river flow (originating from high mountains) along with elevation differences. Then moving to the plains there is considerable human influence on hydrologic patterns with the diversion and use of water, effluent flows, groundwater pumping and reuse of water. Hydrologic changes are due to a number of factors including climate change, and it is more difficult to separate out the climate change signal.

12.7 Societal Impacts in Mountains

The mountains are dominated by alpine ecosystems, including steep slopes, snow- and ice-covered lands, grasslands and forests. Human systems include villages, small-scale agriculture and pastoral systems. Snow and ice melt and permafrost thaw can significantly influence water systems used by people for pastoral activities and crop cultivation. Changes in melt patterns can disrupt these systems. Mountain agriculture and pastoralism are impacted by changes in both temperature and water regimes (Rasul and Molden 2019). In high mountain areas in Afghanistan, northern Pakistan and western and central India, snow and glacial melt water is used for irrigation and helps retain soil moisture on pastures and grasslands. People living in high mountain areas of the HKH region also depend on glacial melt water for drinking and other domestic uses (Rasul and Molden 2019). It is reported that people in such areas have been facing a shortage of water for drinking and domestic uses, partly due to reduced glacier and snow melt (McDowell et al. 2013; Rasul and Molden 2019; Rasul et al. 2020). In some cases, communities have lost their water source, and in extreme cases, this has led to displacement of communities. Interestingly, some areas previously covered by glaciers, and snow are growing new vegetation and have become potential grazing grounds.

A number of field-based studies indicated that subsistence agriculture, the main source of livelihood in high mountain areas of Bhutan, India, Nepal, and Pakistan, is being impacted by melting glaciers, thinning snow cover and changing precipitation patterns (Dame and Nüsser 2011; Rasul et al. 2020; Rasul 2021). Agriculture and agro-pastoralism in high mountain areas of the HKH depends heavily on cryospheric resources, and in particular, the spring and summer melt runoff is a vital source of water for plant growth. People in those areas are thus highly vulnerable to changes in the cryosphere (Rasul 2021). In northwestern India, for example, farmers have been impacted by a shortage of irrigation water and soil moisture due to the gradual recession of low-lying glaciers and changing patterns of snowfall (Grossman 2015). In Ladakh, India, agricultural production almost entirely relies on meltwater for irrigation. Farmers divert snow meltwater and channel glacial melt-water to farm fields and settlements where topography allows. It is reported that faced with water shortage, farmers are taking many adaptation measures including changing cropping patterns across the western Himalaya region (Clouse 2016; Dame and Nüsser 2011; Rasul et al. 2019).

Livestock is an integral part of farming systems and livelihoods in high mountain areas of the HKH. Animal husbandry and agropastoral livelihoods are also impacted by cryosphere change in the HKH region, though less visibly. Changes in snowfall patterns and overall decline in snow cover result in fodder and water shortages that affect livestock production, food security and livelihoods. For instance, it is reported that plant density in meadows has declined in Nagqu Prefecture in the Tibet Autonomous Region of China, in part due to global warming and cryospheric change, as well as increasing intensity of grazing, which affects yak and sheep herders who depend on it for their livelihood (He and Richards 2015; Rasul et al. 2020).

In mountains, life has always been risky, but now it is even more so as avalanches, landslides, floods and glacier lake outburst floods have increased, in part due to climate change. When glaciers melt, lakes often form downstream, dammed by moraines which are often unstable. As the lakes grow with increased melt, the danger of glacial lake outburst floods (GLOF) increases (Byers et al. 2019; ICIMOD 2011; Rounce et al. 2017). In the Koshi Basin, there have been six GLOF events since 1990, and four of these occurred since 2015. In addition, 42 lakes in the Koshi, three in the Gandaki and two in the Karnali basins were identified as potentially dangerous lakes (Bajracharya et al. 2020). Likewise, permafrost thaw tends to destabilise steep slopes actually held together by frozen water, leading to an increase in hazards from landslides.

In the mountains, disaster and its cascading events are common and dangerous. For instance, the Kedarnath flood of 2013 was caused by a rainfall triggered landslide into a glacier lake. The avalanche that destroyed the Langtang village in Nepal in 2015 was the result of an earthquake in combination with a 1:100-year spring snow pack. The recent Chamoli disaster led to over 70 deaths, with 134 missing. A few days before the disaster a strong westerly disturbance caused heavy rain, leading to rising water levels. Then, a massive rockfall mixed with ice and snow occurred on Ronti peak. This melted the ice below, causing a flood wave to surge down the already swelled river and mix with previously deposited sediments and debris. Infrastructure in the flood path, in particular hydropower plants, exacerbated the damage. The unfinished Tapovan Vishnugad hydropower project was destroyed, and the flood inflicted substantial damage on the Rishi Ganga hydropower project (Shrestha et al. 2021a). Disaster can be considered one reason why people migrate away from their homes, a topic that will be discussed in detail in another section.

Cryosphere change has significant impacts on local economy and society, and a few studies have demonstrated this. The IPCC Special Report on Ocean and Cryosphere suggest that economic losses are incurred through two pathways—due to climate-induced disasters, and through additional risks and loss of potential opportunities brought about by cryopsheric changes (Hock et al. 2019). Hock et al. (2019) reported that cryosphere-induced disasters such as flood, landslides and avalanches are increasing in the HKH region and projected to increase in the future, and additional costs will be required for risk reduction measures. These have already incurred huge economic, social and environmental cost to society. It is reported that over the period 1985–2014, economic losses in mountain regions from all flood and mass movements (including those not directly linked to the cryosphere) were highest in the HKH region (USD 45 billion), followed by the European Alps (USD 7 billion) and the Andes (USD 3 billion) (Stäubli et al. 2018). For instance, the Zhangzangbo glacier outburst flood of 1981 in Tibet killed 200 people and caused extensive damage to infrastructure and property, with an estimated economic loss of USD 456 million (Mool et al. 2001). Similarly, the Dig Tsho flood in the Khumbu region of eastern Nepal in 1985 damaged a hydropower plant and other properties, with estimated economic losses of USD 500 million (Shrestha et al. 2010). As an adaptation measure, a channel was dug in Tsho Rolpa glacier in Nepal in 2002 to lower a glacial lake, an initiatve that costs USD 3 million (Bajracharya 2010).

12.8   Societal Impacts in Hills

People living in hills, especially those in rural areas, have a very close relation with ecosystems. They often rely directly on forests for food, fodder and medicinal goods. Changing rainfall patterns impact forest and other ecosystems and threaten livelihoods. Traditionally, most hill and mountain communities reside on hill slopes or hill tops, not on river valley bottoms. This is changing as more roads are being built and more towns and urban centres are springing up. People are moving to valley bottoms to take advantage of increased connectivity. Changing river flow patterns including discharge and timing of water will have an impact on the amount and timing of hydropower production. This movement to valley bottoms makes people and infrastructure vulnerable in different ways. People on hill slopes are quite susceptible to landslides, and this risk gets aggravated with increasing intensity of rain driven by climate change. There are many causes of landslides besides climate change, such as road building and land cover change including deforestation, but the hazard of landslides will increase in the future.

Of rising concern is the situation of springs in the hills, which are a major contributor to river flows and a source of domestic water and small-scale irrigation (Scott et al. 2019). Springs are drying up all across the countries of the HKH. There are several site-specific reasons, such as changes in land use reducing recharge capacity; increased extraction of water from springs with pipes and pumps and possibly changes in rainfall, runoff and recharge patterns due to climate change, though the latter is not well established. Climate change research and development activities should pay more attention to springs given their important role in rural and urban water supply (Scott et al. 2019).

Rivers grow in size as they move downstream, receiving contributions from the cryosphere and increasingly rainfall-driven flows from the land. Changes in land use such as changes in forest cover influence these patterns. Artisanal fisheries and floodplain agriculture are common across the hills, and both are impacted by changing flows and flood flows.

The HKH is now witnessing a hydropower boom, with an increasing number of hydropower dams being built in the hills on the tributaries of the great HKH rivers (Vaidya et al. 2021). Hydropower could significantly help countries move towards less carbon intensive energy, reduce GHG emissions and increase energy security. Hydropower already produces almost 100% of Bhutan's electricity, 93% in Nepal and 33% in Pakistan (Shrestha et al. 2016; Vaidya et al. 2021) and contributes considerably in India. All the countries of the HKH region have taken the initiative to increase hydropower generation. While river runoff is not expected to reduce until the second half of this century, changes in glacier and snow melt patterns will affect hydropower production in the HKH due to changes in the seasonality of river flows to increased variability of flows. The exact impact of cryospheric change is difficult to determine because the effects vary greatly across the region and even within each country. The impact is likely to be greatest on small-scale 'run of the river' hydropower plants with little or no storage, of the type common in the Himalayan region (Boehlert et al. 2016; Turner et al. 2017). Reduction in dry season runoff can reduce or even halt hydropower production from these plants. Even if there is an initial increase in annual flow from glaciers as they recede (Rees et al. 2004), such increase will occur in the warm wet season and will not compensate for decreased water availability during the dry season. For instance, hydropower generation in the Khulekhani project in Nepal is projected to decrease in future by 0.5–13% (Shrestha et al. 2020).

Hydropower plants in the mountains are susceptible to disasters such as floods, GLOFs and landslides. The disastrous cryosphere-related flood of February 2021 in Uttarakhand, which destroyed two hydropower installations, was a stark reminder of the power of floods. Bajracharya et al. (2020) show that the number of potentially dangerous glacier lakes is rising, which has increased risks for hydropwer plants and other infrastructure. For instance, a hydropower plant in Nepal was destroyed by the Dig Tsho GLOF in Nepal in 1985. Similarly, a hydropower project in Uttarakhand, India, was damaged by a cryosphere-related disaster linked with a cloudburst near Chorabari glacier in June 2013 (Schwanghart et al. 2016). A hydropower plant on the Bhote Koshi near the Nepal–China border was severely damaged in 2016 by a GLOF (Liu et al. 2020), and in 2014, a landslide in the Jure area of central Nepal destroyed hydropower projects (Bhatt 2017).

12.9 Climate-Induced Migration in Hills and Mountains

Human mobility and migration are integral to mountain livelihoods. For centuries, mountain people have moved from one place to another to avoid extreme winters, access more productive land and respond to agricultural seasonality (Macfarlane 1976; Pathak et al. 2017). By successfully adopting multi-local livelihoods, mountain communities have flourished in harsh climatic and topographic conditions. There are three major types of migration in the HKH region—transhumance mobility, labour migration and permanent migration. Human mobility and migration in the region are impacted by climate change and cryospheric changes as they have changed the availability of water and increased the frequency of hazards (Carey et al. 2017; Rasul and Molden 2019; van der Geest and Schindler 2016). In high mountains, transhumance mobility and associated herding activities are affected by the changes in snow and glaciers (Gentle and Thwaites 2016; Namgay et al. 2014; Nyima and Hopping 2019). Prolonged winter snow resulting in food and water shortages and increasing hazard events like avalanche resulting in death of livestock in large numbers makes herding a highly risky livelihood option (Shaoliang et al. 2012; Tuladhar et al. 2021).

For a long time, communities in the mountains, hills and plains have adopted labour migration as a livelihood diversifying strategy and also as a response to agricultural seasonality (Adger et al. 2015; Tuladhar et al. 2021). Climate change influences labour migration through its adverse impact on local livelihoods (Foresight 2011). There is a growing body of evidence that attributes migration to the impacts of droughts, floods, landslides, erratic precipitation and its impact on agriculture and other nature-based livelihoods (Hugo 1996; Viswanathan and Kavi Kumar 2015). Apart from climate change and cryospheric changes, other major drivers of migration are economic and employment opportunities, better access to education and health facilities (Gioli et al. 2014; Hugo 1996; Maharjan et al. 2020; Rigaud et al. 2018; Warner and Afifi 2014) and the younger generation's desire to move away from farm-based livelihoods (Carling and Schewel 2018). Labour migration, in turn, helps households to better adapt to climate change impacts through spatially diversifying household livelihood sources and spreading risks (Gemenne and Blocher 2017; Le De et al. 2013; Maharjan et al. 2021). In the last two decades, there has been a growing trend of permanent out-migration of entire households, leading to a decline in mountain population. There is anecdotal evidence that drying up of springs in the mid-hills, as a result of climate change and development interventions, is resulting in out-migration of entire villages in the HKH region. In the Upper Ganga basin, drying springs and rising floods have disrupted local agriculture-based livelihoods resulting in large-scale migration, particularly of youth, to urban centres in search of alternative livelihoods (Bhadwal et al. 2017). In the last population census, 36 out of 55 mountain and hill districts in Nepal reported a negative population growth rate (CBS 2012). Similar trend of depopulation and a rise in 'ghost villages' (villages without inhabitants) is also seen in Uttarakhand, India, that has led to a loss of traditional livelihoods and cultures (Pathak et al. 2017).

However, while there are many intertwined factors for migration, climate change could serve as a tipping point in people's decision to move. There is new evidence that often people's migration decisions are influenced by their perception of changes in climate (Koubi et al. 2016a, b). As the HKH region is highly vulnerable to the impacts of climate change due to its high dependence on nature-based livelihoods, this region is likely to see an increase in migration in future. If no action is taken, climate change is expected to result in 40 million internal climate migrants in South Asia (Rigaud et al. 2018).

12.10 Societal Impacts in the Plains

The plains of the great rivers of Asia, which emanate from the HKH mountains, are home to dense populations, major cities, industries, high economic activity and intense agricultural production. The plains depend on mountains for fertile soil, abundant water supplies and other mountain resources like forests. The Indo-Gangetic Plains is the food basket of India and Pakistan, similarly, the Yangtze is known as China's rice bowl, and the Yellow River Basin also significantly contributes to China's food production. China Water Risk (Hu and Tan 2018) reported that in 2015 the ten HKH river basins generated a total GDP of USD 4 trillion. However, there is growing concern that climate change impacts might threaten economic activities in the plains.

It is not easy to disentangle climate change impacts from other impacts. For example, dams placed at the mouth of rivers where rivers meet the plains (e.g. Tarbela and Tehri dams) exert significant influence on flow regimes. Diversions for cities and agriculture influence river flows, and groundwater dependence has made a huge difference to water supply over the last decades. Water management itself makes a tremendous difference. Where water is poorly managed, even with ample supplies, people can experience water scarcity; in contrast, where it is well managed, people can get enough water even with minimal supply (Molden 2020). It is hard to single out the impact of climate change as it is mixed with many other factors. Nevertheless, changes in flow patterns will present an additional water management challenge in areas that are already stressed by a situation where demand outstrips supply.

With climate change and increased rainfall, monsoon season river flows are likely to increase further, and the likelihood of floods will increase. However, as glaciers shrink and snow melt diminishes, water supply will gradually reduce, especially in the dry season when more water is needed for irrigation. This may affect agriculture and food security in large parts of South Asia, including the Indo-Gangetic Plains, where the lion's share of water is used for food production. Declining runoff, due to a combination of decreasing ice mass and early snow melt, is expected to reduce the productivity of irrigated agriculture in the Indo-Gangetic Plains and other plain areas unless preventive measures are taken (Siderius et al. 2013; Biemans et al. 2019).

Using a coupled cryosphere-hydrology-crop modelling approach, Biemans et al. (2019) analysed the impacts of glacier and snow melt on agricultural production in the the Indus, Ganges and Brahmaputra River basins. The study showed that there is strong spatial and temporal variability of impacts with more dependence on meltwater in the arid Indus Basin and with meltwater being more critical during the pre-monsoon dry season. The contribution of glacier and snow melt to river flows is very high in the Indus. Overall 37% and in the pre-monsoon season up to 60% of total irrigation withdrawals originate from mountain snow and glacier melt, and it contributes an additional 11% to total crop production. In contrast, in the Ganges plains, there is significant contribution in the pre-monsoon from March to May, where snow and glacier melts contribute 20% of supply, but negligible amounts during the monsoon, and in the Brahmaputra Basin, the contribution is much smaller. The authors estimated that meltwater contributes to wheat production for 64 million people and rice production for 52 million people. With climate change, it is expected that the modulating role of meltwater will diminish, and in the long term, an important source of supply would diminish.

The combined effects of a growing population and increasing food demand, declining per capita land availability and crop yield, and reduced water availability threaten future food security for a large population in the HKH region (Aggarwal 2008; Immerzeel et al. 2010; Rasul 2010, 2014). People in the Indus Basin will likely be impacted more because of their high dependence on irrigated agriculture. More than 90% of all crops in the Indus Basin are irrigated, with glacier and snow melt a major source of irrigation water.

Cryospheric change and changes in rainfall patterns may also threaten urban water supply in the future. In the Himalayan region, many of the world's mega cities are situated along the banks and in the catchments of rivers that originate in the mountains, and a large number of major cities partly depend on glacier- and snow-fed surface water for their drinking water supply. A number of large cities in India such as Haridwar, Varanasi, Patna, Kanpur, Allahabad, Munger, Bhagalpur, Delhi, Agra, Mathura and Kolkata are situated along the mighty river Ganges and its tributaries and depend partly on glacier and snow meltwater that feeds the river. In Nepal, the government is developing an inter-basin water transfer project to supply 3.5 million litres of water per day to Kathmandu city from the glacier-fed Melamchi river (Khadka and Khanal 2008). Similarly, the Bhagirathi and Yamuna rivers are a major source of water for Delhi. In Pakistan, the Soan river supplies water to Simly Dam, a major reservoir of Islamabad, and in southern Punjab, water seepage from the glacier-fed Indus Basin Irrigation System is an important source of water (Jehangir et al. 1998). While the demand for water in urban areas is increasing with the growth of population, urbanisation and industrialisation, the reduction in water from glacier melt and snow melt is likely to exacerbate the challenge of urban water supply in many towns and cities in the HKH region.

Looking to the future, with rising populations and increased economic activity, there will be more demand for water from cities, agriculture and industries. At the same time, there is already significant ecosystem degradation. All of the great HKH rivers, except the Yellow and the Yangtze, are transboundary; yet cooperation between countries on rivers remains limited. Water sharing between states in India is also an issue. Even without climate change, the management of these river basins is an urgent challenge. Climate change will add more challenges including changes in hydrology and the threat of more floods and droughts.

12.11 Flooding in the Plains

The HKH Assessment Report provided a grim picture of disasters in the HKH and associated river basins, with 36% of all disaster events in Asia between 1980 and 2015 occurring in the HKH and 21% of events globally (Vaidya et al. 2019). The report stated that the number of disaster events, people killed and affected and economic losses increased by 143% from the ten-year period of 1990–2000 to the period of 2000–2010. Floods are the most common disaster in the HKH (Shrestha et al. 2015) accounting for 17% of people killed and 51% of damage.

Plains adjacent to mountains are prone to severe flooding due to increased climate and rainfall variability. The 2010 floods in Pakistan killed more than 2000 people with a loss of USD 10 billion; the 2013 Uttarakhand flood killed more than 5000 people; and Bangladesh, which lies at the intersection of three HKH rivers, is most vulnerable to floods (Vaidya et al. 2019). The increased intensity of rain in the mountains and plains plays a role, although the climate change attribution has not been adequately assessed. Wijngaard et al. (2017) projected that the 50-year return period flood is expected to increase by up to 305% relative to the current level with the largest increase in the upstream headwaters of the Upper Brahmaputra Basin. In contrast, in the Upper Indus Basin, the 50-year return level is expected to decrease by up to 25%. These changes are attributed to changes in rainfall and melt runoff patterns.

Disasters are a combination of the interplay between hazard, exposure and vulnerability (IPCC 2012) with climate change in the mix, so it is quite difficult to separate out climate change impacts from other human or environmental causes. Climate change will increase hazards, and unless preventative measures are taken to reduce vulnerabilities and exposure, the region is likely to experience more and larger disasters.

To reduce disaster risk, it is important to consider the links between mountains and plains and install flood early warning systems in downstream areas. It is also necessary to protect forests, empower communities and take collective action that transcends national boundaries.

12.12 Key Actions

The HKH mountains are a global resource, important for human survival and well-being and for biodiversity. The first global action required is to immediately reduce greenhouse gas emissions and slow the rise of temperatures in the mountains. While HKH mountain communities and downstream populations already feel the impact of climate change, the global community needs to become more aware of what is happening in the HKH. Located in a strategically important geopolitical region, countries that share the HKH mountains are prone to conflict. The global community has a responsibility to help HKH mountain communities, some of the poorest in the world, to adapt to the changing environment and build resilience. Even if the most drastic greenhouse mitigation measures are taken, changes will continue, and thus, adaptation is essential.

Adaptation and resilience building will be key for mountain communities and downstream communities. Local communities in different parts of the HKH region have already adopted a range of measures, including migration and mobility, in response to challenges posed by changes in the cryosphere. To cope with increasing water stress and uncertainty resulting from cryosphere shrinkage, farmers are increasing water storage and modifying livestock grazing patterns, constructing new water channels, storing water by creating artificial glaciers, shifting away from water intensive crops and growing new crops suited to water stress conditions. In some cases, communities have abandoned agriculture and livestock practices and undertaken new sources of livelihood such as tourism and labour migration. In cases where the impact of cryospheric changes is extreme, communities have been compelled to take the decision to relocate to more favourable areas. Many response measures are site-specific and carried out at a local scale and represent autonomous initiatives of local communities. With the changes in hydrology, downstream communities must adapt their water systems to cope with too much or too little water. Where possible, water augmentation will have to be combined with water demand management, and at all levels, institutions have to become much more adaptable and responsive to change.

Some responses are beyond the financial means of local communities—for example, building the infrastructure needed to deliver water to villages or fields. A key to building adaptation measures is to co-develop solutions with communities and local governments, taking advantage of indigenous knowledge and also strengthening institutions needed to sustain adaptation. Resilience building provides an important framework, as it is a forward-looking approach that also sees the inherent vulnerability of the system beyond climatic drivers and thus helps identify interventions necessary for positive transformation (Mishra et al. 2017).

Many of the impacts will be regional in the sense that impacts of climate change cross-national boundaries, and cooperation is required to deliver solutions. Countries share areas that are important to biodiversity, they share rivers and forested areas, and often disasters extend across borders. Moreover, there is a huge opportunity for countries to share knowledge to help in the adaptation process. However, the region is rife with political tension between countries. While this certainly hampers efforts to strengthen cooperation, the threat of climate change itself could be turned into an opportunity to promote cooperation. ICIMOD has taken important steps through its programmes on river basins and transboundary landscapes (Molden et al. 2017), river basin networks like the Upper Indus Basin Network (Shrestha et al. 2021b) and most recently through a Ministerial Declaration agreeing to work together on mountains (ICIMOD 2020).

A regional adaptation plan is required for the HKH region. A regional action plan supported by regional funding could deliver big impact and also provide measures of both adaptation and mitigation. Key elements of such a plan would include

- Efforts for increased cooperation on river basins, so that water supply, storage, transportation and energy generation and transmission could be developed more optimally, sharing benefits and costs.
- Managing cross-border disasters, improving flow of information and setting up regional responses to disasters when they happen.
- Efforts to build cooperation on important biodiversity hot spots; supporting local communities who live in these areas.

- Better linking of upstream and downstream activities, including compensating upstream communities for conservation efforts and supporting downstream communities in adaptive management of resources impacted by climate change. Focus should be on agriculture as it is the biggest user of water.

Infrastructure is a key concern, considering the need for development and also the need to do things differently in light of climate change. For example, hydropower project impacts the environment, but they are also impacted by the environment. Proper environmental risk assessment and implementation of environmental mitigation measures can ensure the sustainability of hydropower (Vaidya et al. 2021). Many hydropower plants were constructed based on historical meteorological and hydrological data and may need considerable modification to operate under a different hydrological regime.

Governments have a critical role to play in raising awareness about present and future impacts and vulnerabilities, building adequate capacity to cope with impacts, and to help build from the inherent adaptive capacity of local communities. Governments of the HKH region have to work closely with mountain and downstream communities to understand the changing situation and develop solutions. This will require financial and human resources, and these must be mobilised in different ways than before. Much more foresight is required, and there is a need to deal with a a high level of uncertainties. Plans and programs that worked in the past will not be sufficient now. It is necessary to improve research capacity to understand, assess and predict so that appropriate response measures can be developed. Now is also the time for governments to reach out beyond their boundaries to their neighbouring countries for ideas and inspiration, but also to address real and growing transboundary issues such as floods. The International Community has set up instruments like the Green Climate Fund, and this is a positive step, but more could be done to support regional and transboundary issues.

We need more robust science to back policy formulation, but we also need to do science in a different way, using more inter and transdisciplinary approaches involving many disciplines including social sciences and connecting research with communities and policy makers. More measurements and observatories are required in the mountains. Upstream and downstream linkages, including downstream impacts on the plains, need to be better quantified. More attention should be paid to key linkages such as the water–food–energy nexus and the role of air quality. Experts in the social sciences and physical sciences should work together to find ways to alleviate poverty, build institutions and build awareness of the complex political ecology in the region.

There is a lot to be done, and it is not too soon to start. This chapter has brought out some of the existing knowledge on climate change in the mountains and its impacts downstream. We feel that much more needs to be done to understand this critical link. The chapter also shows that we know enough to take action now— through investments, strengthening cooperation and institutions and expanding scientific knowledge and practice to improve our responses.

Acknowledgements and Disclaimer This study was partially supported by core funds of ICIMOD contributed by the governments of Afghanistan, Australia, Austria, Bangladesh, Bhutan, China, India, Myanmar, Nepal, Norway, Pakistan, Sweden and Switzerland.

The views and interpretations in this publication are those of the authors. They are not necessarily attributable to their institution and do not imply the expression of any opinion by their institutions concerning the legal status of any country, territory, city or area of its authority or concerning the delimitation of its frontiers or boundaries or the endorsement of any product.

References

Adger W, Arnell N, Black R, Dercon S, Geddes A, Thomas D (2015). Focus on environmental risks and migration: causes and consequences. Environ Res Lett 10(60201). https://doi.org/10.1088/1748-9326/10/6/060201

Aggarwal PK (2008) Global climate change and Indian agriculture: impacts, adaptation and mitigation. India J Agric Stud 78(10):911–919

Bajracharya SR (2010) Glacial lake outburst flood disaster risk reduction activities in Nepal. Int J Erosion Control Eng 3(1):92–101. https://www.jstage.jst.go.jp/article/ijece/3/1/3_1_92/_pdf

Bajracharya SR, Shrestha B (2011) The status of glaciers in the Hindu Kush-Himalayan region. International Center for Integrated Mountain Development, Kathmandu

Bajracharya SR, Maharjan SB, Shrestha F, Sherpa TC, Wagle N, Shrestha AB (2020) Inventory of glacial lakes and identification of potentially dangerous glacial lakes in the Koshi, Gandaki, and Karnali River Basins of Nepal, the Tibet Autonomous Region of China, and India. Research report, ICIMOD and UNDP

Bansod SD, Yin ZY, Lin Z, Zhang X (2003) Thermal field over Tibetan Plateau and Indian summer monsoon rainfall. Int J Climatol 23(13):1589–1605. https://doi.org/10.1002/joc.953

Bhadwal S, Ghosh S, Gorti G, Govindan M, Mohan D, Singh P, Singh S, Yogya Y (2017) The Upper Ganga Basin—will drying springs and rising floods affect agriculture? HI-AWARE working paper 8. HI-AWARE, Kathmandu

Bhambri R, Bolch T, Kawishwar P, Dobhal DP, Srivastava D, Pratap B (2013) Heterogeneity in glacier response in the upper Shyok valley, northeast Karakoram. Cryosphere 7(5):1385–1398. https://doi.org/10.5194/tc-7-1385-2013

Bhatt RP (2017) Hydropower development in Nepal—climate change, impacts and implications. Renew Hydropower Technol. https://doi.org/10.5772/66253

Biemans H, Siderius C, Lutz AF, Nepal S, Ahmad B, Hassan T, von Bloh W, Wijngaard RR, Wester P, Shrestha AB, Immerzeel WW (2019) Importance of snow and glacier meltwater for agriculture on the Indo-Gangetic Plain. Nat Sustain 2:594–601. https://doi.org/10.1038/s41893-019-0305-3

Boehlert B, Strzepek KM, Gebretsadik Y, Swanson R, McCluskey A, Neumann JE, McFarland J, Martinich J (2016) Climate change impacts and greenhouse gas mitigation effects on U.S. hydropower generation. Appl Energy 183:1511–1519. https://doi.org/10.1016/j.apenergy.2016.09.054

Bolch T et al (2019) Status and change of the cryosphere in the extended Hindu Kush Himalaya Region. In: Wester P, Mishra A, Mukherji A, Shrestha AB (eds) The Hindu Kush Himalaya assessment—mountains, climate change, sustainability and people. Springer, Cham, pp 209–255. https://doi.org/10.1007/978-3-319-92288-1_7

Bookhagen B, Burbank DW (2010) Toward a complete Himalayan hydrological budget: spatiotemporal distribution of snowmelt and rainfall and their impact on river discharge. J Geophys Res Earth Surf 115(3):1–25. https://doi.org/10.1029/2009JF001426

Bräuning A (2006) Tree-ring evidence of "Little Ice Age" glacier advances in southern Tibet. Holocene 16(3):369–380. https://doi.org/10.1191/0959683606hl922rp

Byers AC, Rounce DR, Shugar DH, Lala JM, Byers EA, Regmi D (2019) A rockfall-induced glacial lake outburst flood, Upper Barun Valley Nepal. Landslides 16:533–549. https://doi.org/10.1007/s10346-018-1079-9

Carey M, Molden OC, Rasmussen MB, Jackson M, Nolin AW, Mark BG (2017) Impacts of glacier recession and declining meltwater on mountain societies. Ann Am Assoc Geogr 107 (2):350–359. https://doi.org/10.1080/24694452.2016.1243039

Carling J, Schewel K (2018) Revisiting aspiration and ability in international migration. J Ethn Migr Stud 44(6):945–963. https://doi.org/10.1080/08941920.2019.1590667

CBS (2012) National population and housing census 2011. National report. National Central Bureau of Statistics, Government of Nepal, Kathmandu

Choudhary A, Dimri AP (2017) Assessment of CORDEX-South Asia experiments for monsoonal precipitation over Himalayan region for future climate. Clim Dyn 50:3009–3030. https://doi.org/10.1007/s00382-017-3789-4

Clouse C (2016) Frozen landscapes: climate-adaptive design interventions in Ladakh and Zanskar. Landsc Res 41:821–837. https://doi.org/10.1080/01426397.2016.1172559

Dame J, Nüsser M (2011) Food security in high mountain regions: agricultural production and the impact of food subsidies in Ladakh, Northern India. Food Secur 3:179–194. https://doi.org/10.1007/s12571-011-0127-2

De Kok RJ, Tuinenburg OA, Bonekamp PNJ, Immerzeel WW (2018) Irrigation as a potential driver for anomalous glacier behavior in High Mountain Asia. Geophys Res Lett 45:2047–2054. https://doi.org/10.1002/2017GL076158

De Kok RJ, Kraaijenbrink PDA, Tuinenburg OA, Bonekamp PNJ, Immerzeel WW (2020) Towards understanding the pattern of glacier mass balances in High Mountain Asia using regional climatic modelling. Cryosphere 14(9):3215–3234. https://doi.org/10.5194/tc-14-3215-2020

DHM (2017) Observed climate trend analysis in the districts and physiographic regions of Nepal (1971–2014). Department of Hydrology and Meteorology, Ministry of Population and Environment, Government of Nepal. Kathmandu, Nepal. http://www.dhm.gov.np/uploads/climatic/1935165359NAP_TrendReport_DHM_2017.pdf. Accessed 15 Apr 2021

Diaz HF, Bradley RS (1997) Temperature variations during the last century at high elevation sites. Clim Change 36:253–279

Farinotti D, Immerzeel WW, de Kok RJ, Quincey DJ, Dehecq A (2020) Manifestations and mechanisms of the Karakoram glacier anomaly. Nat Geosci 13:8–16. https://doi.org/10.1038/s41561-019-0513-5

Foresight (2011) Migration and global environmental change: future challenges and opportunities. Final project report. The Government Office for Science, London. https://assets.publishing.service.gov.uk/government/uploads/system/uploads/attachment_data/file/287717/11-1116-migration-and-global-environmental-change.pdf. Accessed 15 Apr 2021

Forsythe N, Fowler HJ, Li XF, Blenkinsop S, Pritchard D (2017) Karakoram temperature and glacial melt driven by regional atmospheric circulation variability. Nat Clim Chang 7:664–670. https://doi.org/10.1038/nclimate3361

Gemenne F, Blocher J (2017) How can migration serve adaptation to climate change? Challenges to fleshing out a policy ideal. Geogr J 183:336–347. https://doi.org/10.1111/geoj.12205

Gentle P, Thwaites R (2016) Transhumant pastoralism in the context of socioeconomic and climate change in the mountains of Nepal. Mt Res Dev 36(2):173–182. https://doi.org/10.1659/mrd-journal-d-15-00011.1

Gioli G, Khan T, Bisht S, Scheffran J (2014) Migration as an adaptation strategy and its gendered implications: a case study from the Upper Indus Basin. Mt Res Dev 34:255–265. https://doi.org/10.1659/MRD-JOURNAL-D-13-00089.1

Grossman D (2015) As Himalayan Glacier melts, two towns face the fall out. Yale Environment 360. Yale school of Forestry and Environment Studies. http://e360.yale.edu/features/as_himalayan_glaciers_melt_two_towns_face_the_fallout. Accessed 15 Apr 2021

Gruber S, Haeberli W (2007) Permafrost in steep bedrock slopes and its temperature-related destabilization following climate change. J Geophys Res 112. https://doi.org/10.1029/2006JF000547

Gruber S, Fleiner R, Guegan E, Panday P, Schmid M-O, Stumm D, Wester P, Zhang Y, Zhao L (2017) Review article: Inferring permafrost and permafrost thaw in the mountains of the Hindu Kush Himalaya region. Cryosphere 11:81–99. https://doi.org/10.5194/tc11-81-2017

Gurung DR, Giriraj A, Aung KS, Shrestha B, Kulkarni AV (2011) Snow-cover mapping and monitoring in the HinduKush-Himalayas. International Center for Integrated Mountain Development, Kathmanud

Hasson S, Lucarini V, Pascale S (2013) Hydrological cycle over South and Southeast Asian river basins as simulated by PCMDI/CMIP3 experiments. Earth Syst Dyn 4(2):199–217. https://doi.org/10.5194/esd-4-199-2013

Hasson S, Böhner J, Lucarini V (2015) Prevailing climatic trends and runoff response from Hindukush–Karakoram–Himalaya, upper Indus basin. Earth Syst Dyn Discuss 6:579–653. https://doi.org/10.5194/esdd-6-579-2015

He S, Richards K (2015) Impact of meadow degradation on soil water status and pasture management: a case study in Tibet. Land Degrad Dev 26:468–479. https://doi.org/10.1002/ldr.2358

Hewitt K (2005) The Karakoram anomaly? Glacier expansion and the 'elevation effect', Karakoram Himalaya. Mt Res Dev 25(4):332–340. https://doi.org/10.1659/0276-4741(2005)025[0332:TKAGEA]2.0.CO;2

Hewitt K (2011) Glacier change, concentration, and elevation effects in the Karakoram Himalaya, Upper Indus Basin. Mt Res Dev 31(3):188–200. https://doi.org/10.1659/MRD-JOURNAL-D-11-00020.1

Hock R, Rasul G, Adler C, Cáceres B, Gruber S, Hirabayashi Y, Jackson M et al (2019) High mountain areas. In: Pörtner H-O, Roberts DC, Masson-Delmotte V, Zhai P, Tignor M, Poloczanska E, Mintenbeck K, Alegría A, Nicolai M, Okem A, Petzold J, Rama B, Weyer NM (eds) IPCC special report on the ocean and cryosphere in a changing climate. IPCC, Geneva, pp 131–202. https://www.ipcc.ch/srocc. Accessed 20 Apr 2021

Hu F, Tan D (2018) No water, no growth. Does Asia have enough water to develop? China Water Risk, Hong Kong. -https://www.chinawaterrisk.org/resources/analysis-reviews/no-water-no-growth-does-asia-have-enough-water-to-develop/. Accessed 20 Apr 2021

Hugo G (1996) Environmental concerns and international migration. Int Migr Rev 30(1):105–131. https://www.jstor.org/stable/2547462

Huss M, Hock R (2018) Global-scale hydrological response to future glacier mass loss. Nat Clim Chang 8:135–140. https://doi.org/10.1038/s41558-017-0049-x

ICIMOD (2011) Glacial lakes and glacial lake outburst floods in Nepal. International Center for Integrated Mountain Development (ICIMOD), Kathmandu, Nepal

ICIMOD (2020) The Hindu Kush Himalaya ministerial mountain summit 2020. https://www.icimod.org/hkhmms. Accessed 20 Apr 2021

Immerzeel WW, van Beek LPH, Bierkens MFP (2010) Climate change will affect the Asian water towers. Science 328(5984):1382–1385. https://doi.org/10.1126/science.1183188

Immerzeel WW, Lutz AF, Andrade M, Bahl A, Biemans H, Bolch T, Hyde S et al (2020) Importance and vulnerability of the world's water towers. Nature 577(7790):364–369. https://doi.org/10.1038/s41586-019-1822-y

IPCC (2012) Managing the risks of extreme events and disasters to advance climate change adaptation. Special report of the intergovernmental panel on climate change. https://www.ipcc.ch/site/assets/uploads/2018/03/SREX_Full_Report-1.pdf. Accessed 20 Apr 2021

IPCC (2018) Summary for policymakers. In: Masson-Delmotte V, Zhai P, Pörtner H-O, Roberts D, Skea J, Shukla PR, Pirani A, Moufouma-Okia W, Péan C, Pidcock R, Connors S, Matthews JBR, Chen Y, Zhou X, Gomis MI, Lonnoy E, Maycock T, Tignor M, Waterfield T (eds) Global warming of 1.5°C. An IPCC special report on the impacts of global warming of 1.5°C above pre-industrial levels and related global greenhouse gas emission pathways, in the context of strengthening the global response to the threat of climate change, sustainable development, and efforts to eradicate poverty. In Press

IPCC (2019) IPCC special report on the ocean and cryosphere in a changing climate [Pörtner H-O, Roberts DC, Masson-Delmotte V, Zhai P, Tignor M, Poloczanska E, Mintenbeck K, Alegría A, Nicolai M, Okem A, Petzold J, Rama B, Weyer NM (eds)]. In press

Jehangir WA, Mudasser M, Hassan M, Ali Z (1998) Multiple uses of irrigation water in the Hakra-6/R distributary command area, Punjab, Pakistan. Pakistan National Program, International Irrigation Management Institute, Lahore. https://www.researchgate.net/publication/254425808_Multiple_Uses_of_Irrigation_Water_in_the_Hakra_6-R_Distributary_Command_Area. Accessed 25 Apr 2021

Kang S, Wang F, Morgenstern U, Zhang Y, Grigholm B, Kaspari S, Schwikowshi M et al (2015) Dramatic loss of glacier accumulation area on the Tibetan Plateau revealed by ice core tritium and mercury records. Cryosphere 9(3):1213–1222

Kayastha RB, Harrison SP (2008) Changes of the equilibrium-line altitude since the Little Ice Age in the Nepalese Himalaya. Ann Glaciol 48:93–99

Khadka RB, Khanal AB (2008) Environmental Management Plan (EMP) for Melamchi water supply project, Nepal. Environ Monit Assess 146(1–3):225–234. https://doi.org/10.1007/s10661-007-0074-8

Khanal S, Lutz AF, Kraaijenbrink PDA, van den Hurk B, Yao T, Immerzeel WW (2021) Variable 21st century climate change response for rivers in High Mountain Asia at seasonal to decadal time scales. Water Resour Res 57:e2020WR029266. https://doi.org/10.1029/2020WR029266

Kick W (1989) The decline of the Little Ice Age in high Asia compared with that in Alps. In: Oerlemans J (ed) Glacier fluctuation and climate change. Springer, Dordrecht, pp 129–140. https://doi.org/10.1007/978-94-015-7823-3_8

Koubi V, Spilker G, Schaffer L, Bernauer T (2016a) Environmental stressors and migration: evidence from Vietnam. World Dev 79:197–210. https://doi.org/10.1016/j.worlddev.2015.11.016

Koubi V, Spilker G, Schaffer L, Böhmelt T (2016b) The role of environmental perceptions in migration decision-making: Evidence from both migrants and non-migrants in five developing countries. Popul Environ 38(2):134–163. https://doi.org/10.1007/s11111-016-0258-7

Kraaijenbrink PDA, Bierkens MFP, Lutz AF, Immerzeel WW (2017) Impact of a global temperature rise of 1.5 degrees Celsius on Asia's glaciers. Nature 549:257–260

Krishnan R, Shrestha AB, Ren G, Rajbhandari R, Saeed S, Sanjay S, Syed MA, Vellore R, Xu Y, You Q, Ren Y (2019) Unravelling climate change in the Hindu Kush Himalaya: rapid warming in the mountains and increasing extremes. In: Wester P, Mishra A, Mukherji A, Shrestha AB (eds) The Hindu Kush Himalaya assessment—mountains, climate change, sustainability and people. Springer, Cham, pp 57–97

Krishnan R, Sanjay J, Gnanaseelan C, Mujumdar M, Kulkarni A, Chakraborty S (2020) Assessment of climate change over the Indian Region. A Report of the Ministry of Earth Sciences (MoES), Government of India. SpringerOpen, 242 pp. https://doi.org/10.1007/978-981-15-4327-2

Le De L, Gaillard JC, Friesen W (2013) Remittances and disaster: a review. Int J Disaster Risk Reduction 4:34–43. https://doi.org/10.1016/j.ijdrr.2013.03.007

Liu X, Chen B (2000) Climatic warming in the Tibetan Plateau during recent decades. Int J Climatol 20(14):1729–1742. https://doi.org/10.1002/1097-0088(20001130)20:14%3c1729::AID-JOC556%3e3.0.CO;2-Y

Liu M, Chen N, Zhang Y, Deng M (2020) Glacial lake inventory and lake outburst flood/debris flow hazard assessment after the Gorkha earthquake in the Bhote Koshi Basin. Water 12:1–21. https://doi.org/10.3390/w12020464

Lutz AF, Immerzeel WW, Shrestha AB, Bierkens MFP (2014) Consistent increase in High Asia's runoff due to increasing glacier melt and precipitation. Nat Clim Chang 4:587–592. https://doi.org/10.1038/nclimate2237

Macfarlane A (1976) Resources and population: a study of the gurungs of Nepal. Cambridge University Press, Cambridge

Maharjan A, de Campos RS, Singh C, Das D, Srinivas A, Bhuiyan MRA, Ishaq S et al (2020) Migration and household adaptation in climate-sensitive hotspots in South Asia. Curr Clim Change Rep 6:1–16. https://doi.org/10.1007/s40641-020-00153-z

Maharjan A, Tuladhar S, Hussain A, Bhadwal S, Ishaq S, Saeed BA, Sachdeva I, Ahmad B, Ferdous J, Hassan SMT (2021) Can labour migration help households adapt to climate change? Evidence from four river basins in South Asia. Clim Dev. https://doi.org/10.1080/17565529.2020.1867044

Maurer JM, Schaefer JM, Rupper S, Corley A (2019) Acceleration of ice loss across the Himalayas over the last 40 years. Sci Adv 5(6):eaav7266. https://doi.org/10.1126/sciadv.aav7266

Mayewski PA, Jeschke PA (1979) Himalayan and trans-Himalayan glacier luctuations since AD 1812. Arct Alp Res 11(3):267. https://doi.org/10.2307/1550417

McDowell G, Ford JD, Lehner B, Berrang-Ford L, Sherpa A (2013) Climate-related hydrological change and human vulnerability in remote mountain regions: a case study from Khumbu, Nepal. Reg Environ Change 13:299–310. https://doi.org/10.1007/s10113-012-0333-2

Mishra V (2015) Climatic uncertainty in Himalayan water towers. J Geophys Res Atmos 120 (7):2680–2705. https://doi.org/10.1002/2014JD022650

Mishra A, Ghate R, Maharjan A, Gurung J, Gurung CG, Dorji T, Wester P (2017) Building resilient solutions for mountain communities in the HKH region. ICIMOD position paper, Kathmandu

Molden D (2020) Scarcity of water or scarcity of management?. Int J Water Resour Dev 36:2–3, 258–268. https://doi.org/10.1080/07900627.2019.1676204

Molden DJ, Shrestha AB, Nepal S, Immerzeel WW (2016) Downstream implications of climate change in the Himalayas. In: Biswas AK, Tortajada C (eds) Water security, climate change and sustainable development. Springer, Singapore, pp 65–82

Molden D, Sharma E, Shrestha AB, Chettri N, Pradhan NS, Kotru R (2017) Advancing regional and transboundary cooperation in the conflict-prone Hindu Kush-Himalaya. Mt Res Dev 37 (4):502–508. https://doi.org/10.1659/MRD-JOURNAL-D-17-00108.1

Mool PK, Bajracharya SR, Joshi SP (2001) Inventory of glaciers, glacial lakes and glacial lake outburst floods, monitoring and early warning systems in the Hindu Kush Himalaya region. ICIMOD/UNEP, Kathmandu

Namgay K, Millar JE, Black R, Samdup T (2014) Changes in transhumant agro-pastoralism in Bhutan: a disappearing livelihood? Hum Ecol 42(5):779–792. https://doi.org/10.1007/s10745-014-9684-2

Nepal S, Pandey A, Shrestha AB, Mukherji A (2018) Revisiting key questions regarding upstream–downstream linkages of land and water management in the Hindu Kush Himalaya (HKH) region. HI-AWARE working paper 21. Himalayan Adaptation, Water and Resilience (Hi-AWARE) Research. Kathmandu, Nepal

Nie Y, Pritchard HD, Liu Q, Hennig T, Wang W, Wang X, Liu S et al (2021) Glacial change and hydrological implications in the Himalaya and Karakoram. Nat Rev Earth Environ 2:91–106. https://doi.org/10.1038/s43017-020-00124-w

NRDC (2017) The Paris agreement of climate change. Issue Brief. https://assets.nrdc.org/sites/default/files/paris-agreement-climate-change-2017-ib.pdf. Accessed 25 Apr 2021

Nyima Y, Hopping KA (2019) Tibetan lake expansion from a pastoral perspective: local observations and coping strategies for a changing environment. Soc Nat Resour 32(9):965–982. https://doi.org/10.1080/08941920.2019.1590667

Panday PK, Thibeault J, Frey KE (2014) Changing temperature and precipitation extremes in the Hindu Kush-Himalayan region: an analysis of CMIP3 and CMIP5 simulations and projections. Int J Climatol 35(10):3058–3077. https://doi.org/10.1002/joc.4192

Pathak S, Pant L, Maharjan A (2017) De-population trends, patterns and effects in Uttarakhand, India—a gateway to Kailash Mansarovar. ICIMOD working paper 2017/22. International Center for Integrated Mountain Development (ICIMOD), Kathmandu, Nepal.

Paul F (2015) Revealing glacier flow and surge dynamics from animated satellite image sequences: examples from the Karakoram. Cryosphere 9(6):2201–2214. https://doi.org/10.5194/tc-9-2201-2015

Pepin N, Bradley RS, Diaz HF, Baraer M, Caceres EB, Forsyhe N, Greenood G (2015) Elevation-dependent warming in mountain regions of the world. Nat Clim Chang 5:424–430. https://doi.org/10.1038/nclimate2563

Qian Y, Flanner MG, Leung LR, Wang W (2011) Sensitivity studies on the impacts of Tibetan Plateau snowpack pollution on the Asian hydrological cycle and monsoon climate. Atmos Chem Phys 11(5):1929–1948. https://doi.org/10.5194/acp-11-1929-2011

Rasul G (2010) The role of the Himalayan mountain systems in food security and agricultural sustainability in South Asia. Int J Rural Manag 6(1):95–116. https://doi.org/10.1177/097300521100600105

Rasul G (2014) Food, water, and energy security in South Asia: a nexus perspective from the Hindu Kush Himalayan region. Environ Sci Policy 39:35–48. https://doi.org/10.1016/j.envsci.2014.01.010

Rasul G (2021) Twin challenges of COVID-19 pandemic and climate change for agriculture and food security in South Asia. Environ Challenges 2:100027. https://doi.org/10.1016/j.envc.2021.100027

Rasul G, Molden D (2019) The global social and economic consequences of mountain cryopsheric change. Front Environ Sci 7:91. https://doi.org/10.3389/fenvs.2019.00091

Rasul G, Saboor A, Tiwari PC, Hussain A, Ghosh N, Chettri GB (2019) Food and nutrition security in the Hindu Kush Himalaya: unique challenges and niche opportunities. In: Wester P, Mishra A, Mukherji A, Shrestha A (eds) The Hindu Kush Himalaya assessment—mountains, climate change, sustainability and people. Springer, Cham, pp 301–338

Rasul G, Pasakhala B, Mishra A, Pant S (2020) Adaptation to mountain cryosphere change: issues and challenges. Clim Dev 12(4):297–309. https://doi.org/10.1080/17565529.2019.1617099

Rees HG, Holmes MGR, Young AR, Kansaker SR (2004) Recession-based hydrological models for estimating low flows in ungauged catchments in the Himalayas. Hydrol Earth Syst Sci 8:891–902. https://doi.org/10.5194/hess-8-891-2004

Ren Y-Y, Ren G-Y, Sun X-B, Shrestha AB, You Q-L, Zhan Y-J et al (2017) Observed changes in surface air temperature and precipitation in the Hindu Kush Himalayan region during 1901–2014. Adv Clim Change Res 8(3). https://doi.org/10.1016/j.accre.2017.08.001

Rigaud KK, de Sherbinin A, Jones B, Bergmann J, Clement V, Ober, K, Schewe J et al (2018) Groundswell: preparing for internal climate migration. World Bank, Washington, DC. https://openknowledge.worldbank.org/handle/10986/29461. Accessed 26 Apr 2021

Rounce DR, Byers AC, Byers EA, McKinney DC (2017) Brief communication: observations of a glacier outburst flood from Lhotse Glacier, Everest area, Nepal. Cryosphere 11:443–449. https://doi.org/10.5194/tc-11-443-2017

Rounce DR, Hock R, Shean DE (2020) glacier mass change in high mountain Asia through 2100 using the open-source python glacier evolution model (PyGEM). Front Earth 7:331. https://doi.org/10.3389/feart.2019.00331

Saikawa E, Panday A, Kang S, Gautam R, Zusman E, Cong Z, Somanathan E (2019) Air pollution in the Hindu Kush Himalaya. In: Wester P, Mishra A, Mukherji A, Shrestha A (eds) The Hindu Kush Himalaya assessment—mountains, climate change, sustainability and people. Springer, Cham, pp 339–377. https://doi.org/10.1007/978-3-319-92288-1_10

Sanjay J, Krishnan R, Shrestha AB, Rajbhandari R, Ren GY (2017) Downscaled climate change projections for the Hindu Kush Himalayan region using CORDEX South Asia regional climate models. Adv Clim Chang Res 8(3):185–198. https://doi.org/10.1016/j.accre.2017.08.003

Schwanghart W, Worni R, Huggel C, Stoffel M, Korup O (2016) Uncertainty in the Himalayan energy–water nexus: Estimating regional exposure to glacial lake outburst floods. Environ Res Lett 11(074005). https://doi.org/10.1088/1748-9326/11/7/074005

Scott CA, Zhang F, Mukherji A, Immerzeel W, Mustafa D, Bharati L (2019) Water in the Hindu Kush Himalaya. In: Wester P, Mishra A, Mukherji A, Shrestha A (eds) The Hindu Kush Himalaya assessment—mountains, climate change, sustainability and people. Springer, Cham, pp 257–292. https://doi.org/10.1007/978-3-319-92288-1_8

Shaoliang Y, Ismail M, Zhaoli Y (2012) Pastoral communities' perspectives on climate change and their adaptation strategies in the Hindukush-Karakoram-Himalaya. In: Kreutzmann H (ed) Pastoral practices in High Asia. Advances in Asian human-environmental research. Springer, Dordrecht, pp 307–322. https://doi.org/10.1007/978-94-007-3846-1

Shean DE, Bhusan S, Montesano P, Rounce DR, Arendt A, Osmangolu B (2020) A systematic, regional assessment of high mountain Asia glacier mass balance. Front Earth Sci 7(363):1–19. https://doi.org/10.3389/feart.2019.00363

Shrestha AB, Wake CP, Mayewski PA, Dibb JE (1999) Maximum temperature trends in the Himalaya and its vicinity: an analysis based on temperature records from Nepal for the period 1971–94. J Clim 12:2775–2786

Shrestha AB, Eriksson M, Mool P, Ghimire P, Mishra B, Khanal NR (2010) Glacial lake outburst flood risk assessment of Sun Koshi basin, Nepal. Geomat Nat Haz Risk 1(2):157–169. https://doi.org/10.1080/19475701003668968

Shrestha MS, Grabs WE, Khadgi VR (2015) The establishment of a regional flood information system in the Hindu Kush Himalayas: challenges and opportunities. Int J Water Resour Dev 31:238–252. https://doi.org/10.1080/07900627.2015.1023891

Shrestha P, Lord A, Mukherji A, Shrestha RK, Yadav L, Rai N (2016) Benefit sharing and sustainable hydropower: lessons from Nepal. International Center for Integrated Mountain Development (ICIMOD), Kathmandu, Nepal

Shrestha A, Shrestha S, Tingsanchalia T, Budhathoki A, Ninsawat S (2020) Adapting hydropower production to climate change: a case study of Kulekhani Hydropower Project in Nepal. J Clean Prod 279:1–14

Shrestha AB, Shukla D, Pradhan NS, Dhungana S, Azizi F, Memon N, Mohtadullah K, Lotia H, Ali A, Molden D, Daming H, Dimri AP, Huggel C (2021b) Developing a science-based policy network over the Upper Indus Basin. Sci Total Environ 784:147067. https://doi.org/10.1016/j.scitotenv.2021.147067

Shrestha AB, Steiner J, Nepal S, Maharjan SB, Jackson M, Rasul G, Bajracharya B (2021a) Understanding the Chamoli flood: cause, process, impacts and context of rapid infrastructure development. International Center for Integrated Mountain Development (ICIMOD), Kathmandu, Nepal. https://www.icimod.org/article/understanding-the-chamoli-flood-cause-process-impacts-and-context-of-rapid-infrastructure-development/. Accessed 26 Apr 2021

Siderius C, Biemans H, Wiltshire A, Rao S, Franssen WHP, Kumar P, Gosain AK et al (2013) Snowmelt contributions to discharge of the Ganges. Sci Total Environ 468–469:S93–S101. https://doi.org/10.1016/j.scitotenv.2013.05.084

Stäubli A, Nussbaumer SU, Allen SK, Huggel C, Arguello M, Costa F et al (2018) Analysis of weather-and climate-related disasters in mountain regions using different disaster databases. In: Mal S, Singh R, Huggel C (eds) Climate change, extreme events, and disaster risk reduction. Sustainable development goals series. Springer, Cham, pp 17–41. https://doi.org/10.1007/978-3-319-56469-2_2

Thompson LG, Yao T, Mosley-Thompson E, Davis ME, Henderson KA, Lin PN (2000) A high-resolution millennial record of the South Asian monsoon from himalayan ice cores. Science 289(5486):1916–1920. https://doi.org/10.1126/science.289.5486.1916

Tuinenburg OA, Hutjes RWA, Kabat P (2012) The fate of evaporated water from the Ganges basin. J Geophys Res 117(D1):D01107. https://doi.org/10.1029/2011jd016221

Tuladhar S, Pasakhala B, Maharjan A, Mishra A (2021) Unravelling the linkages of cryosphere and mountain livelihood systems: a case study of Langtang, Nepal. Adv Clim Chang Res 12(1):119–131. https://doi.org/10.1016/j.accre.2020.12.004

Turner SWD, Hejazi M, Kin SH, Clarke L, Edmonds J (2017) Climate impacts on hydropower and consequences for global electricity supply investment needs. Energy 141:2081–2090. https://doi.org/10.1016/j.energy.2017.11.089

Vaidya RA, Shrestha MS, Nasab N, Gurung DR, Kozo N, Pradhan NS, Wasson RJ (2019) Disaster risk reduction and building resilience in the Hindu Kush Himalaya. In: Wester P, Mishra A, Mukherji A, Shrestha A (eds) The Hindu Kush Himalaya assessment—mountains, climate change, sustainability and people. Springer, Cham, pp 389–419. https://doi.org/10.1007/978-3-319-92288-1_11

Vaidya RA, Molden DJ, Shresthat AB, Wagle N, Tortajada C (2021) The role of hydropower in South Asia's energy future. Int J Water Resour Dev 37(3):367–391. https://doi.org/10.1080/07900627.2021.1875809

van der Geest K, Schindler M (2016) Brief communication: loss and damage from a catastrophic landslide in Nepal. Nat Hazard 16(11):2347–2350. https://doi.org/10.5194/nhess-16-2347-2016

Viswanathan B, Kavi Kumar KS (2015) Weather, agriculture and rural migration: evidence from state and district level migration in India. Environ Dev Econ 20(4):469–492. https://doi.org/10.1017/S1355770X1500008X

Wang G, Yuanshou Li, Yibo W, Qingbo W (2008) Effects of permafrost thawing on vegetation and soil carbon pool losses on the Qinghai-Tibet Plateau, China. Geoderma 143(1–2):143–152. https://doi.org/10.1016/j.geoderma.2007.10.023

Warner K, Afifi T (2014) Where the rain falls: Evidence from 8 countries on how vulnerable households use migration to manage the risk of rainfall variability and food insecurity. Climate Dev 6(1):1–17. https://doi.org/10.1080/17565529.2013.835707

Wester P, Mishra A, Mukherji A, Shrestha AB (eds) (2019) The Hindu Kush Himalaya assessment —mountains, climate change, sustainability and people. Springer, Cham

Wijngaard RR, Lutz AF, Nepal S, Khanal S, Pradhananga S, Shrestha AB, Immerzeel WW (2017) Future changes in hydro-climatic extremes in the Upper Indus, Ganges, and Brahmaputra River basins. PLOS ONE 12(12):e0190224. s

Wiltshire AJ (2014) Climate change implications for the glaciers of the Hindu Kush, Karakoram and Himalayan region. Cryosphere 8:941–958. https://doi.org/10.5194/tc-8-941-2014

Wu G, Zhang Y (1998) Tibetan Plateau Forcing and the timing of the Monsoon onset over South Asia and the South China Sea. Mon Weather Rev 126(4):913–927. https://doi.org/10.1175/1520-0493(1998)126%3c0913:TPFATT%3e2.0.CO;2

Yang M, Frederick E, Nikolay IS, Guo D, Wan G (2010) Permafrost degradation and its environmental effects on the Tibetan Plateau: a review of recent research. Earth-Sci Rev 103(1–2):31–44. https://doi.org/10.1016/j.earscirev.2010.07.002

Zemp M, Huss M, Thibert E, Eckert N, McNabb R, Huber J, Barandun M, Machguth H, Nussbaumer SU, Gärtner-Roer I, Thomsan L, Paul F, Maussion F, Kutuzov S, Cogley JG (2019) Global glacier mass balances and their contributions to sea-level rise from 1961 to 2016. Nature 568:382–386. https://doi.org/10.1038/s41586-019-1071-0

Zhang T, Wang T, Krinner G, Wang X, Gasser T, Peng S, Piao S, Yao T (2019) The weakening relationship between Eurasian spring snow cover and Indian summer monsoon rainfall. Climatology 5(3):eaau8932. https://doi.org/10.1126/sciadv.aau8932

Zhao L, Wu Q, Marchenko SS, Sharkhuu N (2010) Thermal state of permafrost and active layer in Central Asia during the international polar year. Permafrost Piglacial Process 21(2):198–207. https://doi.org/10.1002/ppp.688

Chapter 13
Assessment of and Adaptation Measures to the Impacts of Climate Change on Water Resources in China

Aifeng Lv and Shaofeng Jia

Abstract China is experiencing severe water stress, and this will be exacerbated by climate change. The systematic study of the effects of climate change on water resources in China and the formulation of appropriate and feasible adaptation countermeasures are critical for China's sustainable development. This chapter assesses past and projected future climate change in China and their effects on water resources and additionally proposes climate change adaptation measures. Between 1961 and 2019, China experienced an overall warming trend; however, the rate of warming varied among river basins, with the highest rate being observed in the northwestern river basins. During this period, precipitation in China showed a non-significant upward trend, with only the northwestern river basins showing a significant upward trend. From 2021 to 2100, the average temperature in China is predicted to increase, with a significantly higher warming rate expected under the high-emission "RCP8.5" scenario than under the medium-emission "RCP4.5" scenario. It is predicted that precipitation in China will show a fluctuating upward trend in the future, but that the increasing trend will slow after 2070. It is predicted that, between 2021 and 2100, the overall water resources in China will show a slight increase due to climate change. The Huai River basin, Hai River basin, Yangtze River basin, Pearl River basin, and the southeastern river basins are predicted to show different rates of water resource increase, while the water resources of the Songhua River basin, the northwestern river basins, and the Yellow River basin are predicted to gradually decrease under different climate change scenarios. To deal with the impacts of future climate change on China's water resources, efforts in basic research, planning under changing environments, institutional innovation, and investment are needed.

Keywords Climate change · Water resources · Assessment · Adaptation · China

A. Lv · S. Jia (✉)
Water Resources Research Department, Institute of Geographic Sciences and Natural Resources Research, Chinese Academy of Sciences, Beijing, China
e-mail: jiasf@igsnrr.ac.cn

A. Lv
e-mail: lvaf@igsnrr.ac.cn

© The Author(s), under exclusive license to Springer Nature Singapore Pte Ltd. 2022
A. K. Biswas and C. Tortajada (eds.), *Water Security Under Climate Change*,
Water Resources Development and Management,
https://doi.org/10.1007/978-981-16-5493-0_13

13.1 Introduction

Climate change, as well as its impacts on ecosystems, human societies, and economies, has been observed globally. Water resources are the main way that these impacts are felt (Bates et al. 2008; IPCC 2014; Del Buono 2021). Climate change is believed to change the precipitation regime (amount, timing, and distribution) and increase the rates of evapotranspiration due to increasing temperature (Li et al. 2020). These changes will affect the runoff production and the accessibility of water resources for human consumption and ecosystems (Idrizovic et al. 2020). As well as affecting the water cycle, climate change may alter water demand and how water is used (Wada et al. 2013; Wang et al. 2014). Higher temperatures and evaporation rates could increase the water demand in many areas. Climate change has brought much uncertainty to regional water supply and water demand, especially in water-stressed regions, introducing a new challenge to regional water resource management.

With its brisk socio-economic development, China is experiencing ever more serious water scarcity (Jiang 2009, 2015). Even though the overall amount of water resources in China is around 2800 billion m^3, the amount per capita is only about 2185 m^3, which is less than 25% of the global average. Furthermore, since the country's water resources have an uneven spatiotemporal distribution, water resources have become an important aspect limiting the country's sustainable socio-economic development.

China has been greatly affected by global climate change. The National Assessment Report on Climate Change (III) found that the average warming rate of China's land area between 1909 and 2011 was higher than the global average, reaching 0.9–1.5°C (He et al. 2007). This trend is predicted to persist in the future. In this context, there is an urgent need to assess climate change and its effect on China's water resources and make scientific judgments on future change trends to provide scientific validation for water resource management under environmental changes.

The primary aim of this chapter is to systematically investigate climate change and its effect on China's water resources in China. Section 13.2 analyses the effect of climate change and its effect on water resources in China and its ten main river basins (Fig. 13.1) between 1961 and 2019; it mainly focuses on variations in temperature, precipitation, and water quantity in these ten basins over this period and explores the effect of climate change on water resources. Section 13.3 analyses the projected future climate change in China and its likely effect on water resources. Climate change is mainly analysed by using the future climate data output from global climate model (GCM) (Zhang et al. 2017) to study the predicted future variations of temperature and precipitation in the ten main river basins. The evaluation of the effect of future climate change on China's water resources is mainly performed through a literature review. In Sect. 13.4, adaptation strategies for the effects of climate change on water resources in China are introduced.

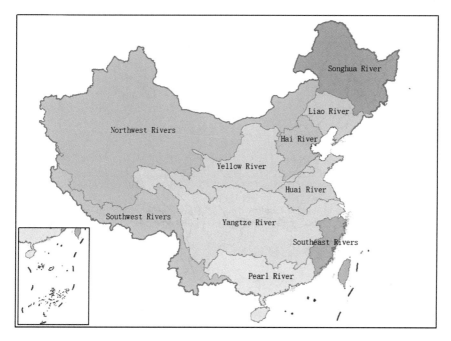

Fig. 13.1 Main river basins in China

13.2   The Effect of Climate Change on Water Resources in China in 1961–2019

13.2.1   Temperature Change in 1961–2019 by Basin

Figure 13.2 demonstrates the changes in the annual mean temperature in China during 1961–2019. In this period, the average temperatures across China showed a warming trend, increasing by about 1.75 °C. From 1960 to 1990, the country experienced a slow warming trend, while after 1990, the warming rate accelerated significantly. At the seasonal scale, similar warming trends were observed for spring, summer, autumn, and winter, with warming rates of 0.38, 0.25, 0.24, and 0.3 °C per decade, respectively (Fig. 13.3).

Between 1961 and 2019, the temperature change in the ten main river basins of China was similar to the national temperature change trend. During this period, the most significant temperature rise was observed in the northwestern river basins, with an increase of about 2°C, followed by the Liao River basin and the Yellow River basin. The temperature increase in the Yangtze River basin was lower than that in the other nine river basins. By season, the warming trend was the largest in spring for all basins. The rate of warming increased with basin latitude, with the rate of warming in the northern basins being significantly higher than that in the southern basins (Fig. 13.4).

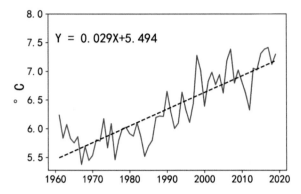

Fig. 13.2 Temperature change in China during 1961–2019. *Data source* 0.5° × 0.5° gridded monthly mean temperature dataset (V2.0) developed by the National Meteorological Information Center (NMIC) of China

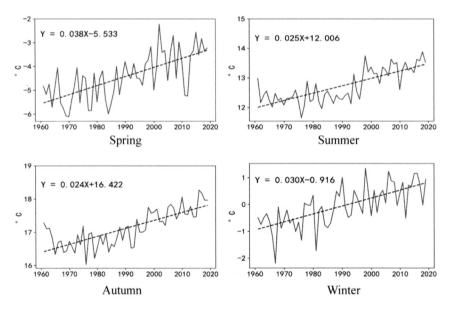

Fig. 13.3 Temperature change in China during 1961–2019 by season. *Data source* 0.5° × 0.5° gridded monthly mean temperature dataset (V2.0) developed by the National Meteorological Information Center (NMIC) of China

13.2.2 Precipitation Changes in 1961–2019 by Basin

During 1961–2019, the average precipitation in China showed an increasing but statistically insignificant trend. Between 1961 and 1990, precipitation in China fluctuated relatively gently around a value of 590 mm, and after the 1990s, the

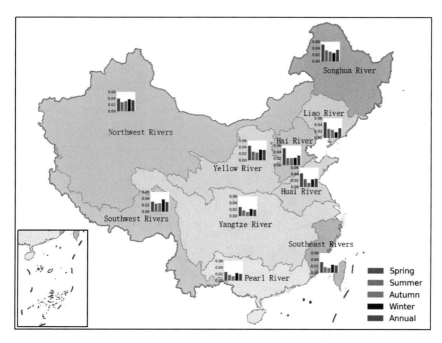

Fig. 13.4 Temperature trends in China's major river basins during 1961–2019. *Data source* 0.5° × 0.5° gridded monthly mean temperature dataset (V2.0) developed by the National Meteorological Information Center (NMIC) of China

Fig. 13.5 Precipitation in China between 1961 and 2019. *Data source* 0.5° × 0.5° gridded monthly precipitation dataset (V2.0) developed by the National Meteorological Information Center (NMIC) of China

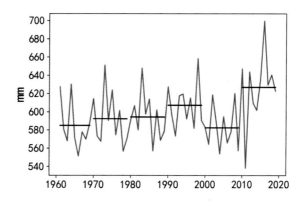

frequency of precipitation fluctuations increased. During 2000–2010, precipitation reached a minimum value of about 580 mm, and after 2010, precipitation increased, with an average value of 630 mm between 2010 and 2020 (Figs. 13.5 and 13.6).

Between 1961 and 2019, the precipitation trend in China showed variability between seasons, and the rate of change varied from season to season. The highest precipitation and the largest precipitation fluctuations were observed in autumn,

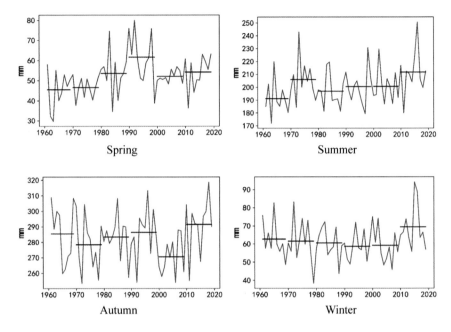

Fig. 13.6 Precipitation trends in China during 1959–2019 by season. *Data source* $0.5° \times 0.5°$ gridded monthly precipitation dataset (V2.0) developed by the National Meteorological Information Center (NMIC) of China

while the second-highest precipitation and the second-highest precipitation fluctuations were observed in summer.

The trends of precipitation change between 1961 and 2019 in the main river basins of China are different. In the Songhua River basin, precipitation reduced from 1960 to 1980, decreased sharply in the late 1990s, and began to fluctuate and increase after the year 2000. In the Huai River basin, the variability of precipitation cycles from 1960 to 1990 was higher than that after 1990. The Yangtze River basin and the southeastern river basins had more frequent precipitation fluctuations, and all experienced the maximum precipitation around 2015. The Pearl River basin had low precipitation around 1990 and 2010, while for the rest of the study period, the precipitation remained at a high level. The southwestern river basins had a high precipitation level throughout the whole of 1961–2019. The precipitation in the southwestern river basins fluctuated roughly in decadal cycles. Only the northwestern river basins had a significant upward trend, with a precipitation increase of about 80 mm between 1961 and 2019. The percentage variation of precipitation per decade is greatest in the spring of the main river basins in China, with the highest value in the spring of 1990–2000 (Fig. 13.7).

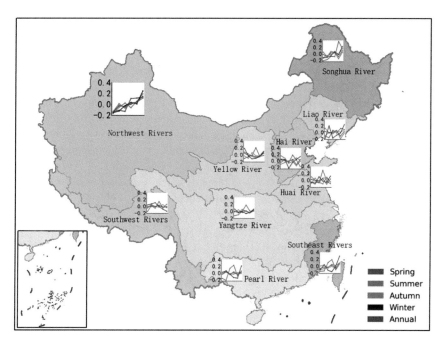

Fig. 13.7 Percentage changes in precipitation per decade during 1961–2019 in the main river basins in China. *Data source* 0.5° × 0.5° gridded monthly precipitation dataset (V2.0) developed by the National Meteorological Information Center (NMIC) of China

13.2.3 Changes in Water Resources in 1961–2019 by Basin

Data from the first and second national water resource assessments were used to analyse the changes in water resources in China's ten main river basins in 1956–1979 and 1980–2000. China's brisk economic development and the intensification of its anthropogenic actions have significantly impacted the exploitation and preservation of regional water resources in the past two decades. At present, the third water resources assessment has been completed; however, the specific results have not been announced. Therefore, the data from the China Water Resources Bulletin (2001–2019) were used to analyse the changes in China's water resources between 2001 and 2019 (Table 13.1).

Table 13.1 Data sources that were used to analyse the changes in China's water resources

Source	Period covered	Data used	Timescale
The first water resource assessment	1956–1979	Surface water resources and total water resources	Annual average
The second water resource assessment	1956–2000		
The China water Resources bulletin	2001–2019		Yearly

Source Li et al. (2012); China Water Resources Bulletin (2001–2019)

A comparison between the annual average water resource change rate of the first water resource assessment (1956–1979) with that of the second water resource assessment (1956–2000) suggested that the amount of water resources in China slightly increased between 1956 and 2000 and that there were different trends in Southern and Northern China (Fig. 13.8). In Southern China, the quantities of surface and total water resources rose by 1.78% and 1.70%, respectively. However, in Northern China, both surface and total water resources fell, notably in the Hai River basin, where the surface and total water resources fell by 24.95% and 12.13%, respectively. The quantities of surface and total water resources in other river basins in Northern China, such as the Liao River basin, the Yellow River basin, and the Huai River basin, also exhibited decreasing trends at different rates.

Using data from the Water Resources Bulletin, the rates of change of the annual surface and total water resources in each basin from 2001–2019 relative to the multi-year average surface and total water resources between 1956 and 2000 were calculated. The results are shown in Fig. 13.2. The findings indicate that there are large differences between the changes in water resources in Southern and Northern China. In the south, the water resource changes were relatively low, except in the southeastern basins. The variability of water resources in the northern basins was greater than that in the southern basins, and the variability of water resources varies among the northern basins. There was a decreasing trend of water resources in most of the northern basins. The rate of change in the Haihe River basin was negative in all years except 2012, indicating a decrease in water resources in the Haihe River basin compared to the mean volume of water from 1956 to 2000. In the Huaihe River basin, the water resources showed a relatively large change between 2002 and

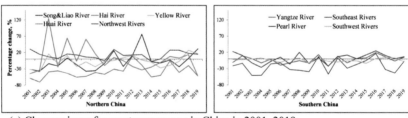

(a) Changes in surface water resources in China in 2001–2019

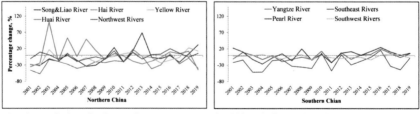

(b) The variations in the total water resources in China during 2001–2019

Fig. 13.8 Variations in the surface and total water resources in China during 2001–2019. *Source* China Water Resources Bulletin (2001–2019)

2009. Meanwhile, the variations in water resources in the Yellow River basin and the northwestern basins were relatively low.

13.2.4 The Effect of Climate Change on Water Resources

Since the start of the twentieth century, China's climate has changed significantly (Chen et al. 2005). Many scholars have summarised changes in the amount of water resources in the country based on research on hydrological cycle processes in different regions under global climate change. Under climate change, the total amount of water resources in China has presented a slightly increasing trend in the twentieth century. In recent decades, with climate warming, the annual precipitation has tended to decrease in Northern China and increase in Southern China, in particular in the Yangtze River, where the annual precipitation and concentration have significantly increased (Zhang et al. 2010). Furthermore, Wu et al. (2009) suggested that the precipitation in Southern China increased between 1990 and 2000. Moreover, in most areas of Western China, the annual precipitation demonstrated an increasing trend between 1990 and 1999 (Ren et al. 2000). Global warming is constantly accelerating hydrological cycle processes, which has variable effects on regional water resources.

The Yellow River basin, one of the largest river basins in China, has recently been experiencing decreasing precipitation and increasing evaporation due to rising temperatures. Since the 1980s, the water resources of the basin have been declining (Liu et al. 2009). Additionally, in the Weihe River basin, the water resources in 1980–2000 were lower by 2.0% compared to 1956–1979 (Zhou et al. 2009). Moreover, a study of water resource variations in the Yangtze River basin found that the total precipitation in the entire basin has not changed much compared to the multi-year average during 1970–2000, except for a decreasing tendency in the Qingjiang and Minjiang basins, and the water resources in the entire Yangtze River basin have not changed considerably since the 1970s (Hu et al. 2008; Zhang et al. 2010). The characteristics of precipitation changes in the Huaihe River basin and the southwestern river basins since the 1970s are consistent with those of the Yangtze and Yellow River basins, with decreasing precipitation and yearly runoff (Wang and Dai 2008; Liu et al. 2020). Most of the runoff in the northwestern basins of China originates from glacial snowmelt. With increasing global temperature, the precipitation in the northwestern river basins has not decreased significantly, whereas snowmelt has increased, increasing annual runoff (Wang and Zhang 2006; Li et al. 2008; Ekegemu-Abra et al. 2019). In an analysis of annual runoff trends in the Liaohe River basin, Hu et al. found that the runoff has been decreasing in recent decades and the intra-annual distribution is very uneven (Ying and Jiang 1996; Sun et al. 2015). Climate change also affects water vapour transport. Gao and Feng (2019) found that precipitation in the Haihe River basin tended to decrease during 1956–2019 due to the decrease in the amount of water vapour reaching the basin, and the decrease was extremely pronounced in summer, causing a reduction in the total water resources of about 8 billion m³ since the year 2000 (Bao et al. 2014; Gao and Feng 2019). In the Songhua River basin, the annual precipitation showed a non-significant decrease from 1958 to 2009. However, in the main stream of Songhua River and Nenjiang River Basin, the decreasing trend of water resources was obvious (Lu et al. 2012; Wang et al. 2017a, b, c). In the southeastern river basins, take the Pearl River basin for example, precipitation was higher in the 1990s than in the 1980s, with an increase of about 50–150 mm (Chen et al. 2005).

13.3 The Effect of Climate Change on Water Resources Under Future Climate Scenarios

13.3.1 Future Temperature Change

From 2021 to 2050, the average temperature in China is predicted to increase under both the RCP4.5 and RCP8.5 scenarios, with warming of 0.6°C and 1°C, respectively. The temperature projections for the ten river basins for 2021–2050 also show a significant upward trend, with the largest warming rate in the northwestern river basins, followed by the Songhua River basin. The Liao River, Hai River, Yellow River, Huai River, Yangtze River, and southwestern river basins have similar projected temperature trends under both the high- and medium-emission scenarios. In general, the predicted temperatures in each basin for RCP8.5 are significantly higher than those for RCP4.5. The projected temperature increases in the Pearl River basin, and the southeastern river basins are lower than those in the other basins (Fig. 13.9).

From 2050 to 2100, the overall temperature projection China has a rising trend under both RCP4.5 and RCP8.5, and the difference between the temperature projections under the two emission scenarios gradually increases, with a significantly higher temperature increase projected for RCP8.5 than for RCP4.5. Under RCP8.5, the temperature maintains an upward trend until 2100, while for RCP4.5, the

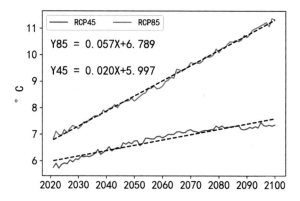

Fig. 13.9 Projected temperature trends in China during 2021–2100. *Source* Zhang et al. (2017)

projected temperature begins to level off and remains stable after 2070. The major watersheds in China still maintain an increasing trend under RCP8.5. Under RCP4.5, the predicted temperature variations in China's ten main river basins are similar to the national temperature changes, with the temperature in all of the basins levelling-off and remaining stable after 2070.

The seasonal variation of the predicted temperature in the ten river basins differs under the medium- and high-emission scenarios. The predicted temperature trends under the two scenarios are generally consistent for all seasons, with the predicted temperature values for the RCP8.5 scenario being higher than those for RCP4.5. In the RCP8.5 scenario, the temperature is projected to rise in all seasons, and the warming rate is increasing. For RCP4.5, the temperature variations in different basins differ in different seasons. In spring, the temperature in the Yangtze River basin and the southwestern river basin is predicted to decrease, while the temperature in the rest of the basins is projected to increase, although with a decreasing warming trend; in summer, the temperature in the Songhua River basin and the northwestern river basins is predicted to decrease, while temperature increases are projected in the rest of the basins with this decreasing warming trend except for the Yangtze River basin, where there is projected to be an increasing warming trend. In autumn, the temperature is projected to rise in all basins, although the warming trend in the Songhua River basin, Liaohe River basin, Yellow River basin, and southwestern and northwestern river basins is projected to weaken. In winter, the temperature in the Haihe, Yangtze, and southwestern river basins is projected to decrease, while the temperature in the Liaohe River basin is projected to increase (Tables 13.2 and 13.3).

13.3.2 Future Precipitation Change

From 2020 to 2050, precipitation in China shows a fluctuating upward trend under both RCP4.5 and RCP8.5. The precipitation trends of the two emission scenarios are generally consistent, with an average increase in precipitation of 40 mm projected for China in RCP8.5 and an 18 mm increase projected for RCP4.5. Additionally, both scenarios project an increase in precipitation in the ten main river basins in China, although the magnitude of the increase varies among the basins. The projected increase in precipitation is larger in the southwestern and northwestern river basins, and the predicted precipitation for RCP8.5 is significantly higher than that for RCP4.5. The predicted precipitation in the Yangtze River basin, the Pearl River basin and the southeastern river basins are not projected to increase significantly under either scenario; the two scenarios give similar precipitation values for these basins, although the predicted values are generally higher under RCP8.5.

From 2050 to 2100, the precipitation in China is projected to rise for both RCP4.5 and RCP8.5, with increases of up to about 50 mm and up to about 10 mm, respectively. For RCP8.5, the projected precipitation in China during this period

Table 13.2 Differences between the seasonal and annual mean temperatures for 1961–2019 and the projected mean temperatures for 2021–2100 for RCP4.5 and RCP8.5 in China's ten main river basins

Basin	Spring		Summer		Autumn		Winter		Annual mean	
	RCP4.5	RCP8.5	RCP4.5	RCP8.5	RCP4.5	RCP8.5	RCP4.5	RCP8.5	RCP4.5	RCP8.5
Songhua River	1.41	4.14	−0.14	1.84	1.31	3.64	1.59	4.45	1.04	3.52
Liao River	0.28	2.82	0.01	2.03	0.86	3.13	0.63	3.13	0.45	2.78
Hai River	0.01	2.27	0.05	2.03	0.86	3.13	-0.04	2.21	0.22	2.41
Yellow River	0.31	2.54	0.29	2.31	1.13	3.46	0.41	2.67	0.54	2.74
Huai River	0.65	2.71	0.50	2.33	1.16	3.33	0.19	2.31	0.63	2.67
Yangtze River	−0.22	1.85	0.55	2.45	0.57	2.74	−0.68	1.45	0.05	2.12
Southeastern Rivers	3.21	4.90	2.77	3.50	1.61	3.48	2.40	4.19	2.50	4.27
Pearl River	1.78	3.43	1.88	3.67	1.08	2.99	0.04	1.87	1.20	2.99
Southwestern Rivers	−1.26	1.12	0.80	2.94	0.79	2.76	−1.09	1.26	−0.19	2.02
Northwestern Rivers	0.36	2.74	−0.49	1.68	0.89	3.45	0.64	3.16	0.35	2.76
China	0.38	2.67	0.17	2.2	0.92	3.24	0.33	2.71	0.45	2.71

Unit: °C

Source Zhang et al. (2017)

Note The values in the table were obtained by subtracting the 1961–2019 average temperature from the projected 2021–2100 average temperature. Thus, positive values mean that the average temperature is projected to increase, and vice versa for negative values

Table 13.3 Differences between the seasonal and annual mean temperature trends for 1961–2019 and the projected mean temperature trends for 2021–2100 for the ten main river basins in China

Basin	Spring		Summer		Autumn		Winter		Annual mean	
	RCP4.5	RCP8.5	RCP4.5	RCP8.5	RCP4.5	RCP8.5	RCP4.5	RCP8.5	RCP4.5	RCP8.5
Songhua River	−0.031	0.015	−0.015	0.018	−0.01	0.028	−0.001	0.045	−0.015	0.026
Liao River	−0.026	0.016	−0.009	0.024	−0.004	0.032	0.005	0.045	−0.009	0.029
Hai River	−0.031	0.005	−0.004	0.029	0	0.035	−0.003	0.034	−0.01	0.025
Yellow River	−0.022	0.012	−0.009	0.024	−0.002	0.035	−0.013	0.024	−0.011	0.024
Huai River	−0.022	0.012	−0.006	0.023	0.01	0.045	−0.005	0.03	−0.006	0.028
Yangtze River	−0.006	0.029	0.001	0.032	0.009	0.044	−0.001	0.033	0.001	0.035
Southeastern Rivers	−0.014	0.013	−0.002	0.028	0.004	0.034	−0.007	0.02	−0.005	0.024
Pearl River	−0.008	0.02	−0.001	0.03	0.005	0.036	−0.006	0.021	−0.002	0.027
Southwestern rivers	−0.006	0.034	−0.003	0.033	−0.006	0.027	−0.015	0.023	−0.007	0.029
Northwestern rivers	−0.016	0.022	−0.01	0.027	−0.009	0.033	−0.017	0.025	−0.013	0.027
China	−0.016	0.021	−0.007	0.027	−0.003	0.035	−0.009	0.029	−0.009	0.028

Unit: °C/a

Source Zhang et al. (2017)

Note The values were obtained by subtracting the slope of the fitted straight line for the projected 2021–2100 temperature trend from the slope of the fitted straight line for the 1961–2019 trend. Thus, positive values indicate that the average warming trend is projected to increase, and vice versa for negative values

shows an increasing trend, while for RCP4.5, the projected precipitation tends to flatten out after 2070, with smaller increases. The predicted precipitation in RCP8.5 shows an increasing trend for all river basins in China, with significant increases in the Songhua River basin, the Liao River basin, the southwestern river basins, and the northwestern river basins; the predicted precipitation in the Pearl River basin and the southeastern river basins shows an increasing trend, while the precipitation in the Hai River, Huai River, Yellow River, and Yangtze River basins shows an increasing but insignificant trend. For the RCP4.5 emission scenario, the predicted precipitation values show a flat trend in all river basins except for the southwestern and northwestern river basins, which show a small increase in precipitation.

Under both RCP4.5 and RCP8.5, the precipitation variation in China shows obvious seasonal characteristics. The predicted seasonal precipitation trends are more consistent for both scenarios in the same season, and the predicted values of precipitation for RCP8.5 are higher than those for RCP4.5. In autumn, the projected national precipitation is the highest of any season, and the increasing trend is the largest. In summer, the projected national precipitation is marginally lower than in autumn and shows a significant increasing trend. In spring, the predicted precipitation values fluctuate widely and have a rising trend. In winter, the national precipitation trend generally tends to be flat, with a small increasing trend (Fig. 13.10; Table 13.4).

For RCP4.5, for the periods 2021–2050 and 2050–2100, spring precipitation is predicted to rise in all of China's ten main river basins except the Pearl River basin; in summer and fall, all basins are projected to have a rising trend of precipitation; and in winter, all basins are projected to experience increasing precipitation, except for the southeastern river basins, which are predicted to have decreasing precipitation. Meanwhile, for RCP8.5, for the periods 2021–2050 and 2050–2100, spring precipitation shows a decreasing trend in the Pearl River basin, the southeastern river basins, and the southwestern river basins and shows a rising trend in the other basins; in summer and autumn, precipitation shows an increasing trend in all basins; and in winter, precipitation is predicted to increase in all basins aside from the Huai River basin, the Yangtze River basin, and the southeastern river basins, where precipitation is predicted to decrease.

Fig. 13.10 Projected precipitation trends in China during 2021–2100. *Source* Zhang et al. (2017)

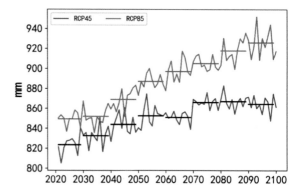

Table 13.4 Differences between the projected average precipitation during 2021–2050 and 2050–2100 for the ten major watersheds in China

Basin	Spring		Summer		Autumn		Winter		Annual mean	
	RCP4.5	RCP8.5	RCP4.5	RCP8.5	RCP4.5	RCP8.5	RCP4.5	RCP8.5	RCP4.5	RCP8.5
Songhua River	3.41	6.81	5.70	12.66	14.44	22.97	3.56	6.16	27.12	48.60
Liao River	3.67	7.51	8.25	11.25	17.84	39.53	2.20	4.75	31.95	63.05
Hai River	4.11	12.39	12.21	15.96	21.14	35.65	4.34	8.80	41.80	72.80
Yellow River	4.59	11.67	14.27	24.92	13.82	16.27	5.10	8.62	37.78	61.47
Huai River	6.26	19.05	13.84	29.52	10.46	11.77	3.87	−0.94	34.42	59.40
Yangtze River	4.78	4.92	16.92	36.02	6.89	19.94	3.51	−3.03	32.09	57.85
Southeastern rivers	1.41	−14.37	10.78	31.33	5.41	28.28	−5.20	−3.59	12.40	41.66
Pearl River	−1.62	−6.58	12.77	18.44	6.11	30.80	2.69	4.52	19.95	47.19
Southwestern rivers	0.70	−0.02	18.30	33.45	28.99	69	4.50	3.12	52.49	105.55
Northwestern rivers	2.59	6.84	3.68	8.94	7.20	7.99	1.80	4.51	15.26	28.29
China	3.04	5.78	9.72	19.42	11.16	21.30	2.85	3.41	26.77	49.90

Unit: mm

Source Zhang et al. (2017)

Note The values were obtained by subtracting the projected average precipitation for 2021–2050 from the projected average precipitation for 2050–2100. Thus, positive values mean that the trend of average precipitation is projected to increase, and negative values mean that the trend of average precipitation is projected to decrease

13.3.3 Assessment of the Effect of Future Climate Change on Water Resources

The application of the outputs of multiple global climate models to drive hydrological models is one of the most effective methods that are currently available to investigate the effect of future climate change on water resources (Liu and Xu 2016; Zhu et al. 2018; Saddique et al. 2019; Hughes and Farinosi 2020; Wang et al. 2020a). This method has been widely used to predict the future water resources change trends in China and has allowed great advancements. Many studies involving projections of future water resources in main river basins in China have found that the future water resource change trends in various river basins vary under different emission scenarios. As the longest river in China, the water resources of the Yangtze River basin are projected to be deeply affected by climate change. Various studies have that the runoff in the Yangtze River basin will show a small decrease of about 2% between 2000 and 2030 and will experience a significant increase after 2060 (Liu et al. 2008; Jin et al. 2009; Ye et al. 2010; Chen et al. 2012; Zhou et al. 2015; Wang et al. 2016, 2017a; b, c). Studies have found that water resources in the Yellow River basin and the Songhua River basin will decline in the future under the impact of projected climate change, and the rate of decrease in the Yellow River basin will increase with time (Li et al. 2016; Liu and Xu 2016; Zhu et al. 2016; Wang et al. 2017a, b, 2020c). Furthermore, other research indicates that, in the future, the hydrological cycle in the Huai River basin will continue to be enhanced and the basin's water resources will tend to increase, with precipitation growth ranging from 5.8 to 8.3% under different typical RCP scenarios (Jin et al. 2017; Chen et al. 2018; Yang et al. 2020). Ding et al. (2010) and Yang et al. (2011), respectively, applied the water and energy transfer processes model in large river basins (WEP-L) and the spatial averaging watershed hydrological model (SAWH) to simulate the future change of water resources in Hai River basin by inputting the results of a downscaling climate model. The results indicated that the runoff in the basin will decline due to global warming and increasing evaporation (Ding et al. 2010; Yang et al. 2011). Furthermore, Jin et al. (2016) analysed the projected changes of water resources in the Hai River basin and found that the water resources in the north of the basin may increase in the next 30 years, but there are large uncertainties in predicting the changes in water resources in river systems in the south of the basin under different climate models. Although there are inconsistencies in predictions of future variation trends in the Hai River basin for different climate scenarios, many research results demonstrate that the extreme precipitation in the Hai River basin has intensified in recent decades. The water resources in Northwestern China, which are dominated by the Tarim River basin, are projected to present a downward trend in the near future under the medium- and high-emission scenarios (Huang et al. 2014; Zhu and Chang 2015; Guo et al. 2016; Yuan et al. 2016; Shi 2020; Sun 2020; Wang et al. 2017c), while the annual runoff in other basins in Western China is projected to slightly increase in the near future (Li et al. 2013; Luo et al. 2017; Dong et al. 2019; Lou et al. 2020). Under RCP2.6

and RCP4.5, the water resources in the Pearl River basin show an increasing trend, but the trend declines in the next 60 years due to the increasing concentration of greenhouse gases (Du et al. 2014; Deng et al. 2015). Other studies show that water resources in the southeastern and southwestern river basins of China will show increasing trends at different rates until 2050 (Shan et al. 2016; Zaman et al. 2018; Wang et al. 2019, 2020b). Under future climate change, water resources in the Liao River basin are not projected to fluctuate to a large degree (Zhu et al. 2018; Shen and Fang 2019). Overall, water scarcity is projected to become more severe in the middle and high latitudes of China, especially in the Yellow River basin and the northwestern river basins, which is consistent with the expected future global trend of water resources change (IPCC 2014). However, water resource stress is projected to decrease in Southern China, for example, in the southeastern river basins and the Pearl River basin (Wang et al. 2013) (Table 13.5).

13.4 Adaptation Measures for the Impact of Climate Change on Water Resources

13.4.1 Active Scientific Research

(1) *Water security.* The issue of water security under a changing environment is of global importance and a critical problem facing China's sustainable development (Xia et al. 2015). Studies on the effect of climate change on water security in China began in the 1980s. In 1988, the "Study of the Impact of Climate Change on Water Resources in Northwestern and Northern China" was regarded as an important topic in China's Seventh Five-Year Plan. In 1991, the Eighth Five-Year Plan project set up a special topic entitled "Impacts of Climate Change on Hydrology and Water Resources and Countermeasures for Adaptation," which revealed the formation and evolution mechanism of several types of climate anomalies affecting water security. Subsequently, in 2003, a special topic entitled "Threshold and Comprehensive Evaluation of the Impact of Climate Change on China's Freshwater Resources" was set up in the National Key Scientific and Technological Projects of the Tenth Five-Year Plan. Then, in 2010, the National 973 Program for Key Basic Research and Development supported three studies on the effect of climate change on water security (Xia and Shi 2016).

(2) *Water cycle.* It is highly significant to propose response strategies to climate change based on the reasonable assessment of the effect of climate change on the water cycle (IPCC 2007). To ensure the sustainability of water resources, adaptive countermeasures should be formulated according to the change of the water cycle under climate change (Liu 2004). Xia et al. (2011) concluded that the effect of climate change on the water cycle is one of the most important interdisciplinary scientific issues in the topics of international global change

Table 13.5 Projected future change trends of water resources in the ten main river basins of China

Basin	Change trend of water resources in the future		References
	2021–2050	2051–2100	
Huai River basin	Water resources show a growth trend under different emission scenarios, with a growth rate of about 10%	The annual precipitation of the basin will increase, with the precipitation increase mainly occurring in the middle of the basin, and the frequency of extreme events will increase.	Jin et al. (2017), Jiang et al. (2020), Yang et al. (2020)
Yellow River basin	Under the scenarios of medium and low discharge, the water resources of the basin will be reduced by about 5%	The water resources in the upper reaches of the river will decrease by 10–20%	Liu and Xu (2016), Zhu et al. (2016), Wang et al. (2020c)
Hai River basin	The overall water resources of the basin will increase by 3–7%, and the precipitation in 2030 will be uncertain	The water resources will increase by 5–39% by the 2080s	Bao et al. (2014), Jin et al. (2016)
Liao River basin	The change of water resources will have no obvious trend.		Zhu et al. (2018), Shen and Fang (2019)
Yangtze River basin	The runoff of the basin will begin to increase with time.	After 2060, water resources will show a significant increasing trend.	Liu et al. (2008), Jin et al. (2009), Liu and Xu (2016)
Songhua River basin	The flow in the lower reaches will decrease by about 25%	–	Li et al. (2016)
Pearl River basin	The annual average precipitation in the basin will increase, with the largest increase in autumn, and the intensity of extreme precipitation will increase continuously		Du et al. (2014), Deng et al. (2015)
Southeastern Rivers	The water resources in the basins will increase.		Zaman et al. (2018), Wang et al. (2020b)
Northwestern Rivers	Water resources will decrease gradually.	The average annual runoff may decrease or increase.	Huang et al. (2014), Yuan et al. (2016), Luo et al. (2017), Sun (2020)
Southwestern Rivers	The average annual precipitation will increase slightly.	The future flow of the basin will decrease gradually.	Shan et al. (2016), Wang et al. (2019, 2020a, b, c)

and water science. The analysis of the evolution rules of hydrological elements in China is at an early stage yet has produced relatively abundant research; studies on the response of hydrological components and water resources to future climate change have been performed in many Chinese river basins, and valuable results have been achieved (Zhang and Wang 2015).

(3) *Water resource vulnerability and adaptive systems.* Since the start of the twenty-first century, the vulnerability and adaptability of water resources under climate change have become an important demand and a hot research topic. China has often stressed the importance of researching water resource vulnerability and adaptive countermeasures in major research fields and via large projects. The National Key Basic Research Development Plan set up a project entitled "Under the Influence of Climate Change Vulnerability and Adaptation Countermeasures of Water Resources." Xia et al. (2015) studied the theory and method of water resource adaptive management countermeasures under climate change, analysed the cost–benefit and restrictive factors of adaptation measures, and discussed the system construction and realisation way.

13.4.2 Planning for Water Security Under Changing Environment

In China's National Adaptation Strategy (NAS), water resource management is highlighted as a priority area. Various plans of resource use and conservation and ecological restoration are being developed to assist the water sector to adapt while managing complex demands (National Development and Reform Commission 2013). Changes to policy and institutions to assimilate climate change into water resources planning and realise IWRM have been the long-term aims of water resources management in China.

(1) *Water supply.* Following the Chinese National Adaptation Strategy to Climate Change, the country has intensified efforts to construct a water-efficient society, built various water storage and diversion projects in accordance with local conditions, improved key water source projects and irrigation projects, and implemented the construction of the eastern and middle route of the South-to-North Water Diversion Project. The South-North Water Transfer Project transfers 45 billion m^3 of water annually—twice the runoff of the Yellow River—from the Yangtze River to Northern and Northwestern China and is expected to lessen local water shortage. Through the compilation of the National Comprehensive Water Resources Plan and the comprehensive river basin plan, the water allocation plan for main river basins has been formulated to improve the allocation pattern of water resources and improve the emergency water supply guarantee capacity.

(2) *Flood control.* China has formulated and implemented the Water Law and Flood Prevention Law and completed flood control plans and other water conservancy plans for main river basins. A flood control and mitigation engineering system for large rivers has been established to enhance the ability to resist floods and waterlogging disasters; this system mainly consists of reservoirs, embankments, flood storage and detention areas, a mountain flood

prevention and control system mainly based on management measures, a national command system for flood control and drought relief, and a flood risk management system. At present, a flood control system for large river basins has been mostly completed. The key points to be improved in the future are the flood defences in small- and medium-sized rivers to cope with the more uncertain hydrological changes caused by climate change in such rivers.

(3) *Sponge city.* It refers to the ability of cities to act like sponges, resilient to environmental changes and natural disasters caused by rain, which can also be called "low impact development" (LID). Incorporating "sponge cities" into water security planning can effectively reduce the risk of urban flooding, especially under situations where climate change and urban construction combine to exacerbate the disaster risk. Many cities in China have carried out specific plans of "sponge city." The State Council's guidance on promoting the construction of sponge cities has specified that, according to the relevant requirements of new-type urbanisation and water security strategies, the pilot construction of sponge cities should be promoted with an emphasis on the prevention and control of urban waterlogging and on guaranteeing urban ecological security. By 2030, over 80% of the urban built-up areas in China are expected to achieve these targets.

(4) *The prevention of other water-related disasters.* To effectively prevent and control secondary disasters (e.g., landslides and debris flows) caused by floods, droughts, and other related disasters (e.g. storm surges and coastal-zone disasters) under climate change, China has strengthened its capacity for the monitoring and early warning of extreme weather and climate events and has set up an emergency response plan for extreme meteorological events and secondary disasters. Major progress has been made in the prevention of extreme weather and climate events such as strong typhoons and regional rainstorms and floods (Information Office of the State Council 2008). Additionally, an integrated climate change observation system has been established.

13.4.3   Innovation of Mechanisms for Adapting to Climate Change

(1) *Ministry of Ecosystem and Environment.* To ensure the overall planning and sustainable development of watershed ecosystems, China established the Ministry of Ecology and Environment in 2018 to reintegrate the departmental responsibilities related to water ecology and environment of the Ministry of Agriculture, Ministry of Environmental Protection, and Ministry of Water Resources. The Ministry of Ecosystem and Environment can integrate and unify the management of all links of the water ecological environment and effectively integrate the protection mechanisms of river and lake ecological environments in response to climate change.

(2) *River Chief System.* The key of the River Chief System is making clear the responsibilities of the local chief in water management. In this system, the principal person in charge of the party committee or the local government is designated as a "River Chief" at different levels of government (provincial-, prefectural-, and county-level). The main responsibility of the River Chief is to undertake the management and protection of rivers and lakes, such as the supervision of water resource conservation, river and lake shoreline management, water ecological space management, water pollution prevention and control planning, and water environment management. As an innovative mechanism that combines the responsibility of the party and government, the River Chief System which has the clear attribution of administrative responsibility can strengthen the administrative efficiency of different departments, including water-related departments, at different levels, and improve the unity of water resource management systems, to strengthen the decision-making and implementation ability of the government regarding issues related to climate change.

(3) *Market mechanisms.* As an important institutional tool for addressing water resource supply and demand constraints under climate change, market mechanisms—for example, water pricing and water rights—can improve water resource utilisation efficiency, resolve water conflict, and promote the sustainable use and management of water resources. In certain locations, the trading of water property rights can be considered as adaptive marketing behaviour (Pan et al. 2011). Water markets can improve the utilisation efficiency of agricultural water use and can induce farmers to save water and promote the orderly flow of water resources between different industries and regions (United Nations 2020). Recently, various water rights transfer projects have been implemented in China (Calow et al. 2009). Water rights have been allocated to the provincial- and county-level and even to rural households in some areas. Many regions have carried out pilot transactions of water rights, including those for major projects, among farmers, between irrigated areas and industrial enterprises, and between provinces.

(4) *Social participation.* Increased public involvement in the management of climate risk is recommended as a means to obtain adaptive capacities at multiple levels, prevent institutional traps, and prioritise risk reduction for socially vulnerable groups (United Nations 2020). China has held 33 consecutive "Water Week" events to inform the public of the importance of the efficient utilisation of water resources. The establishment of the "Citizen River Chief" system has provided an effective channel for the public to engage in water environment supervision. Additionally, the establishment of local water users' associations, especially in water-scarce areas, has helped to encourage farmers to participate in water management.

13.4.4 Investment

(1) *Investment in facility construction*. Water-related extremes aggravated by climate change exacerbate the threats to water, sanitation, and hygiene (WASH) infrastructure (United Nations 2020). China has long stressed the significance of increasing its water storage and improving its water resources and management infrastructure. The nation's water resource policies and management practices, which are supply-driven and engineering-based, have allowed it to increase the available surface water and raise the total water storage through infrastructure construction (Cheng et al. 2009; Cheng and Hu 2011; Jiang 2009). China has invested significant funds in the construction of water resources facilities. Engineering measures to improve adaptation to climate change improve the adaptive capacity of socio-economic systems by improving engineering infrastructure such as water conservancy facilities, environmental infrastructures, inter-basin water transfer projects, disease-monitoring networks, and meteorological monitoring stations. Water reservoirs and inter-basin water transfers (e.g. the Three Gorges Dam and the South-to-North Water Diversion Project), which improve the allocation of water resources, control flooding on main rivers, and alleviate drought, are expected to remain to be important aspects of China's future water resource development (MWR 2008; Cheng et al. 2009).

(2) *Funding Investment*. On 29 January 2011, China revealed a plan to invest the US$600 billion during the next decade to preserve and enhance access to water (Yu 2011; Gong et al. 2011). This huge investment primarily targets the improvement of water infrastructures such as reservoirs, drilling wells, water supply for key irrigation districts, and inter-basin water transfer projects (Yu 2011), which is considered as the most effective means to deal with the effects of climate change, droughts and floods, and food security. Additionally, the central government recently launched the Special Program for Water Pollution Control and Treatment, which invested 35.6 billion yuan between 2008 and 2020 to control water pollution and improve water quality nationwide (MEP 2008). According to the Third National Climate Assessment, from 2010 to 2030, the cost of drought adaptation nationwide will reach 500 billion yuan.

Acknowledgements This research received funding from the Strategic Priority Research Program of the Chinese Academy of Sciences (XDA20010201) and the National Natural Science Foundation of China (41671026). The authors thank Professor Xuezhen Zhang for providing the multi-model ensemble forecasts of Chinese climate change data under different emission scenarios and thank the China Meteorological Data Service Centre for providing historical climate data.

References

Bao XJ, Zhang JY, Yan XL, Wang GQ, He RM (2014) Changes of precipitation in Haihe River Basin in 60 years and analysis of future scenarios. Hydro-Sci Eng 05:8–13

Bates BC, Kundzewicz ZW, Wu S, Palutikof JP (2008) Climate change and water. Technical paper of the intergovernmental panel on climate change. IPCC Secretariat, Geneva

Calow RC, Howarth SE, Wang J (2009) Irrigation development and water rights reform in China. Int J Water Resour Dev 25(2):227–248

Chen Y, Gao G, Ren GY, Liao YM (2005) Spatial and temporal variation of precipitation over ten major basins in China between 1956 and 2000. J Nat Resour 20(005):637–643

Chen H, Xiang T, Zhou XC, Xu Y (2012) Impacts of climate change on the Qingjiang Watershed's runoff change trend in China. Stoch Env Res Risk Assess 26(6):847–858

Chen SC, Huang BS, Wang XG, Hao ZC (2018) Characteristics of precipitation change and CMIP5 climate model evaluation in Yanglou watershed. Pearl River 039(006):58–62, 97

Cheng H, Hu Y, Zhao J (2009) Meeting China's water shortage crisis: current practices and challenges. Environ Sci Technol 43(2):240–244

Cheng H, Hu Y (2011) Economic transformation, technological innovation, and policy and institutional reforms hold keys to relieving China's water shortages. Environ Sci Technol 45 (2):360–361

Del Buono D (2021) Can biostimulants be used to mitigate the effect of anthropogenic climate change on agriculture? It is time to respond. Sci Total Environ 751:141763

Deng XY, Zhang Q, Li JF, Sun P, Chen XH (2015) Prediction of water resources in Dongjiang River Basin based on different climate change scenarios. J Sun Yat Sen Univ 54(02):141–149

Ding XY, Jia YW, Wang H, Niu CW (2010) Impact of climate change on water resources in Haihe River Basin and its countermeasures. J Nat Resour 25(04):604–613

Dong LJ, Dong XH, Zeng Q, Wei C, Yu D, Bo HJ, Guo J (2019) Study on future runoff variation trend of Yalong River Basin under climate change. Adv Clim Chang Res 15(06):596–606

Du YD, Yang HL, Liu WQ (2014) Simulation analysis of precipitation characteristics in Pearl River Basin under future RCPs scenarios. J Trop Meteorol 30(03):495–502

Ekegemu-Abra, Wang YJ, Lin HB, Xu HL, Zhou HY (2019) Analysis of water resources change trend and water use efficiency in Tarim River Basin. J Shihezi Univ 37(01):112–120

Gao JD, Feng D (2019) Analysis on the change trend of water resources in Haihe River Basin from 1998 to 2017. J Irrig Drainage 38(S2):101–105

Gong P, Yin Y, Yu C (2011) China: invest wisely in sustainable water use. Science 331 (6022):1264–1265

Guo J, Su XL, Singh V, Jin JM (2016) Impacts of climate and land use/cover change on streamflow using SWAT and a separation method for the Xiying River Basin in Northwestern China. Water 8(5):192

He JK, Liu B, Chen Y, Xu HQ, Guo Y, Hu XL, Zhang X, Li Y, Zhang A, Chen W (2007) China's National assessment report on climate change (III): integrated evaluation on policies of China responding to climate change. Advances in Climate Change Research

Hu DL, Yan DH, Song XS, Zhang MZ, Yu H, Yang SY (2008) Analysis on the change trend of water resources in the Yangtze River Basin above Yibin. South North Water Diversion Water Conservancy Sci Technol 02:53–56

Huang JL, Tao H, Su BD, Gemmer M, Wang YJ (2014) Simulation of extreme climate events in Tarim River Basin and prediction under RCP4.5 scenario. Arid Land Geogr 37(3):490–498

Hughes DA, Farinosi F (2020) Assessing development and climate variability impacts on water resources in the Zambezi River basin. Simulating future scenarios of climate and development. J Hydrol: Reg Stud 32:100763

Idrizovic D, Pocuca V, Mandic M, Djurovic N, Matovic G, Gregoric E (2020) Impact of climate change on water resource availability in a mountainous catchment: a case study of the Toplica River catchment, Serbia. J Hydrol 587:124992

Information Office of the State Council (2008) The white paper: China's policies and actions for addressing climate change. Beijing, China

IPCC (2007) Climate change: impacts, adaptation and vulnerability: contribution of working group II to the fourth assessment report of the intergovernmental panel on climate change. Cambridge University Press, Cambridge

IPCC (2014) Climate change 2014. Synthesis report

Jiang Y (2009) China's water scarcity. J Environ Manage 90:3185–3196

Jiang Y (2015) China's water security: current status, emerging challenges and future prospects. Environ Sci Policy 54:106–125

Jiang T, Lv YR, Huang JL, Wang YJ, Su BD, Tao H (2020) Overview of new scenario of CMIP6 model (SSP-RCP) and its application in Huaihe River Basin. Prog Meteorol Sci Technol 10 (05):102–109

Jin XP, Huang Y, Yang WF, Chen L (2009) Impact of future climate change on water resources in the Yangtze River Basin. Yangtze River 40(08):35–38

Jin JL, Wang GQ, Liu CS, Liu YL, Bao ZX (2016) Evolution trend of water resources in Haihe River Basin under climate change. J North Chin Univ Water Resour Hydropower 37(05):1–6

Jin JL, He J, He RM, Liu CS, Zhang JY, Wang GQ, Bao ZX (2017) Impacts of climate change on water resources and extreme flood events in Haihe River Basin. Sci Geogr Sin 08:103–110

Li JB, Wang G, Li XH, Ma JZ (2008) Impacts of climate change and human activities on water resources in Shiyang River Basin in recent 50 years. J Arid Land Resour Environ 02:75–80

Li YY, Wen K, Shen FX, Zhang SF, Wang JQ (2012) Impacts and adaption of climate change in China. China Water Conservancy and Hydropower Press, Beijing, p 433

Li L, Shen HY, Dai S, Li HM, Xiao JS (2013) Response and trend prediction of surface water resources to climate change in the source region of the Yangtze River. J Geog Sci 23(02):208–218

Li F, Zhang G, Xu YJ, Loukas A (2016) Assessing climate change impacts on water resources in the Songhua River Basin. Water 8(10):420

Li LJ, Song XY, Xia L, Fu N, Feng D, Li HY, Li YL (2020) Modelling the effects of climate change on transpiration and evaporation in natural and constructed grasslands in the semi-arid Loess Plateau, China. Agric Ecosyst Environ 302:107077

Liu CM (2004) Study of some problems in water cycle changes of the Yellow River Basin. Adv Water Sci 15(5):608–614 (in Chinese)

Liu B, Jiang T, Ren GY, Fraedrich K (2008) Change trend of surface water resources in the Yangtze River Basin before 2050. Adv Clim Chang Res 03:145–150

Liu P, Xu ZS, Wang L, Liu JF (2009) Research progress on the impact of climate change on water resources in the Yellow River Basin. Meteorol Environ Sci 32(S1):275–278

Liu L, Xu ZX (2016) Hydrological implications of climate change on river basin water cycle: Case studies of the Yangtze River and Yellow River basins, China. Appl Ecol Environ Res 15 (4):683–704

Liu XL, Hong L, Feng D (2020) Analysis on the change trend of water resources in Huaihe River Basin from 1998 to 2017. Anhui Agric Sci 48(13):207–210

Lou W, Li ZJ, Liu YH (2020) Prediction of future precipitation change in the upper reaches of Jinghe River Basin under Multi Model. South North Water Diversion Water Conservancy Sci Technol 18(06):1–16

Lu ZH, Xia ZQ, Yu LL, Wang JC (2012) Spatiotemporal evolution characteristics of precipitation in Songhua River Basin from 1958 to 2009. J Nat Resour 27(06):990–1000

Luo M, Meng F, Liu T, Duan Y, Frankl A, Kurban A, De Maeyer P (2017) Multi–model ensemble approaches to assessment of effects of local climate change on water resources of the Hotan River Basin in Xinjiang, China. Water 9(8):584

MEP (Ministry of Environmental Protection) (2008) Special programme on water pollution control and treatment (2008–2020). Beijing, China

MWR (Ministry of Water Resources) (2008) China's water agenda 21. Resources & Hydropower Press, Beijing

National Development and Reform Commission (2013) The National plan for addressing climate change (2013–2020). Beijing, China

Pan JH, Zheng Y, Markandya A (2011) Adaptation approaches to climate change in China: an operational framework. Economía Agraria y Recursos Naturales 11(1):99–112

Ren GY, Wu H, Chen ZH (2000) Spatial characteristics of precipitation change trend in China. J Appl Meteorol 03:322–330

Saddique N, Usman M, Bernhofer C (2019) Simulating the impact of climate change on the hydrological regimes of a sparsely gauged mountainous basin, Northern Pakistan. Water 11(10):2141

Shan HC, Yuan F, Sheng D, Zou L, Liu YP (2016) Analysis of climate change characteristics in Xijiang River Basin under AIB Scenario using PRECIS. China Rural Water Hydropower 12:84–87

Shen J, Fang HC (2019) Study on water resources evolution of Hunhe River Basin based on different discharge scenarios. Ground Water 41(03):121–124

Shi YJ (2020) Study on future runoff change of Tarim River basin based on GCM model. Dev Manage Water Resour 05:7–15

Sun FH, Li LG, Yuan J, Dai P (2015) Analysis on the impact of climate change on water resources in Liaohe River Basin. J Meteorol Environ 31(06):147–152

Sun YX (2020) Study on future precipitation and temperature variation in Tarim River Basin based on GCM Model. Ground Water 42(204):172–175

United Nations (2020) Launch of UN World Water development report 2020: water and climate change. Thai News Service Group, Bangkok

Wada Y, Wisser D, Eisner S, Flörke M, Gerten D et al (2013) Multimodel projections and uncertainties of irrigation water demand under climate change. Geophys Res Lett 40:4626–4632

Wang JN, Zhang LC (2006) Water resources change and regional response in Ebinur Lake Basin. J Arid Land Resour Environ 04:157–161

Wang GY, Dai SB (2008) Changes of water resources and water environment in Huaihe River Basin in recent 50 years. J Anhui Normal Univ 01(75–78):87

Wang JX, Huang JK, Yan TT (2013) Impacts of climate change on water and agricultural production in ten large river basins in China. J Integr Agric 12(007):1267–1278

Wang XJ, Zhang JY, Shahid S, Guan EH, Wu YX, Gao J, He RM (2014) Adaptation to climate change impacts on water demand. Mitig Adapt Strat Glob Change 21(1):81–99

Wang M, Liu M, Xia ZH, Wang K, Xiang H, Qin PC, Ren YJ (2016) Impact of future climate change on water resources in Honghu Lake Basin based on SWAT Model. J Meteorol Environ 32(04):39–47

Wang XG, Hu J, Lv J, Liu HC, Wei CF, Zhang Z, Zhang Y (2017) Analysis on runoff variation characteristics of Songhua River Basin from 1956 to 2014. Soil Water Conserv China 10(61–65):72

Wang G, Zhang J, Jin J, Weinberg J, Bao Z, Liu C, Liu Y, Yan X, Song X, Zhai R (2017) Impacts of climate change on water resources in the Yellow River Basin and identification of global adaptation strategies. Mitig Adapt Strat Glob Change 22:67–83

Wang GQ, Zhang JY, Xu YP, Bao ZX, Yang XY (2017) Estimation of future water resources of Xiangjiang River Basin with VIC Model under multiple climate scenarios. Water Sci Eng 10(02):87–96

Wang SX, Zhang LP, Li Y, She DX (2019) Extreme flood events in Lancang River Basin under climate change scenarios. Adv Clim Chang Res 15(01):23–32

Wang D, Liu MB, Chen XW, Gao L (2020) Spatiotemporal variation characteristics of future blue-green water in Shanmei Reservoir Basin based on CMIP5 and SWAT. South North Water Diversion Water Conservancy Sci Technol. https://kns.cnki.net/kcms/detail/13.1430.TV.20201026.1329.004.html. Accessed 13 Apr 2021

Wang KY, Niu J, Li TJ, Zhou Y (2020) Facing water stress in a changing climate: a case study of drought risk analysis under future climate projections in the Xi River Basin, China. Front Earth Sci 8:86

Wang GQ, Qiao CP, Liu ML, Du FR, Ye TF, Wang J (2020) Analysis on the future trend of water resources in the Yellow River Basin under climate change. Hydro-Sci Eng 02:1–8

Wu L, Qin ZR, Huang DZ, Shen H, Li JN (2009) Difference analysis of regional seasonal precipitation in South China. Meteorol Res Appl 30(03):5–7, 11

Xia J, Liu CZ, Ren GY (2011) Opportunity and challenge of the climate change impact on the water resource of China. Adv Earth Sci 26(1):1–12 (in Chinese)

Xia J, Shi W, Luo XP, Hong S, Ning LK, Christopher JG (2015) Revisions on water resources vulnerability and adaption measures under climate change. Adv Water Sci 26(2):279–286 (in Chinese)

Xia J, Shi W (2016) Perspective on water security issue of changing environment in China. J Hydraul Eng 47(03):292–301 (in Chinese)

Yang ZY, Yu YD, Wang JH, Yan DH (2011) Impact of climate change on water resources in the Ethan River Basin. Adv Water Sci 22(02):175–181

Yang QQ, Gao C, Zha QY, Zhang PJ (2020) Changes of climate and runoff in the upper reaches of Huaihe River under RCP Scenario. Anhui Agric Sci 48(3):217–222

Ye X, Zhang Q, Bai L, Hu Q (2010) A modeling study of catchment discharge to Poyang Lake under future climate in China. Quatern Int 244(2):221–229

Ying AW, Jiang GB (1996) Response of water resources to climate change in Liaohe River Basin. Adv Water Sci S1:67–72

Yu CQ (2011) China's water crisis needs more than words. Nature 470:307

Yuan F, Ma M, Ren L, Shen H, Li Y, Jiang S, Yang X, Zhao C, Kong H (2016) Possible future climate change impacts on the hydrological drought events in the Weihe River Basin, China. Adv Meteorol 2905198

Zaman M, Naveed Anjum M, Usman M, Ahmad I, Saifullah M, Yuan S, Liu S (2018) Enumerating the effects of climate change on water resources using GCM scenarios at the Xin'anjiang watershed, China. Water 10(10):1296

Zhang ZX, Zhang JC, Sheng RF (2010) Influence of seasonal variation of precipitation on water resources in the Yangtze River Basin. J Qingdao Technol Univ 31(01):67–72

Zhang XQ, Wang YH (2015) A review of adaptive management research on China's water resources management under climate change. Resour Environ Yangtze Basin 24(12):2061–2068 (in Chinese)

Zhang XZ, Li XX, Xu XC, Zhang LJ (2017) Ensemble projection of climate change scenarios of China in the 21st century based on the preferred climate models. Acta Geogr Sin 72(9):1555–1568

Zhou ZH, Qiu YQ, Jia YW, Wang H, Wang JH, Qin DY (2009) Analysis of water resources evolution in Weihe River Basin under changing environment. Hydrology 29(01):21–25

Zhou J, He D, Xie Y, Liu Y, Yang Y, Sheng H, Guo H, Zhao L, Zou R (2015) Integrated SWAT model and statistical downscaling for estimating streamflow response to climate change in the Lake Dianchi Watershed, China. Stoch Env Res Risk Assess 29:1193–1210

Zhu YL, Chang JX (2015) Runoff prediction of Weihe River based on climate model and hydrological model. J xi'an Univ Technol 31(4):400–408

Zhu Y, Lin Z, Wang J, Zhao Y, He F (2016) Impacts of climate changes on water resources in Yellow River Basin, China. Proc Eng 154:687–695

Zhu X, Zhang C, Qi W, Cai W, Zhao X, Wang X (2018) Multiple climate change scenarios and runoff response in Biliu River. Water 10(2):126

Chapter 14
Using Waternomics to Develop and Avoid Systemic Shocks to the Economy

Debra Tan

Abstract Water is essential for economic development. For a country with limited water resources, how to balance trade-offs between economic development, water resource availability and quality is key. Policy decisions should thus wed economic planning to water resources and pollution management; this concept is called "waternomics". In developing Asia, there is a powerful case for countries to curate a waternomic roadmap, especially since the locational nature of water and climate risks points to a significant clustering of economic risk exposure in vulnerable regions—specifically, major river basins and coastal economic hubs.

Keywords Asia · Waternomics · Economic development · River basins

Governments and central banks are starting to take action to alleviate such threats because if left unattended, chronic and acute water risks (amplified by climate change) could trigger systemic shocks and collapse in local and national economies as well as the global financial system. Water, climate challenges, economic planning and financial resilience, therefore, need to be managed comprehensively to ensure socio-economic and water security. This chapter sets out key waternomic challenges in Asia and the rise of region-based waternomic policies in China to address such risks. An overview of action in the financial sector to manage such chronic threats is also included to help multidisciplinary decision-makers make better decisions today for a water and economic secure tomorrow.

14.1 Key Waternomic Challenges in Asia

Asia faces serious and urgent water challenges. Already today, two of the world's most populous countries, India and China, are water stressed (WRI 2013), and rampant water pollution from decades of rapid development has only further

D. Tan (✉)
China Water Risk (CWR), Hong Kong, China
e-mail: dt@chinawaterrisk.org

exacerbated water challenges. Meanwhile, across the continent, although water infrastructure is improving, hundreds of millions are still left without access to improved water sources (World Bank 2015a).

Yet, many countries in Asia facing similar challenges still need to develop. As the economy runs on water, no water means no growth. Beyond quenching thirst, water is also used to grow food, generate power, mine resources, make clothes, electronics and other consumables so multiple sectors, trade and employment will be affected if increasingly precious water resources are mismanaged. Asia's tight liquidity constraint means that a parallel conversation on waternomics should also be progressed alongside the traditional conversation of Water, Sanitation and Hygiene (WASH) to ensure long-term social, economic and water security in Asia. Climate change and its impact on water resources only lends urgency.

A compelling reason for a waternomics approach in Asia is because the region faces a triple threat in the water–climate nexus that could put one in two Asians and US\$4.3 trillion of annual Gross Domestic Product (GDP) at risk (CWR 2018; Wester et al. 2019; CWR, Manulife Asset Management and AIGCC 2019).

Limited water resources to support development under the current economic model—Water is essential for growth across agriculture, industry and power generation (thermal cooling and hydropower). By 2050, as demand for water rises by 30–40% across Asia Pacific, an estimated 3.4 billion people will be living in water-stressed regions according to the Asian Development Bank (ADB) which also warned it will likely constrain economic growth in a number of countries (ADB 2016).

Waternomic performance analysis which benchmarks per capita GDP against per capita water use of the G20 countries shows that key Asian economies like China and India not only lag but also face liquidity constraints (HSBC 2015; FECO and CWR 2016; Yang et al. 2016). To achieve a per capita GDP of over USD50,000, the USA uses at least 1543 m^3 of water per capita, which is around 16% of its total annual renewable water resources of 2018 m^3 per capita (CWR 2018). However, China and India only have total renewable water resources of 2018 m^3/pax and 1458 m^3/pax, respectively; see Fig. 14.1 (CWR 2018).

The harsh reality is that both China and India do not have sufficient water to pursue "development as usual"; their current water-intensive export-led economic growth model is not sustainable and will not help them ensure food, energy, economic and water security (CWR 2018). They will have no choice but to maximise waternomic performance. This includes revamping agriculture and currently highly polluting and water-intensive industries as well as careful planning of energy expansion and future industries so as to build a collective roadmap towards more GDP on less water and less pollution (HSBC 2015; FECO and CWR 2016; Yang et al. 2016; CWR 2018). Such strategies are discussed later in the next section "cross-cutting waternomic strategies".

China, in recognising its environmental constraints, has already started to de-prioritise GDP to rebalance its economy and environment in its move to strive towards an "ecological civilization", a concept that has now been entrenched in its

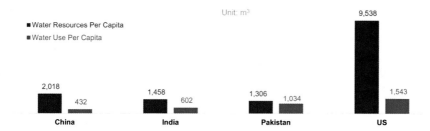

Fig. 14.1 Limited water resources per capita constraint development choices in Asia. CWR based on FAO AQUASTAT (2010). *Note* This chart is extracted from the report "No Water, No Growth —Does Asia have enough water to develop?" (CWR 2018) for use in this chapter with permission from China Water Risk; © China Water Risk 2021, all rights reserved

constitution (China Environment News 2018). The country has also set waternomic targets (including water use and wastewater discharged per unit of GDP) at various levels—nationally, provincially, regionally and by river basin as a step towards holistic management of water and business unusual innovations and circular economies should also be encouraged for green development (CWR 2019a; Yang et al. 2019).

People, cities and GDP are clustered in ten major river basins—Rivers are important to Asia. The Amu Darya, Brahmaputra, Ganges, Indus, Irrawaddy, Mekong, Salween, Tarim, Yangtze and Yellow are the continent's cradles of civilisation. Just under two billion people (Wester et al. 2019) across 16 countries (Afghanistan, Bangladesh, Bhutan, Cambodia, China, India, Kyrgyzstan, Laos, Myanmar, Nepal, Pakistan Tajikistan, Thailand, Turkmenistan, Uzbekistan and Vietnam) are clustered in these ten basins which are estimated to generate US$4.3 trillion of GDP per annum (CWR 2018). Already, material portions (>50%) of half of these ten river basins are facing "high" to "extremely high" water stress (CWR 2018).

These ten rivers basins support over 280 large cities, each with a population of over 300,000; many of these cities are Asia's capitals and economic powerhouses including mega-cities (more than 14 million people) of Delhi, Shanghai, Dhaka, Kolkata, Chongqing, Lahore and Chengdu (CWR 2018). Such clustering of cities in the river basins leads to high concentrations of national GDP generated in these river basins.

For example, the Ganges River Basin alone supports over 600 million people and generates around a third of India's GDP annually; yet the water from this river at the maximum average annual flow will not even fill up one Lake Erie, the smallest of the Great Lakes (CWR 2018). Meanwhile, the Indus River Basin is home to almost 90% of its population and accounts for 92% of its annual GDP; and the Yellow and Yangtze River Basins together account for 42% of China's population and 28% of its annual GDP; see Fig. 14.2 (CWR 2018).

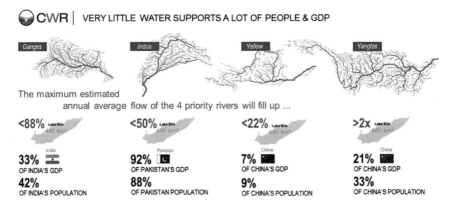

Fig. 14.2 Very little water supports a lot of people and GDP. *Source* CWR based on "No water, no growth—Does Asia have enough water to develop?" (CWR 2018). *Note* This infographic is extracted from the report "Yangtze water risks, hotspots and growth—Avoiding regulatory shocks from the march to a Beautiful China" (CWR 2019a) for use in this chapter with permission from China Water Risk; infographic © China Water Risk 2021, all rights reserved

The Asian Investor Group for Climate Change (AIGCC), Manulife Asset Management (Manulife) and CWR warned that the magnitude and concentration of water and economic risks illustrated above have grave implications and clearly warrant attention from governments, central banks, financial institutions and corporates (CWR, Manulife Asset Management and AIGCC 2019). Looking forward, population growth coupled with rising urbanisation will only increase water stress as more people flock to cities in the river basins. By 2050, over 60% of the population in Asia Pacific region (APAC) are expected to live in cities (ADB 2016).

Climate change will exacerbate water scarcity—These ten rivers share a common source region: the Hindu Kush Himalayas (HKH), where the glacial melt from an estimated 7547 km^3 of ice feed into the rivers (Bajracharya et al. 2015; Mukherji et al. 2015; CWR 2018; National Tibetan Plateau Third Pole Environment Data Centre n.d.). Unfortunately, climate change is already evident in the HKH threatening the upper watersheds and further downstream other components of river flow are also impacted—accelerated glacial melt, reduced snowfall and changing monsoon patterns will impact future river flow (Wester et al. 2019). Future projections made by Chinese Academy of Science-Institute of Geographic Sciences and Natural Resources Research (CAS-IGSNRR) using five climate ensemble models under Intergovernmental Panel on Climate Change's (IPCC) RCP4.5 scenario on four key climate and hydrological indicators: (1) temperature, (2) snowfall, (3) rainfall and (4) river flow for each of the ten rivers, are not encouraging; see Fig. 14.3 (CWR 2018).

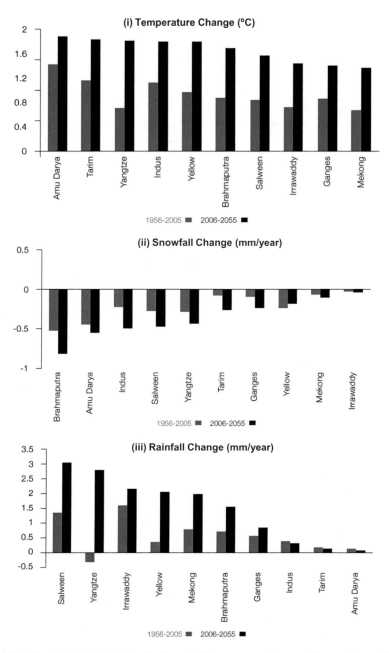

Fig. 14.3 Changes in key indicators for the past 50 years versus the next 50 years. *Source* CWR based on data calculated by CAS-IGSNNER from five ensemble models (BCC-CSM1.1, CanESM2, CCSM4, MIROC5, MPI-ESM-LR). *Notes* Rainfall, snowfall and run-off change are expressed in equivalent water eight. These charts are extracted from the report "No water, no growth—Does Asia have enough water to develop?" (CWR 2018) for use in this chapter with permission from China Water Risk; © China Water Risk 2021, all rights reserved

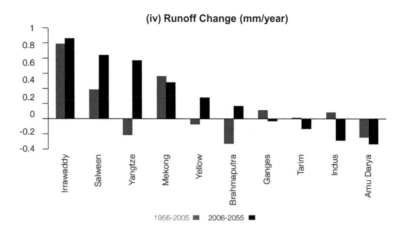

Fig. 14.3 (continued)

- **Temperatures are expected to continue rising across the ten basins**: changes in temperature will double for the next 50 years (2006–2055) as compared to the past 50 years (1956–2005) in six out of ten basins (CWR 2018). Rising temperatures will have direct impacts on the water cycle and the availability of water resources, especially in snow-dominant or glacier-fed river basins (Oki 2016). Moreover, rise in water temperature could reduce raw water quality and threaten drinking water sources (Oki 2016; Döll et al. 2015).
- **Snowfall is projected to continue declining**: losses are likely more than double for Indus, Tarim and Ganges basins (CWR 2018). Glaciers are also expected to retreat, particularly those below 5700 m above sea level (masl) which are more sensitive to climate change (Bajracharya et al. 2015). Over 60% of the total glacier area of the HKH rregion lie within 5000–6000 masl making them more vulnerable to climate change (Bajracharya et al. 2015). Glacier and snow melt are important contributors to river flow at 62–79% for the upper Indus, 20% for the upper Ganges and 42% for the upper Tarim (Wang et al. 2007; Gao et al. 2010; Chen 2013; Lutz and Immerzeel 2013; Zhang et al. 2013; Lutz et al. 2014).
- **Rainfall is on the rise except for three basins**: Indus, Tarim and Amu Darya. The Yangtze and Yellow face the largest upticks in rainfall (CWR 2018). Indeed in 2020, the Yangtze faced the worst flood since 1961 in terms of the precipitation recorded with more than 37 million people affected and economic losses of USD11.7 billion incurred (BBC News 2020).
- **Mixed impact on runoffs with four rivers seeing shrinkages in flow**: The Ganges, Tarim, Indus and Amu Darya are expected to see overall shrinkages in flow in the future versus the past (CWR 2018). This does not bode well for the 925+ million people who reside in these basins nor the US$1.3 trillion of annual GDP generated there (CWR 2018).

There are clear socio-economic risks ahead accelerated by climate change, and here, it is important to note the following:

- Climate projections are full of uncertainties, and while the results above show changes over 50 years, it is important to prepare for annual seasonal variability as climate change will push out the extremes in rainfall and river flow and current water infrastructure may not be able to deal with widening extremes of seasonal droughts and floods. These are not shown in the charts above; their purpose is to illustrate broad trends.
- More rain and run-off may not translate to more water supply as infrastructure will need to be built for increased drainage and greater water storage to avoid floods. Moreover, increases in run-off are often seasonal and happen within a short time period which could lead to floods. This uneven temporal distribution of precipitation has led to some serious floods during the rainy seasons, especially in mountain regions (ICIMOD 2014).
- Changes in monsoon patterns can also be worrying. Not only could food production be impacted but for countries like India, monsoons have been used to predict economic trends—poor monsoon in 2014 and 2015 led to a 2% drop in agricultural output growth (World Bank 2015b).
- The RCP4.5 scenario above assumes that global temperatures will likely not rise more than 2°C from the pre-industrial period by 2100. Unfortunately, the World Meteorological Organisation (WMO) announced in January 2020 that global temperatures are now at least 1.1°C above the pre-industrial period (WMO 2020a); it also projected that there is a 20% chance that one of the next five years (2020–2024) will be at least 1.5°C warmer (WMO 2020b). At this rate, the IPCC has warned that 1.5°C of warming can be reached as early as 2030 (IPCC 2018). Reaching the intended Paris Agreement target of 1.5°C seventy years ahead of the original target of 2100 will no doubt bring forward the impacts of increased seasonal variability.
- COVID-19 has brought some relief to carbon emissions in 2020, but experts warn that a decade of coronavirus is required to steer the world back on track for 1.5°C (Le Quéré et al. 2020). Carbon neutrality pledges so far (including China) indicate a median temperature rise of around 2.5°C by 2100 (Climate Action Tracker 2020). The USA rejoining the Paris Agreement and achieving carbon neutrality by 2050 will reduce warming by another 0.1°C. (Climate Action Tracker 2021). However, saying is not doing, and there is broad consensus in the financial sector that we are heading on a climate path of 3–4°C (CWR 2020a).

Taken together, the triple threat above throws out monumental and daunting challenges ahead. The fact that eight out of the ten rivers discussed are transboundary lend further complexity. The current arrangements of people, resources and economies shaped by previous conditions will likely have to change; given what is at stake, we have to adapt. No country will be immune from such risks.

Asia must chart a path of "development unusual" as decisions made across the continent and the world, be they in economic, industrial, agricultural and power expansion will not only put pressure on already limited water resources, and they could accelerate climate change which in turn exacerbates scarcity. Rampant water pollution from decades of rapid development only further intensifies the problem. Going forward, water and climate challenges will therefore need to be managed comprehensively alongside development to ensure socio-economic and water security; waternomic strategies can provide a way forward.

14.2 Cross-Cutting Waternomic Strategies for "Development Unusual"

A new paradigm of "development and business unusual" in Asia must be created if prosperity is to be ensured. Multiple actions will have to be taken, some of which are fundamental: a change in mindset, governance, business practices as well as consumption habits. Because water is a cross-cutting issue, a waternomic roadmap for development will also cut across multiple sectors. Summarised below are eight strategies recommended in CWR's report *"No Water No Growth—Does Asia have enough water to develop"* to rethink economic and development models through a waternomic lens (CWR 2018). These actions should be concurrent, cohesive and urgent as follows:

1. **Protect Asia's rivers to ensure water resources for one in two Asians**: by ensuring proper and efficient use of water from Asia's rivers and limiting as well as preventing water pollution. Investment in infrastructure to protect against extreme weather events must be made to avoid "water refugees" and migration due to the lack of water. Improved water quantity and quality could help increase the ecological carrying capacity of the river. These need to be carried out concurrently with improving access to clean water, health and sanitation in rural areas, and industrial expansion plans must also be considered alongside the health of the river.

2. **Rethink economic and development models through a waternomic lens**: This includes a pantheon of actions; at the core of these is de-emphasising economic growth and prioritising the environment. Water use quotas plus water use and discharge rights trading could be used. These could be set nationally, regionally or by sector to tighten water use and encourage tech upgrades of irrigation and industrial equipment, as well as to rein in pollution. Also, optimising GDP, industry and crop mix as well as managing virtual water trade, could also help save water and reduce pollution. Policies could be set to favour more GDP and less polluting and water-intensive industries in the future, while encouraging dirty thirsty industries to go circular. A combination of such strategies is already in practice in China; these are discussed in more detail in "Waternomic lessons from China".

3. **Control agricultural water use and pollution while ensuring food security**: As the largest employer, water user and generally the largest polluter in Asia, agriculture practices need to be rehauled. Strategies in agricultural water savings and fertiliser/chemical use must therefore play an integral part in resolving water management issues. However, increasing irrigated land area may offset water savings from improving irrigation efficiency so crop mix and yield optimisation are also necessary by region. Balancing employment rates, controlling agricultural water use and ensuring food security are a monumental challenge.

4. **Choose the right type of power in the water–energy–climate nexus**: Water is used to generate power, but power is also used to clean and supply water. Moreover, the type of power used today could accelerate climate change, and water solutions like desalination may also be limited given the amount of power it currently requires. As Asia is still power hungry with per capita power generation at 7 MWh/pax for India, a quarter that of China's 26 MWh/pax compared to 40 MWh/pax for Japan and 81 MWh/pax for the USA (2015), smart energy choices (including energy mix and cooling technologies) will have to be made for both climate and water today. With millions of people still lacking access to electricity and abundant hydropower resources, dams are likely here to stay, so stronger hydro/transboundary governance will be required to assuage geopolitical tensions.

5. **Resolve transboundary issues through better waternomic cooperation**: Better cooperation in managing shared rivers can benefit regional relationships, hydropower generation and climate risks (floods and droughts). Although transboundary agreements between the 16 countries lag those in Europe, water agreements do exist between them and provide a base from which to build on. India and China as upper riparians of the ten major rivers must lead both transboundary and regional economic cooperation; innovative mechanisms and cooperation platforms are needed. A way forward could be the Lancang–Mekong Cooperation Mechanism, led by China which moves away from traditional transboundary water management to include regional economic and environmental cooperation. This "waternomic" approach could help countries achieve a more holistic approach to shared development.

6. **Reorganise to focus on basin-level development planning and innovations**: Across the 16 countries discussed above, water stress is more acute at the basin level than at a national level, putting not only people but assets at risk. For example, water use to generate a dollar of GDP can be 0.3 m^3 for India nationally but 0.08 m^3 for both the Ganges and the Brahmaputra. Governments, financial institutions and corporates should therefore assess and quantify exposure to water and climate risks at a basin level. Moreover, given the magnitude of the socio-economic exposure discussed above, it makes sense to focus development and economic planning (including executing recommendations 1–5 above) by river basins across Asia. In China, there are multiple river basins/watersheds focused on regional holistic development pilots including the

Yangtze River Economic Belt (YREB) (along the Yangtze River), the Capital Two Zones Plan (the Zhangjiakou watershed) and the Greater Bay Area (GBA) (around the Pearl River Delta)—these are expanded in the next section "Waternomic lessons from China".

7. **Collaborate to close multidisciplinary research gaps and standardise data across basins**: To assess and mitigate clustered financial exposure to water and climate risks across basins require multidisciplinary research. This requires scientists, policy-makers, businesses, engineers and financiers to collaborate, but these specialised disciplines tend to operate "in their own box". A banker or a business owner is unlikely to crawl over research papers to search for natural risks that may impact their assets; but neither are scientists or engineers expected to know what would be considered a business risk and how water and climate risks could be factored into government/corporate strategy or credit policy. Also, data gaps need to be closed: there is limited comparable data across countries and rivers and worse still, river basin boundaries and regional borders are also not well defined or standardised, making comparative analysis difficult. Such gaps in data/analysis must be identified so that they can be plugged with well-funded multidisciplinary research.

8. **Close the funding gap and drive financial reform**: Despite the gravity of the triple threat, there is still no estimate available globally for the 16 countries regarding the amount of money required to ensure global/regional future water security in a changing climate. Assessing and closing the funding gap are a minimum. As competition for water and extreme weather events become more frequent and intensive, the financial industry will have to adapt. Environmental regulations, waternomic policies and climate change will impact investment portfolios and loan books. Rising systemic exposure at the basin-level means that banks will have to rethink credit policy to factor in environmental risks from a basin perspective. An evolution in financial risk assessment to include chronic and acute water and climate risks is inevitable and has started—such actions by the financial sector to build resilience to systemic shocks are discussed in more detail later in "Magnitude to threats pose systemic risks to countries and global financial systems".

Asian leaders, businesses, financiers, entrepreneurs and scientists have an important role to play across all these eight areas. Multiple assets and significant parts of supply chains lie along the ten rivers and are at risk. The stakes are high, but so are opportunities; private investment to mitigate and adapt business risks and good water stewardship can help alleviate basin-level risks and supplement government action. Business leaders, entrepreneurs and bank CEOs can steer Asia into a new way of doing old things, and governments can set policies to lead the way.

China has taken actions across all eight strategies discussed above, and some strategies adopted may prove useful for the region.

14.3   Waternomic Lessons from China

As part of its push to build a "Beautiful China" where "the sky is blue, the land is green and the water runs clear", multi-pronged strategies have been adopted. The spend on water infrastructure and conservancy of CNY3.58 trillion in its 13th Five Year Plan 2016–2020 (13FYP) period was already 57% more than the last five-year plan period (MWR 2021). In addition, waternomic strategies are afoot; this section sets out key aspects of national and regional strategies to illustrate the concept of "waternomic" development to ensure water and economic security:

- **Top-down vision and buy-in**: A clear vision and top-down buy-in from China's central government reinforces the importance of mindset shifts required to deliver development in tandem with environmental protection:

  "We will make China a beautiful country with blue sky, green vegetation and clear rivers, so that the people will enjoy life in a liveable environment and the ecological benefits created by economic development."

  Source Keynote Speech by H.E. Xi Jinping, President of the People's Republic of China, Opening Ceremony of the 2016 B20 Summit.

China is prioritising its ecological protection and green development through a waternomic lens by striking a balance between water use and allocation, water pollution prevention and economic development; usage of innovative technologies, policies and finance; and coordinated decision-making on water, energy, food and climate change (FECO and CWR 2016; Yang et al. 2016; NDRC 2016a; MIIT et al. 2017; MEE et al. 2017). Governance frameworks have also been improved: including updating the environmental law, reforming government ministries to deliver holistic management of natural resources and ecological environment from mountaintops-to-the-sea for air, water and soil and stepping up the supervision and enforcement of the natural environment (The State Council of the People's Republic of China 2018).

- **National and provincial waternomic targets to transition away from business as usual**: Projected water demand (818 billion m^3) was expected to exceed projected supply (501 billion m^3) if business as usual was to continue by 2030 (The 2030 Water Resources Group 2009). So, during the 12th Five Year Plan 2011–2015 (12FYP), China charted a course of "development unusual" with new strategic emerging industries and circular economies to drive growth plus imposed "stringent water management" through "3 Red Lines" to (1) control total water use, (2) improve water use efficiencies and (3) prevent and control pollution; see Fig. 14.4.

To rein in water use, national water use caps as well as water efficiency targets tied to output values (waternomic targets) were set (The State Council of the People's Republic of China 2012, 2015a; Qin et al. 2015). These targets have economic implications: indicating that unless China beats its own water targets,

12FYP Strategic Emerging Industries (2011)	Circular Economy Development Strategies & Action Plan (2013)	Three Red Lines
1. Energy Saving & Environmental Protection 2. New Energy 3. Bio-technology 4. New Materials 5. Next Generation IT 6. Clean Energy Vehicles 7. High-end Manufacturing	1. Coal 2. Power 3. Steel 4. Non-ferrous Metals 5. Petroleum & Petrochem 6. Chemical 7. Building Materials 8. Paper 9. Food 10. Textile	1. Control total water use 2. Improve water use efficiencies (which included water use per unit of GDP and water discharged per unit of GDP generated) 3. Prevent & control water pollution

Fig. 14.4 Industries to deliver development in tandem with water management and protection. *Source* Various plans/policies of the People's Republic of China: China State Council 12th Five Year Plan 2011–2015, Circular Economy Development Strategies and Action Plan 2013, Most Stringent Water Management Systems Methods 2013, Water Prevention and Control Pollution Plan 2015. *Note* This table is republished in this chapter with permission from China Water Risk; © China Water Risk 2021, all rights reserved

GDP growth will be no more than 7.6% by 2020 and 5.7% by 2030 (HSBC 2015). At that time, giving up growth for the environment appeared like an outlandish claim, as were the seven emerging industries in the 12FYP; but today, China is clearly committed to this new path.

Industries identified in Fig. 14.4 indicate that China is cognizant of its liquidity constraints as well as climate change implications: energy and water-intensive industries that are also polluting with low GDP contributions are included in the circular economy lists, whereas higher GDP industries with lesser climate and water impacts are in the emerging industry list. Holistic thinking in the water–energy–climate nexus is also evident in the 13FYP (2016–2020) which stated that the renewable expansion envisaged would save the nation 3.8 billion m^3 of water (NDRC 2016b).

The national targets are allocated by province and by sector to push provinces to implement policies beyond traditional water efficiency gains (irrigation improvements and industrial equipment upgrades) to strategies that include rethinking crop, energy and industry mixes through a water lens. As provinces were at different stages of development with different sectors contributing to growth, they each faced unique water stress and pollution challenges and unique targets and strategies per province were thus set; see Fig. 14.5 for examples (CWR 2019a; Yang et al. 2019).

Moreover, the upstream/downstream location of the province within the watershed was also considered so that economic planning can be based on the ecological boundary rather than the administrative provincial boundary. Having recognised that there is no one-size-fits-all solution, Beijing has tailored waternomic policies to ensure water and economic security by regional watersheds—this waternomics approach is currently piloted in various regions (MEP et al. 2017; NDRC 2019a, b). Below are examples of three distinct regions to illustrate such actions: (1) the waternomic development of an entire river basin from mountains-to-ocean (Yangtze River) (MEP et al. 2017); (2) waternomic management of an upper watershed (Zhangjiakou) (NDRC 2019a) and (3) waternomics at a key delta region (GBA) (NDRC 2019b). These regions and watersheds are important to China as significant population and economies are located there, including the major cities of Shanghai, Chongqing, Beijing, Shenzhen and Guangzhou.

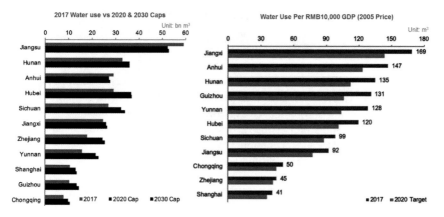

Fig. 14.5 Examples of provincial waternomic targets along the Yangtze River. *Source* CWR based on National Bureau of Statistics of China, YREB Ecological Environment Protection Plan, 13FYP provincial targets by National Development and Reform Commission and Ministry of Water Resources of the People's Republic of China. *Notes* These charts are extracted from the report "Yangtze water risks, hotspots and growth—Avoiding regulatory shocks from the march to a Beautiful China" (CWR 2019a) for use in this chapter with permission from China Water Risk; © China Water Risk 2021, all rights reserved

- **Yangtze River Economic Belt (YREB) waternomic targets and policy innovations to protect 42% of China's population and 45% of China's GDP.** The YREB is a key pilot region designated by President Xi for holistic ecological protection and green development (MEP et al. 2017). As China's industrial heartland (with sizeable shares of national production across polluting sectors (e.g. cloth (57%); cement (51%); auto (46%), crude steel (33%), chemical pesticides (77%), chemical fibre (78%)), the YREB faces key challenges including the lack of environmental infrastructure, dense heavy chemical industries with great environmental risk, rural non-point source pollution and impaired ecological systems (China Government News 2019; CWR 2019a). *"We must not allow the ecological environment of the Yangtze River to continue deteriorating in the hands of our generation, and we must leave our descendants a clean and beautiful Yangtze River"* (Xinhua News 2018).

The YREB is not to be confused with the Yangtze River Basin (YRB).[1] The YREB comprises 11 provinces and municipalities along the Yangtze River: Yunnan, Guizhou, Sichuan, Chongqing, Hunan, Hubei, Jiangxi, Anhui, Zhejiang, Jiangsu and Shanghai—it is home to 595 million people or 43% of China's population. The YREB also generated CNY37.4 trillion (USD5.3 trillion) in 2017 or

---

[1] Note that the YREB is not the Yangtze River Basin (YRB). In terms of economic policy, the focus is on the YREB, whereas water management is focused at the ecological boundary of the YRB. Although China is moving to a holistic waternomic approach with the YREB, the different boundaries still present a challenge to the management of waternomics at both central and provincial government levels. Ideally, an umbrella body is needed to coordinate and manage the waternomics future of the entire YREB (CWR 2019a).

45% of China's GDP; this means that if the YREB was treated as a country, it would be the third-largest economy in the world (CWR 2019a). The YREB is also important for national food and energy security, producing almost two-thirds of China's rice and over three-quarters of the nation's hydropower generation, thus balancing trade-offs between water and economic growth, industrial pollution plus food and energy security is thus a monumental challenge for the government and failure is clearly not an option; see Fig. 14.6 (CWR 2019a).

The economic and pollution disparities along the river lend further complexity—the Yangtze River Delta is much more developed than the upper reaches of the river and a broad strategy of "protect, upgrade and advance" was proposed for the three regions: Protect the Upper Reaches, Upgrade Middle Reaches and the Advance Yangtze River Delta (YRD) (Yang et al. 2016). Not only were provincial

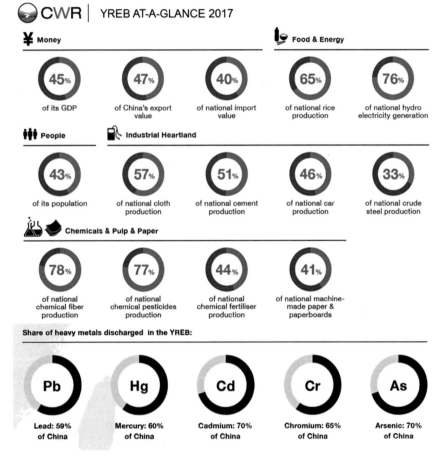

Fig. 14.6 YREB at a glance 2017. *Source* CWR based on "Yangtze water risks, hotspots and growth—Avoiding regulatory shocks from the march to a Beautiful China" (CWR 2019a). *Note* This infographic is extracted from the above-mentioned report for use in this chapter with permission from China Water Risk; infographic © China Water Risk 2021, all rights reserved

waternomic targets set to manage these, but the 195 industrial parks and development zones in the YREB will also be reorganised into five world-class manufacturing clusters, YREB specific "favoured or not" lists have also been created with polluting industries feeling more pressure (CWR 2019a)—see table below in Fig. 14.7 (note that these lists expand on key industries in Fig. 14.4).

Provinces not meeting targets will likely face more scrutiny and laggard provinces, and industries based there will likely face aggressive action; and while this benefits China in the long run, it could disrupt global supply chains as over half of the world's medium and heavy rare earths and chemical fibres are produced in the YREB, as are around two-thirds of global antimony and tungsten (CWR 2019a). Already, hundreds of chemical factories located within 1 km from the river have been or will be either moved or shut down—in Jiangsu province alone, 195 factories and 37 chemical industrial parks were affected and in Hubei province, another 105 factories were impacted (Xinhua News 2019). Ecological protection zones and city clusters and other policy innovations have also been enacted (see box below) plus tighter regulations have ensued with the "Yangtze River Protection Law" passed in 2019 (MOF and NDRC 2015; The State Council of the People's Republic of China 2015b; CPCCC and The State Council of the People's Republic of China 2017a, b; MOF 2018).

The Yangtze River is simply too big to fail and so in addition to the above, trillions of yuan have been set aside to clean up the river. Specifically, in the YREB:

CWR | FAVOURED OR NOT? DIFFERENT LISTS FOR MULTIPLE WATERNOMIC ACTIONS

8 strictly controlled industries along the river	8 highly water-intensive industries to equip water-saving measures	10 key industries with special pollution control actions	6 water intensive industries that will face higher tariff	5 world-class manufacturing clusters
"Guiding Opinions on Strengthening YREB Industrial Green Development" (2017)	*"YREB Ecological & Environmental Protection Plan" (2017)*	*"Action Plan for the War to Protect & Rehabilitate the Yangtze River'" (2018)*		*"Guiding Opinions on Strengthening YREB Industrial Green Development" (2017)*
• Petroleum processing • Chemical raw materials • Chemical manufacturing • Pharmaceutical manufacturing • Chemical fibre manufacturing • Non-ferrous metals • Printing & dyeing • Papermaking	• Power • Steel • Papermaking • Petroleum chemicals • Chemicals • Printing & dyeing • Chemical fibre manufacturing • Food fermentation	• Papermaking • Coking • Nitrogen fertilizer • non-ferrous metals • Printing & dyeing • Agricultural & sideline food processing • Raw material medicine manufacturing, • Leather making, pesticides • Electroplating	• Thermal power • Steel • Textiles • Paper making • Chemicals • Food fermentation	• Electronic information • High-end equipment • Automobiles • Home appliances • Textiles & apparels

Fig. 14.7 Favoured or not? Different lists for multiple waternomic actions in the YREB. *Source* CWR based on "Yangtze water risks, hotspots and growth—Avoiding regulatory shocks from the march to a Beautiful China" (CWR 2019a). *Note* This infographic is extracted from the above-mentioned report for use in this chapter with permission from China Water Risk; infographic © China Water Risk 2021, all rights reserved

CNY252 billion was spent on ecological and environmental protection (2016–2017); CNY380 billion invested in the treatment of environmental pollution (2016); CNY5 billion allocated from the central budget for YREB eco-compensation (2018); and CNY15 billion set aside to incentivise and promote YREB ecological protection and recovery (2018–2020) (CWR 2019a). These together with private, provincial and multilateral funding in the YREB to foster green development, meant that at least CNY2.1 trillion of green investment is deployed in the region; for perspective, this is 1.35x the total defence expenditure of the EU in 2016 (CWR 2019a).

Box 1 Pilot policies and initiatives to green the YREB

These have been summarised from CWR's report "Yangtze water risks, hotstpots and growth—Avoiding regulatory shocks from the march to a Beautiful China" (CWR 2019a)

- **Ecological protection redlines (Eco-Redline)** are drawn to protect watersheds. They limit industrial development in important designated ecological zones. The YREB's total Eco-Redline area is larger than the land area of Thailand. National parks are also being created to protect source regions of key rivers. The core protection zone in the Yangtze River source region is the size of the Czech Republic.
- **Sponge cities** are pilots to increase flood resilience. Given that more rain is expected by the mid-century in the Yangtze River Basin, 12 of the 30 national sponge city pilots are in the YREB.
- **River chiefs** are government officials assigned to take charge of rivers and lakes in their jurisdiction. The concept was first developed in the Yangtze River Delta where deteriorating surface water quality triggered the appointment of the first river chief in 2003. The concept was expanded by the central government in 2016; there are now over 1.2 million river chiefs across China, mainly at village levels to help monitor tributaries.
- **Water use and wastewater discharge permits** are used to allocate water and manage pollution across sectors. China has been piloting trading markets with regard to these permits which allow companies to sell unused quota of the permits, thereby incentivising water savings and pollution control. In Zhejiang, the market value of wastewater discharge permits traded was as high as CNY2.5 billion (2009–2014).
- **Eco-compensation**: to manage upstream/downstream pollution challenges. There are YREB inter-provincial eco-compensation schemes (each CNY100–200 million) in place. Further top-down support was received in 2018 through a YREB specific eco-compensation plan from the central government with CNY5 billion allocated from the central budget for YREB eco-compensation.

- **Capital Two Zones Plan to protect Zhangjiakou, Beijing's upper watershed**. Beijing faces extremely high water stress and Zhangjiakou, the upstream region and the water resource conservation area of Beijing, also faces serious water problems with more than half of its area also extremely highly water stressed. Besides, Zhangjiakou also has other outstanding issues such as water quality, groundwater over-extraction, grassland degeneration and poverty (CWR 2019b). Yet, Zhangjiakou has a long way to develop; see Fig. 14.8.

Both Beijing and Zhangjiakou are within the same ecological boundary: both relying on the Haihe River Basin—a highly stressed and polluted basin (WRI 2013; MEE 2019). Although the south–north water diversion project and the use of reclaimed water have alleviated Beijing's heavy reliance on Zhangjiakou's water supply, Zhangjiakou still plays an important role as Beijing upper watershed and to ensure that development in the Zhangjiakou (which is to host the 2022 winter Olympics) does not impair the capital's water resources, the "Construction Plan for the Capital Water Source Conservation Functional Zone and Ecological Environmental Supporting Zone (the Capital Two Zones) in Zhangjiakou" was released in 2019 (NDRC 2019a). The plan includes various waternomic actions:

- Agricultural water: restrict farming of water-intensive crops, restore farmland to grassland and limit agricultural water use to 600 million m^3 by 2022. Also, irrigated areas are to be reduced by over 26,000 hectares by 2022, with a further reduction of over 52,000 hectares by 2029.

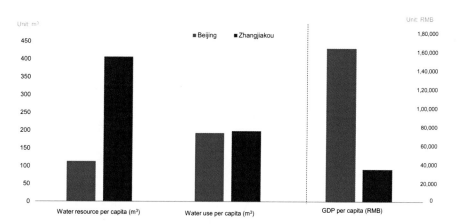

Fig. 14.8 Downstream/upstream water and GDP challenges: Beijing and Zhangjiakou. *Source* CWR based on Beijing Water Authority (2019), Statistical Communiqué of Zhangjiakou on the 2019 Economic and Social Development, Construction Plan for the Capital Water Source Conservation Functional Zone and Ecological Environmental Supporting Zone (the Capital Two Zones) in Zhangjiakou. *Note* This chart is republished in this chapter with permission from China Water Risk; © China Water Risk 2021, all rights reserved

- Industrial water: restrict water-intensive industries, improve wastewater treatment and water reuse, and limit industrial water use to 80 million m^3 by 2022.
- Domestic water: retrofit water supply networks, rein in leakage rates to less than 10%, limit domestic water use to 160 million m^3 (2022) and 200 million m^3 (2035), and groundwater: cap usage at 580 million m^3 by 2022; no further increases thereafter (NDRC 2019a).

The Capital Two Zones Plan also recommended the shutdown of 80% of mines by 2020 with the remaining slated to go green or face upgrades; green industries in Zhangjiakou will also be developed—these include renewable energy (solar and wind), cloud computing and big data (NDRC 2019a). Such actions re-affirm China's commitment to shift beyond traditional water management towards a waternomics approach where economic planning is based on the ecological carrying capacity as well as the boundary of the watershed.

- **Greater Bay Area—ensuring water for growth and dealing with coastal threats in the delta.** China's Greater Bay Area (GBA) brings together Hong Kong, Macao and nine cities in Guangdong to form an "integrated economic and business hub" (NDRC 2019b). It is home to financial services powerhouses like Hong Kong, Shenzhen and Guangzhou; strong manufacturing bases like Foshan and Dongguan as well as the region's main entertainment hub Macao. It is a key growth region in China's 13FYP, accounting for 12% of China's GDP in 2018 (NPC 2016; NBSC 2018).

In 2018, the GBA generated USD 1.6 trillion of GDP and housed a population of around 70 million, and if growth targets are met, the region's GDP is expected to be around USD4.6 trillion by 2030, 2.9x that in 2018 (CLSA 2019). Moreover, the region expects to see an influx of 18 million more people by 2030; this is the equivalent to the population of two more New York cities, thus putting more pressure on already stressed water resources (CLSA 2019). Eight of the 11 GBA cities are as dry as the Middle East—their per capita water resources fall well below the World Bank's Water Poverty Mark, yet they account for 92% of the GBA's 2018 GDP; see Fig. 14.9. The four core GBA cities of Guangzhou, Shenzhen, Hong Kong and Macao are part of this "dry" group, driving 68% of the region's GDP (CLSA 2019).

As climate change only increases the risk of "Day Zero" no water scenarios, planning and execution of strategies to counter and adapt to climate change and potable-water scarcity are difficult enough for one government, let alone three in the GBA for the Guangdong province, Hong Kong and Macao Special Administrative Regions (SARs). While the region has embarked on various strategies, separate governments have meant that actions are not yet cohesive, comprehensive or efficient (CLSA 2019). Disparities can be significant

- Reducing water use: Guangdong managed to save 4.3 billion m^3 of water between 2011 and 2018, equivalent to four times Hong Kong's total water use. But over the same period, the SAR has increased its usage by almost 10%; see

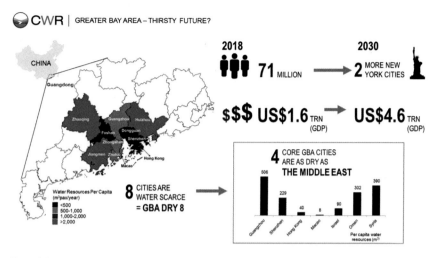

Fig. 14.9 Greater Bay Area—Thirsty future? *Source* CWR based on NBSC (2018), Water Resources Department of Guangdong Province (2018), HKSAR Water Supplies Department (2017/2018), Macao Water (2018), FAO AQUASTAT (2017), HSBC (2019), Sina Finance (2017), HKTDC (2018). *Notes* For Hong Kong and Macao, water resources do not include water imported from mainland China so that it is comparable to other cities. Although Zhuhai's water/pax is above the water poverty mark in 2018, its multi-year average is 985 m^3 and so we have included it in the GBA dry 8. This infographic is republished in this chapter with permission from China Water Risk; infographic © China Water Risk 2021, all rights reserved

Fig. 14.10. Note that Guangdong is the only province in China that faces continuously lower water caps for 2015, 2020 and 2030 (The State Council of the People's Republic of China 2012).

- Improving leakage rates: Given water scarcity levels, leakages should be kept to a minimum—the GBA cities in Guangdong province reduced water leakage rates to 11%, Macao is at 9%, whereas HK is at 25%; since 70–80% of Hong Kong's water is supplied from the water-stressed Dongjiang River, such levels of water losses are extremely wasteful (CLSA 2019).
- Diverting water: The GBA is primarily reliant on surface water sources (96% in 2018), provided by a network of rivers—mainly the Dongjiang, Xijiang and Beijiang Rivers. Of these three, the Dongjiang is the most water stressed with its water resource utilisation rate very close to its exploitation limits (HKSAR Water Supplies Department n.d.). Thus in 2019, Guangdong launched a US $5.14 billion Xijiang water diversion project to alleviate water stress in Dongjiang—the project will provide Guangzhou, Shenzhen and Dongguan with 1.7 billion m^3 annually and will also act as a backup supply to Hong Kong (Water Resources Department of Guangdong Province n.d.). The cost of this diversion project is 45× that of Hong Kong's planned US$1.16 billion desalination plant, but it delivers 35x the water (CLSA 2019).

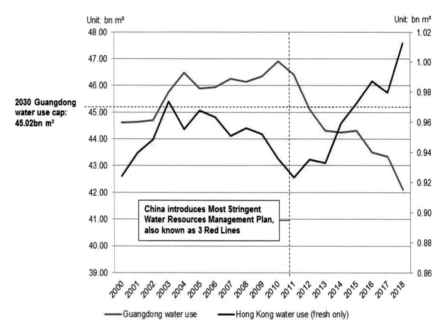

Fig. 14.10 Guangdong lowering water use but Hong Kong keeps using more. *Source* CWR based on HKSAR Water Supplies Department (2000/2018), Water Resources Department of Guangdong Province (2000–2018). *Notes* Hong Kong also uses seawater for flushing purposes, this is not included in the above chart. This chart is republished in this chapter with permission from China Water Risk; © China Water Risk 2021, all rights reserved

- Dealing with floods: The GBA is prone to flooding and already floods in Guangdong have cost CNY200 billion in total damages from 2008 to 2017, and although deaths have declined, over 20 million people can be affected (SFCDRH and MWR 2017, 2018). Efforts to step up flood resilience included CNY62.2 billion investment in disaster mitigation and prevention (Water Resources Department of Guangdong Province 2017).

Besides freshwater challenges, there are also coastal threats to contend with in the GBA, and already, freshwater intake pipes are affected by saltwater intrusion. For example, in Zhongshan, Zhuhai and Macao, active saltwater intrusion from October 2016 to February 2017 led local governments to transfer 133 million m^3 of water from upstream regions to ensure safe water supply (The State Council of the People's Republic of China 2017). As rivers retreat due to over-extraction and the sea encroaches, freshwater intake points may have to be moved further upstream; such upgrades in infrastructure could be extremely costly due to compact urban structures.

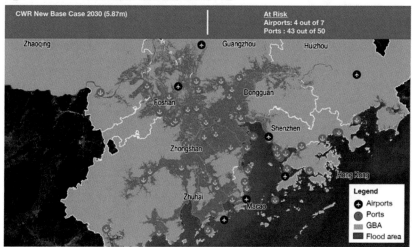

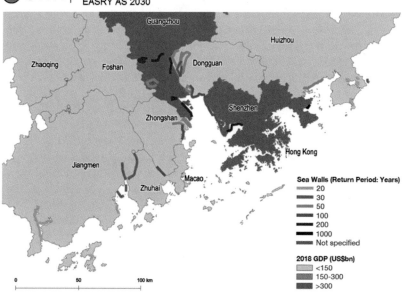

◄Fig. 14.11 Extreme storm tide risk and adaptation efforts in the GBA. *Source* (Top) CWR based on a roughly 30 m horizontal resolution grid derived from NASA's Shuttle Radar Topography Mission (SRTM), Digital Terrain Model (5 m) from the Lands Department of Hong Kong, various GBA government port authority websites, Civil Aviation Administration of China website, Google Maps. (Bottom) CWR. GDP based on Hong Kong Trade Development Council website. Sea walls in the PRD digitised based on "Guangdong Province Sea Walls Plan" map from Guangdong Hydropower Planning and Design Institute website. Sea walls in Shenzhen based on approximate location in Shenzhen Water Bureau report "Shenzhen Flood Prevention and River Remediation Plan 2014–2020". *Notes* The seawalls on the map may not reflect the true length and are only for illustrative purposes. These infographics are extracted from the report "Sovereigns at Risk: APAC Capital Threats—Re-ratings warranted as city capitals and GDP are materially exposed to coastal threats" (CWR 2020b) for use in this chapter with permission from China Water Risk; infographics © China Water Risk 2021, all rights reserved

Worse still, as the GBA is typhoon prone, rising risks from typhoon storm surges as well as sea level rise (SLR) could threaten key sectors that drive the region's GDP of logistics and trade, real estate, finance and entertainment as early as 2030. Unless adaptation actions are taken to protect the cities from coastal threats, extreme storm tides of 5.87 m from a super typhoon will disrupt four out of seven of the GBA's airports; 43 of its 50 ports and half of Macao's casinos by 2030 (CLSA 2019; CWR 2020b); see Fig. 14.11.

While Shenzhen and Guangzhou are implementing adaptation actions to protect their cities, Hong Kong and Macao have been more laissez-faire, leading them to rank differently in the CWR APACCT 20 Index despite being in the same watershed (CWR 2020a). The index designed with input from the financial sector benchmarks vulnerability to coastal threats across 20 key capitals and cities in APAC and takes into consideration physical threats (SLR, storm surge and subsidence) as well as government adaptation efforts to alleviate them (CWR 2020a). As can be seen from the index rankings below in Fig. 14.12, Shenzhen and Guangzhou's rankings improve materially, while Hong Kong and Macao's threat levels remain high at both 1.5°C and 4°C climate scenarios (CWR 2020a).

Note that adaptation efforts are not just limited to Shenzhen and Guangzhou but other mainland China cities of Shanghai and Suzhou (Yangtze River Delta) and Tianjin (Yellow River Delta) have also materially improved their rankings (CWR 2020a).

Water and climate change evidently play key roles in shaping a Beautiful China and is considered in economic planning at national, provincial and basin/watershed levels with clear support from the top. Careful waternomic planning and cohesive resilience strategies can help countries navigate a "thirsty and underwater" future. While the above waternomic policies are tailored for China, they can be used to inspire action across Asia; some successful policies and pilots can be recalibrated for other major river basins across the continent. Given Asia's pressing water and climate challenges, waternomic lessons from China can prove useful.

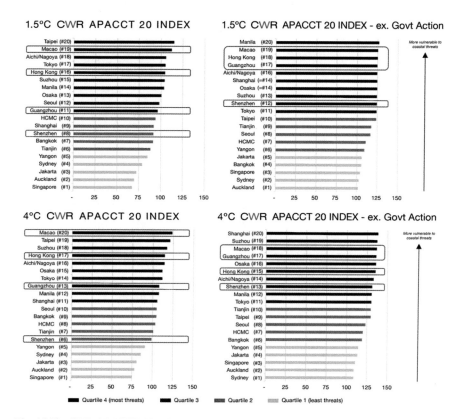

Fig. 14.12 CWR APACCT 20 Index—Rankings change due to adaptation actions. *Source* CWR based on "Avoiding Atlantis: CWR APACCT 20 Index—Benchmarking coastal threats for 20 APAC cities with finance sector input" (CWR 2020a). *Notes* These charts are republished in this chapter with permission from China Water Risk; © China Water Risk 2021, all rights reserved

14.4 Magnitude to Threats Pose Systemic Risks to Countries and Global Financial Systems

The significant clustering of people and economy in river basins and at river deltas point to waternomic threats if water and climate risks vis-à-vis development are mismanaged. To recap, the triple threat on Asia's major river basins discussed above could strand 280+ cities, putting around 1.9 billion people and US$4.3 trillion of annual GDP at risk (CWR 2018). As for 20 key APAC coastal capitals and cities, rising seas will redraw coastlines stranding assets worth trillions of dollars—these cities generate US$5.7trillion of GDP annually (CWR 2020b). For perspective, this is equivalent to the GDP of both France and the UK which totalled US$5.6 trillion in 2018 (World Bank 2018). Thus together, water and climate risks have the ability to strand trillions of dollars' worth of assets across Asia; see Fig. 14.13.

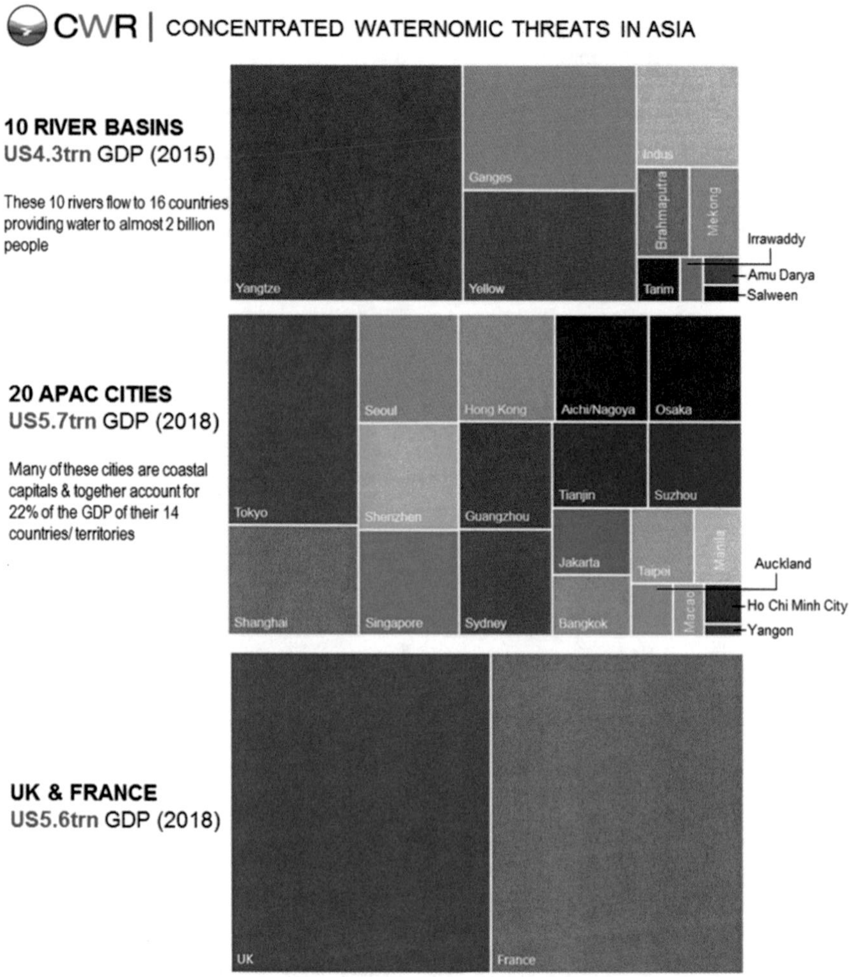

CWR | CONCENTRATED WATERNOMIC THREATS IN ASIA

10 RIVER BASINS
US4.3trn GDP (2015)

These 10 rivers flow to 16 countries providing water to almost 2 billion people

20 APAC CITIES
US5.7trn GDP (2018)

Many of these cities are coastal capitals & together account for 22% of the GDP of their 14 countries/territories

UK & FRANCE
US5.6trn GDP (2018)

Fig. 14.13 Concentrated waternomic threats in Asia and APAC. *Source* CWR based on "No water, no growth—Does Asia have enough water to develop?" (CWR 2018), "Sovereigns at risk: APAC capital threats—Re-ratings warranted as city capitals and GDP are materially exposed to coastal threats" (CWR 2020b), World Bank (2018) GDP data, individual government sources. *Notes* This infographic is republished in this chapter with permission from China Water Risk; infographic © China Water Risk 2021, all rights reserved

It is not just the size of the waternomic threat that should worry governments, central banks and financial institutions but the concentration of risks can also be triggers of financial collapse if they fail to plan for the new risk landscape ahead:

- **Concentration of national GDP along vulnerable river basins and the coastal cities**: In Asia, waternomic threats are concentrated in river basins and coastal cities which generate material shares of the continent's GDP. Just like a significant amount of GDP is generated in the major rivers basins resulting in high financial exposure to water and climate risks, the 20 APAC cities in Fig. 14.13 also account for a sizeable share of their respective country/territory's GDP: 100% in the case of Hong Kong and Singapore; for Japan, the three cities (Tokyo, Aichi/Nagoya and Osaka) account for a third of its GDP; whereas capitals Sydney and Seoul drive around a quarter of their national GDP; and Yangon, Auckland and Manila 38% of their respective GDPs (CWR 2020b).

- **Asia's export-led growth model at risk posing threats to global trade**: Limited water resource may lead countries to rethink export-led growth and use virtual water to manage trade by switching from exporting water-intensive goods to importing them (HSBC 2015; FECO and CWR 2016; Yang et al. 2016; CWR 2018). Moreover, key trade infrastructure such as ports and airports could also be impacted by saltwater threats of storm surges and SLR—globally 80 airports are vulnerable to just 1 m of SLR (Maghsadi and Huang 2020). Unfortunately, on our current climate path, experts warn that SLR can be multi-metre with ice sheet experts projecting a "very likely" range of 2.38–3.29 m by 2100 (Bamber et al. 2019). At 3 m of SLR, it is not just the GBA that is vulnerable, but 20 out of the 23 ports servicing the 20 APAC cities above and 12 of their 25 airports will be permanently submerged (CWR 2020a). It is not just those economies that will suffer, there are global trade implications: just these 23 ports and 25 airports account for 26% of global sea cargo and 23% of global air cargo volumes (CWR 2020b).

- **Expensive urban real estate at risk**. Failure to adapt current water infrastructure to climate change impacts ahead will result in cities running out of water and coastal flooding. Already Manila and Chennai are running short and even London is projecting that it will run out of water by 2030 unless investments are made (CWR, Manulife Asset Management and AIGCC 2019). At a plausible 3 m of SLR by 2100, just for the 20 APAC cities discussed above, urban real estate areas equivalent to 22 Singapores will be underwater and 28 million residents will lose their homes (CWR 2020b). Without adaptation, losses of urban real estate will pose systemic shocks to banking system because when flood risk switches from acute to chronic, insurers will likely stop insuring flood risk, leaving banks to inherit such risks through their mortgage loan books (CWR, Manulife Asset Management and AIGCC 2019; CWR 2020c). Already in New Zealand, insurers have indicated that they will stop insuring homes in vulnerable areas by 2050, with premiums spiking as early as 2030 (Storey et al. 2020). The probability of economic shocks and financial collapse is high when almost 70% of Australia and Japan's total population are clustered by the coast (CWR 2020b).

- **Typhoon prone APAC means extreme storm tides bring forward flooding impacts**. APAC is particularly at risk: five of the top ten countries most affected by "weather-related loss events" from 1998 to 2017 were in Asia (Germanwatch 2019). As the map in Fig. 14.14 shows, Hong Kong has been lucky so far. In 2018, when Super Typhoon Mangkhut, a T10 typhoon, hit the region, it brought storm tides of 3.88 m in Victoria Harbour. Had Mangkhut hit during high tide and taken a slightly different path, storm tides of 5.65 m would have inundated Central, Hong Kong's financial district—the storm tide would have reached past IFC and Exchange Square, all the way to the headquarters of HSBC and Standard Chartered (CLSA 2019). Clearly, this would have been extremely costly and disruptive. Unfortunately, these extreme sea level events to likely occur every year by 2050 in all climate scenarios, especially in tropical regions (IPCC 2019). It is thus imperative that flood drainage systems, sea wall barriers, nature-based solutions as well the exposure of essentials (e.g. underground mass

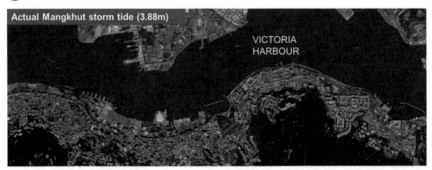

Fig. 14.14 Example of coastal flooding from an extreme weather event in Hong Kong. *Source* CWR based on Digital Terrain Model (5 m) from Lands Department of Hong Kong, Google Maps. Tide gauge location from the HKO website. *Notes* This infographic is republished in this chapter with permission from China Water Risk; infographic © China Water Risk 2021, all rights reserved

transit, cable landing sites, water/wastewater treatment plants, landfills, power plants, food/cold storage) be examined as they will likely have to be rehauled to adapt to the new climate realities. Despite imminent threats ahead, Hong Kong has yet to make public an adaptation plan (CWR 2020b).

14.5 Central Banks and the Financial Sector Are Acting, but Piecemeal Assessment of Water Risks Points to Their Undervaluation

Climate and water risks ahead present grave threats to the people, economy and the financial systems. Global financial systems must thus also evolve and adapt to the new risk landscape. With increasing multi-sectoral and clustered exposure to climate change, global financial regulators through the Network of Central Banks and Supervisors for Greening the Financial System (NGFS) have stated that there is *"a strong risk"* that climate-related financial risks (from extreme weather events) are not fully reflected in asset valuations (NGFS 2019). The NGFS specifically noted that reduced availability of freshwater could amplify extreme weather events causing cascading risks for financial institutions (NGFS 2019).

Now an 80+ member strong coalition of central banks and supervisors, the NGFS has recognised that physical climate risks have micro and macroeconomic impacts that translate into credit, market, underwriting, operational and liquidity risk for all financial institutions. With regard to physical water risks, the NGFS classifies them into two broad categories: (1) Acute—these are event-driven risks and include floods, droughts, tropical cyclones/typhoons, winter storms and hailstorms; (2) Chronic—these are underlying risks and include ecosystems pollution, sea level rise, water scarcity and desertification (NGFS 2020). In addition, water transition risks are also recognised—these are regulatory risks imposed to manage water and include policies on resource conservation, pollution control regulation (NGFS 2020).

But so far, the financial sector has prioritised the assessment of carbon transition risk scenarios, and while some have factored in acute (event-driven) water risks, most have not explored their exposure to chronic water risks (NGFS 2020). Moreover, there is a lack of consensus in valuation methodology across the various types of water and water-related climate risks. That said, there have been piecemeal efforts to build valuation consensus across different types of physical water risks led by CWR, UBS, China's Green Finance Committee, 427, DeNederlandsche Bank, BlackRock, CLSA, S&P Trucost, Moody's, Fitch and McKinsey (CWR 2016; UBS 2016a, b; China's Green Finance Committee 2018; Four Twenty Seven and Geophy 2018; BlackRock 2019; CLSA 2019; CWR, Manulife Asset Management and AIGCC 2019; DNB 2019; Trucost 2019; BlackRock 2020; CWR 2020a, b, c; Fitch Rating 2020; McKinsey Global Institute 2020; Moody's 2020).

Although the water risk conversation has progressed, challenges remain in their holistic assessment. Because acute and chronic risks are interlinked and impact each other, assessing these risks in isolation will likely result in an undervaluation of water risks. Moreover, ignoring chronic risks of water scarcity and sea level rise has resulted in a negative finance feedback loop—capital continues to flow to already water-stressed river basins/vulnerable coastal regions, thereby adding to already concentrated financial risk exposure and accelerating their collapse by adding more strain in the case of water stress; meanwhile, investments also continue in carbon-intensive industries which only serve to amplify and accelerate impacts in vulnerable areas (CWR 2020c).

Traditional means of spreading risks by the financial industry such as sectoral risk spreads will not help as water and climate risks are sector agnostic. Proper holistic pricing of acute as well as chronic water risks will break this negative finance feedback loop. As for regulatory risks brought on by waternomic policies, banks in China such as ICBC have started to assess the impacts of pollution regulations on lending portfolios (ICBC 2016).

With finance waking up to these significant tail risks, valuation adjustments are inevitable. This is not about natural capital accounting but real risk adjustments—the same way leasehold property is afforded a lower valuation compared to freehold assets, SLR and/or no water "Day Zeros" will have a clear downward impact on valuations as the permanent stranding of the region will shorten the life span of all assets located there. Imminent water and climate threats question the going concern status of affected assets and should lead to the re-assessment of discount rates, terminal values, capital adequacy ratios and the cost of capital, which will all affect sovereign and credit ratings as well as equity/project valuations (CWR 2020c).

Such downward adjustments may be significant, and the financial industry may be reluctant to factor in this currently missing chunk of chronic water risks. Unfortunately, the longer we wait, the worse it becomes as climate risks are only rising; when extreme weather events occur annually, acute risks will become chronic risks further widening the existing chronic risk valuation gap and perpetuating the negative finance feedback loop (CWR 2020c). While the financial sector has started assessing acute, chronic and transition risks related to water, there is still a long way to go before financial resiliency to water and climate shocks is achieved. If the sector and/or regulators cannot get on top of these risks, not only will homes be lost but savings will also be at risk as pension funds are location biased; plus, index weightings across Asia favour financials, trade/manufacturing and real estate stocks (CWR, Manulife Asset Management and AIGCC 2019).

Looking forward, basin risks and coastal threats should be holistically assessed from mountains-to-the-oceans; assessing one aspect but not another provides an incomplete picture and increases the likelihood of triggering systemic shocks as comprehensive adaptation solutions are needed. Figure 14.15 attempts to portray the complex web of physical and regulatory water risks as well as their interlinkages; understanding these interlinked risks can facilitate their proper valuation.

Comprehensive physical and transitional water risk scenarios should be prioritised alongside carbon transition scenarios when disclosing climate-related financial

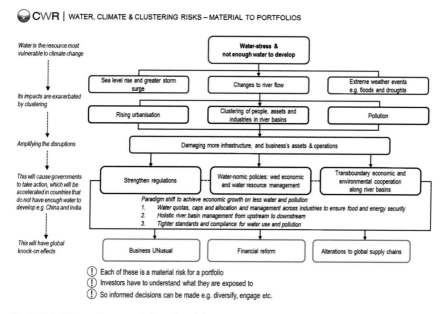

Fig. 14.15 Water, climate and clustering risks, material to portfolios. *Source* CWR (2021). *Notes* This infographic is republished in this chapter with permission from China Water Risk; infographic © China Water Risk 2021, all rights reserved

risk as recommended by The Task Force on Climate Related Financial Disclosures (TCFD), for only then can financial resiliency to water and climate shocks be ensured.

14.6 Waternomics Can Provide a Way Forward to Develop and Avoid Systemic Shocks to the Economy

Water plays a key role in the economy, and given that it is one of the most vulnerable resources to climate change, it should also play a central role in the climate conversation. Assessing waternomic threats as well as developing water-nomic roadmaps can help avoid economic and financial collapse triggered by water and climate risks ahead.

Grave fresh and salt water threats present Asia with a great opportunity to embark on greener development paths to ensure continued water and economic security—fast-tracking decarbonisation, waternomics (achieving more GDP on less water and less pollution) and adaptation to impacts that are already baked-in are a few examples of such paths.

Given the lag in adaptation investments to baked-in impacts, there is an urgent need to properly price in risks (UNEP 2020). Assessing waternomic threats can help build the adaptation business case: given what is at stake, water-related adaptation investments that previously do not make sense when assessed on their own merits may now be bankable. In short, we cannot resolve water security without looking at water's impact on economic and financial security.

Rethinking development through a waternomic lens can also help governments manage economic growth while ensuring water and economic security. Here, China's waternomic roadmap and multiple pilots can provide ideas and ways forward for developing nations. Although such waternomic regulations will bring short-term disruptions and financial risks, they will alleviate clustered financial risks in the long term. The transboundary nature of river basins and water sources also demands deeper regional cooperation to resolve mounting challenges and to protect the only resource we cannot live without.

Clustered and significant financial exposure from water and climate risks are here to stay and rise across Asia. Waternomics can help ensure water for growth as well as facilitate the building of physical and financial resilience to avoid systemic shocks ahead from climate change. Given the current climate path and Asia's vulnerability, there is no time to lose; we must "rethink water" by starting a waternomic conversation in tandem to that of access to water across Asia. We must act now, and it is not just governments but also the private sector—waternomic actions need to be concurrent, cohesive and urgent across multiple sectors (especially the financial sector) to protect our precious water resources and to ensure Asia's long-term prosperity. Failure is not an option as the alternative is unthinkable.

References

ADB (Asian Development Bank) (2016) Asian Water development outlook 2016: strengthening water security in Asia and the Pacific. Asian Development Bank, Philippines

Bajracharya SR, Maharjan SB, Shrestha F, Guo W, Liu S, Immerzeel W, Shrestha B (2015) The glaciers of the Hindu Kush Himalayas: current status and observed changes from the 1980s to 2010. Int J Water Resour Dev 31(2):161–173

Bamber JL, Oppenheimer M, Kopp RE, Aspinall WP, Cooke RM (2019) Ice sheet contributions to future sea-level rise from structured expert judgment. Proc Natl Acad Sci 116(23):11195–11200

BBC (2020) Floods in China: at least 141 people are dead or missing, and the water level of Poyang Lake reaches the highest level in history. https://www.bbc.com/zhongwen/trad/chinese-news-53388985. Accessed 21 Jan 2021

Beijing Water Authority (2019) Beijing water resource bulletin 2019. http://swj.beijing.gov.cn/zwgk/szygb/202009/P020200918627119515926.pdf. Accessed 24 Mar 2021

BlackRock (2019) Getting physical: scenario analysis for assessing climate-related risks. https://www.blackrock.com/ch/individual/en/insights/physical-climate-risks. Accessed 21 Jan 2021

BlackRock (2020) Troubled waters: water stress risks to portfolios. https://www.blackrock.com/us/individual/insights/blackrock-investment-institute/troubled-waters. Accessed 21 Jan 2021

Chen J (2013) Water cycle mechanism in the source region of Yangtze River. J Yangtze River Sci Res Inst 30:1–5 In Chinese

China Environment News (2018) Ecological civilization written into the constitution. http://epaper.cenews.com.cn/html/2018-03/12/content_70526.htm. Accessed 23 January 2021 (In Chinese)

China Government News (2019) Press conference records with Minister of Ecology and Environment Li Ganjie during the 'Two Sessions'. http://www.news.cn/politics/2019lh/zb/zfwzb/20190311d35820/wzsl.htm. Accessed 23 Jan 2021

China's Green Finance Committee (2018) Green finance series—Environmental risk analysis by Financial Institutions. Water Report

Climate Action Tracker (2020) Press release "China going carbon neutral before 2060 would lower warming projections by around 0.2 to 0.3 degrees C". https://climateactiontracker.org/press/china-carbon-neutral-before-2060-would-lower-warming-projections-by-around-2-to-3-tenths-of-a-degree/. Accessed 30 Jan 2021

Climate Action Tracker (2021) Global update: Paris Agreement turning point. https://climateactiontracker.org/press/global-update-paris-agreement-turning-point/. Accessed 30 Jan 2021

CLSA (CLSA Ltd) (2019) Thirsty and underwater: rising risks in the Greater Bay Area. Guest-authored by CWR

CPCCC (Central Committee of the Chinese Communist Party) and The State Council of the People's Republic of China (2017a) Guiding opinions on setting and protecting strict ecological red lines, 7 February 2017. https://www.xinhuanet.com/politics/2017-02/07/c_1120426350.htm. Accessed 23 Jan 2021 (In Chinese)

CPCCC (Central Committee of the Chinese Communist Party) and The State Council of the People's Republic of China (2017b) Opinion on full implementation of the River Chief System, 11 December 2016. http://www.gov.cn/gongbao/content/2017/content_5156731.htm. Accessed 23 Jan 2021 (In Chinese)

CWR (China Water Risk) (2016) Toward water risk valuation—Investor feedback on various methodologies applied to 10 Energy ListCo's

CWR (China Water Risk) (2018) No water, no growth—Does Asia have enough water to develop? (in collaboration with Center for Water Resources Research, Institute of Geographic Sciences and Natural Resources Research, Chinese Academy of Sciences)

CWR (China Water Risk) (2019a) Yangtze water risks, hotspots and growth—Avoiding regulatory shocks from the march to a Beautiful China

CWR (China Water Risk) (2019b) Capital two zones: Protecting Beijing's upper watershed. https://www.chinawaterrisk.org/resources/analysis-reviews/capital-two-zones-protecting-beijings-upper-watershed/. Accessed 30 Jan 2021

CWR (China Water Risk), Manulife Asset Management and AIGCC (Asia Investor Group on Climate Change) (2019) Are Asia's pension funds ready for climate change? Brief on imminent threats to asset owners' portfolios from climate and water risks

CWR (China Water Risk) (2020a) Avoiding Atlantis: CWR APACCT 20 Index—Benchmarking coastal threats for 20 APAC cities with finance sector input

CWR (China Water Risk) (2020b) Sovereigns at risk: APAC capital threats—Re-ratings warranted as city capitals and GDP are materially exposed to coastal threats

CWR (China Water Risk) (2020c) Changing risk landscapes: coastal threats to central banks—everything you need to know about sea level rise, storm surge and financial regulations to recalibrate risks

CWR (China Water Risk) (2021) Big picture: water risk valuation. http://www.chinawaterrisk.org/the-big-picture/water-risk-exposure/. Accessed 30 Jan 2021

DNB (De Nederlandsche Bank) (2019) Values at risk? Sustainability risks and goals in the Dutch financial sector

Döll P, Jiménez-Cisneros B, Oki T, Arnell NW, Benito G, Cogley JG, Jiang T, Kundzewicz ZW, Mwakalila S, Nishijima A (2015) Integrating risks of climate change into water management. Hydrol Sci J 60(1):4–13

FAO AQUASTAT (2010) 2010 statistics from the FAO AQUASTAT database. http://www.fao. org/aquastat/en/databases/. Accessed 24 Mar 2021

FAO AQUASTAT (2017) 2017 statistics the FAO AQUASTAT database. http://www.fao.org/ aquastat/en/databases/. Accessed 24 Mar 2021

FECO (Foreign Economic Cooperation Office) of the Ministry of Environmental Protection of the People's Republic of China (MEP), CWR (China Water Risk) (2016) Water-nomics of the Yangtze river economic belt: strategies and recommendations for green development along the river. Ministry of Environmental Protection of the People's Republic of China

Fitch Rating (2020) Water risks and sovereign ratings. https://www.fitchratings.com/research/ sovereigns/water-risk-relevance-for-sovereign-ratings-to-increase-03-09-2020. Accessed 30 Jan 2021

Four Twenty Seven and GeoPhy (2018) Climate risk, real estate, and the bottom line. http://427mt. com/wp-content/uploads/2018/10/ClimateRiskRealEstateBottomLine_427GeoPhy_Oct2018-4.pdf. Accessed 30 Jan 2021

Gao X, Ye B, Zhang S, Qiao C, Zhang X (2010) Glacier runoff variation and its influence on river runoff during 1961–2006 in the Tarim River Basin, China. Sci China Earth Sci 53(6):880–891

Germanwatch (2019) Global climate risk index 2019. https://germanwatch.org/en/16046

HKSAR Water Supplies Department (2000/2018) Various annual reports from 2000 to 2018. https://www.wsd.gov.hk/en/publications-and-statistics/pr-publications/list-of-publications/ annual-report-water-supplies-department-archive/index.html. Accessed 31 Mar 2021

HKSAR Water Supplies Department (2017/2018) Annual report 2017/18. https://www.wsd.gov. hk/filemanager/common/annual_report/2017_18/en/index.html. Accessed 31 Mar 2021

HKSAR Water Supplies Department (n.d.) Dongjiang water. The government of the Hong Kong special administrative region. https://www.wsd.gov.hk/en/core-businesses/water-resources/ dongjiang-water/index.html. Accessed 30 Jan 2021

HKTDC (Hong Kong Trade Development Council) (2018) Statistics of the Guangdong-Hong Kong-Macao Greater Bay Area. https://research.hktdc.com/en/article/MzYzMDE5NzQ5. Accessed 24 Mar 2021

HSBC (The Hong Kong and Shanghai Banking Corporation Ltd) (2015) No water more trade-offs —Managing China's growth with limited water (Research and analysis by CWR)

HSBC (The Hong Kong and Shanghai Banking Corporation Ltd) (2019) Business talks: Greater Bay Area Bridging the future. https://www.business.hsbc.com.hk/-/media/library/business-hk/ pdfs/en/gba-booklet-english-version.pdf. Accessed 24 Mar 2021

ICBC (Industrial and Commercial Bank of China Limited) (2016) The impact of environmental factors on the credit risk of commercial banks—research and application of ICBC based on stress test. http://upload.xh08.cn/2016/0428/1461824071992.pdf. Accessed 30 Jan 2021 (In Chinese)

ICIMOD (International Centre for Integrated Mountain Development) (2014) Research insights on climate and water in the Hindu Kush Himalayas. International Centre for Integrated Mountain Development, Nepal

IPCC (Intergovernmental Panel on Climate Change) (2018) 2018 IPCC special report on global warming of 1.5 °C

IPCC (Intergovernmental Panel on Climate Change) (2019) Special report on the ocean and cryosphere in a changing climate

Le Quéré C, Jackson RB, Jones MW, Smith AJ, Abernethy S, Andrew RM, De-Gol AJ, Willis DR, Shan Y, Canadell JG, Friedlingstein P (2020) Temporary reduction in daily global CO 2 emissions during the COVID-19 forced confinement. Nat Clim Chang 10(7):647–653

Lutz AF, Immerzee, WW (2013) Water availability analysis for the upper Indus, Ganges, Brahmaputra, Salween and Mekong river basins. Final Report to ICIMOD. Future Report 127

Lutz AF, Immerzeel WW, Shrestha AB, Bierkens MFP (2014) Consistent increase in High Asia's runoff due to increasing glacier melt and precipitation. Nat Clim Chang 4:587–592

Maghsadi N, Huang T (2020) Runways underwater: maps show where rising seas threaten 80 airports around the world. World Resources Institute. https://www.wri.org/blog/2020/02/runways-underwater-maps-show-where-rising-seas-threaten-80-airports-around-world. Accessed 30 Jan 2021

McKinsey Global Institute (2020) Climate risk and response in Asia

MEE (Ministry of Ecology and Environment) (2019) 2019 State of ecology and environment report, People's Republic of China. https://www.mee.gov.cn/hjzl/sthjzk/zghjzkgb/202006/P020200602509464172096.pdf. Accessed 30 Jan 2021 (In Chinese)

MEE (Ministry of Ecology and Environment), NDRC (National Development and Reform Commission) and MWR (Ministry of Water Resources) (2017) Yangtze river economic belt ecological environment protection plan, 17 July, People's Republic of China (In Chinese)

MEP (Ministry of Environmental Protection), MWR (Ministry of Water Resources), and NDRC (National Development and Reform Commission) (2017) Ecological and environmental protection plan of Yangtze river economic belt, People's Republic of China. https://www.mee.gov.cn/gkml/hbb/bwj/201707/t20170718_418053.htm. Accessed 30 Jan 2021 (In Chinese)

MIIT (Ministry of Industry and Information Technology), NDRC (National Development and Reform Commission), MOST (Ministry of Science and Technology), MOF (Ministry of Finance) and MEE (Ministry of Ecology and Environment) (2017) Guiding opinions on strengthening the green development of industry in the Yangtze river economic belt, 27 July, People's Republic of China (In Chinese)

MOF (Ministry of Finance) (2018) Guiding opinion on constructing long term ecological compensation and protection mechanism for Yangtze river economic belt, 24 February, People's Republic of China. http://www.gov.cn/xinwen/2018-02/24/content_5268509.htm. Accessed 30 Jan 2021 (In Chinese)

MOF (Ministry of Finance), NDRC (National Development and Reform Commission) (2015) Interim management measures on pollution discharge permit leasing revenue, 31 July, People's Republic of China. https://szs.mof.gov.cn/zhengwuxinxi/zhengcefabu/201507/t20150731_1397067.html. Accessed 30 Jan 2021 (In Chinese)

Macao Water (2018) Macao water annual report 2017. https://www.macaowater.com/sites/default/files/report/annals/2017%20Macao%20Water%20Annual%20Report%20%28finalized%29.pdf. Accessed 24 Mar 2021

Moody's (2020) Sovereigns—global: sea level rise poses long-term credit threat to a number of sovereigns

Mukherji A, Molden D, Nepal S, Rasul G, Wagnon P (2015) Himalayan waters at the crossroads: issues and challenges. Int J Water Resour Dev 31:151–160

MWR (Ministry of Water Resources) (2021) Press release "The investment in water conservancy construction reached a record high of 770 billion yuan in last year", 12 January, People's Republic of China. http://www.mwr.gov.cn/xw/mtzs/qtmt/202101/t20210112_1495673.html. Accessed 30 Jan 2021 (In Chinese)

National Tibetan Plateau Third Pole Environment Data Centre (n.d.) The second glacier inventory dataset of China (version 1.0) (2006–2011). https://data.tpdc.ac.cn/en/data/f92a4346-a33f-497d-9470-2b357ccb4246/. Accessed 30 Jan 2021

NBSC (National Bureau of Statistics of China) (2018) China statistical year book 2018. http://www.stats.gov.cn/tjsj/ndsj/2018/indexeh.htm. Accessed 30 Jan 2021

NDRC (National Development and Reform Commission) (2016a) China's policies and actions for addressing climate change, People's Republic of China

NDRC (National Development and Reform Commission) (2016b) 13th five year plan for the development of renewable energy, People's Republic of China

NDRC (National Development and Reform Commission) (2019a) Construction plan for the capital water source conservation functional zone and ecological environmental supporting zone (the capital two zones) in Zhangjiakou. People's Republic of China. https://www.ndrc.gov.cn/xxgk/zcfb/ghwb/201908/t20190806_962255.html. Accessed 30 Jan 2021 (In Chinese)

NDRC (National Development and Reform Commission) (2019b) Outline development plan for the Guangdong-Hong Kong-Macao Greater Bay Area, People's Republic of China. http://www.gov.cn/zhengce/2019-02/18/content_5366593.htm#1. Accessed 30 Jan 2021 (In Chinese)

NGFS (Network of Central Banks and Supervisors for Greening the Financial System) (2019) A call for action—climate change as a source of financial risk

NGFS (Network of Central Banks and Supervisors for Greening the Financial System) (2020) Overview of environmental risk analysis by financial institutions

NPC (National People of Congress) (2016) Outline of the 13th five-year plan for the national economic and social development of the People's Republic of China (In Chinese)

Oki T (2016) Water resources management and adaptation to climate change. In: Biswas AK, Tortajada C (eds) Water security, climate change and sustainable development. Springer, Singapore, pp 27–40

Qin Y, Curmi E, Kopec GM, Allwood JM, Richards KS (2015) China's energy-water nexus–assessment of the energy sector's compliance with the "3 Red Lines" industrial water policy. Energy Policy 82:131–143

SFCDRH (State Flood Control and Drought Relief Headquarters) and MWR (Ministry of Water Resources) (2017) Bulletin of flood and drought disasters in China, People's Republic of China (In Chinese)

SFCDRH (State Flood Control and Drought Relief Headquarters) and MWR (Ministry of Water Resources) (2018) Bulletin of flood and drought disasters in China, People's Republic of China (In Chinese)

Sina Finance (2017) Sina Finance article's "Around 2030, the Guangdong-Hong Kong-Macao Greater Bay Area will become the Bay Area with the largest population and the largest economy". http://finance.sina.com.cn/roll/2017-12-21/doc-ifypwzxq4774628.shtml. Accessed 24 Mar 2021 (In Chinese)

Storey B, Owen S, Noy I, Zammit C (2020) Insurance retreat: sea level rise and the withdrawal of residential insurance in Aotearoa New Zealand. Report for the deep south national science challenge

The 2030 Water Resources Group (2009) Charting our water future—economic frameworks to inform decision-making

The State Council of the People's Republic of China (2012) Opinions on the implementation of the most stringent water resources management system, 12 January. http://www.gov.cn/zwgk/2012-02/16/content_2067664.htm. Accessed 30 Jan 2021 (In Chinese)

The State Council of the People's Republic of China (2015a) Water pollution prevention and control action plan, 2 April. http://zfs.mee.gov.cn/fg/gwyw/201504/t20150416_299146.htm. Accessed 30 Jan 2021 (In Chinese)

The State Council of the People's Republic of China (2015b) Guiding opinions on promoting sponge city construction, 16 October. http://www.gov.cn/zhengce/content/2015-10/16/content_10228.htm. Accessed 30 Jan 2021 (In Chinese)

The State Council of the People's Republic of China (2017) Water dispatching during the dry season of the Pearl River was successfully completed, and the safety of water supply in Zhuhai, Macau was strongly guaranteed, 3 March. http://www.gov.cn/xinwen/2017-03/03/content_5172773.htm. Accessed 30 Jan 2021 (In Chinese)

The State Council of the People's Republic of China (2018) State council institutional reform program. http://www.gov.cn/guowuyuan/2018-03/17/content_5275116.htm. Accessed 30 Jan 2021 (In Chinese)

Trucost (2019) Climate change physical risk analytics

UBS (2016a) Are investors pricing in water risk? A geospatial perspective

UBS (2016b) Is China consuming too much water to make?

UNEP (United Nations Environment Programme) (2020) UNEP adaptation gap report 2020

Wang GX, Li YS, Wang YB, Shen YP (2007) Impacts of alpine ecosystem and climate changes on surface runoff in the headwaters of the Yangtze River. J Glaciol Geocryol 29(2):159–168

Water Resources Department of Guangdong Province (2000–2018) Various Guangdong water resources bulletins from 2000 to 2018. http://slt.gd.gov.cn/szygb/. Accessed 24 Mar 2021

Water Resources Department of Guangdong Province (2017) Guangdong 13FYP for water conservancy development. http://www.gpdiwe.com/UploadFile/upi/file/20170220/20170220 112549994999.pdf. Accessed 30 Jan 2021

Water Resources Department of Guangdong Province (2018) Guangdong water resources bulletin 2018. http://slt.gd.gov.cn/gs2018/index.html. Accessed 24 Mar 2021

Water Resources Department of Guangdong Province (n.d). http://slt.gd.gov.cn/zjsjzszypzgc/. Accessed 30 Jan 2021 (In Chinese)

Wester P, Mishra A, Mukherji A, Shrestha AB (2019) The Hindu Kush Himalaya assessment: mountains, climate change, sustainability and people. Springer, Cham

WMO (World Meteorological Organization) (2020a) Press release "WMO confirms 2019 as second hottest year on record", 15 January. https://public.wmo.int/en/media/press-release/ wmo-confirms-2019-second-hottest-year-record. Accessed 30 Jan 2021

WMO (World Meteorological Organization) (2020b) New climate predictions assess global temperatures in coming five years, 8 July. https://public.wmo.int/en/media/press-release/new-climate-predictions-assess-global-temperatures-coming-five-years. Accessed 30 January 2021

World Bank (2015a) World development indicators. https://data.worldbank.org/indicator. Accessed 30 Jan 2021

World Bank (2015b) India development update, April 2015: towards a higher growth path. https:// openknowledge.worldbank.org/handle/10986/21872. Accessed 30 Jan 2021

World Bank (2018) GDP (current US$)—Germany, Canada, United Kingdom, France. https://data.worldbank.org/indicator/NY.GDP.MKTP.CD?end=2018&locations=DE-CA-GB-FR&name_desc=false&start=2018. Accessed 30 Jan 2021

WRI (World Resources Institute) (2013) Aqueduct country and river basin rankings. https://www. wri.org/resources/maps/aqueduct-country-and-river-basin-rankings. Accessed 30 Jan 2021

Xinhua News (2018) Xi calls for high-quality growth through developing Yangtze River economic belt, 27 April. http://www.xinhuanet.com/english/2018-04/27/c_137139635.htm. Accessed 30 Jan 2021

Xinhua News (2019) Yangtze 'Chemical Belt' pollution control should avoid a sweeping approach, 12 February. http://greenfinance.xinhua08.com/a/20190212/1796739.shtml?f=arelated. Accessed 30 Jan 2021

Yang Q, Hu F, Chen YH, Zhang X (2016) Green development strategy and advice for Yangtze River Economic Belt based on Water-nomics theory. Environ Prot 44(15):36–40 In Chinese

Yang Q, Hu F, Zhao Z, Chen YH, Zhang X, Wang H (2019) Evaluation of water resource and water environment in the Yangtze River economic belt and relevant policy strategy. J Beijing Norm Univ 55(6):731–740 In Chinese

Zhang L, Su F, Yang D, Hao Z, Tong K (2013) Discharge regime and simulation for the upstream of major rivers over Tibetan Plateau. JGR Atmos 118(15):8500–8518

Chapter 15
Managing Risks on Egypt Water Resources Security: Climate Change and Grand Ethiopian Renaissance Dam (GERD) as Challenging Aspects

Mohamed Abdel Aty

Abstract Worldwide, availability of water resources is one of the most crucial economic and social concerns of the century. Egypt is considered as one of the most arid countries around the world. The Nile River is the main source of life for the Egyptians since it constitutes more than 97% of Egypt's renewable water resources. Egypt faces great challenges with regard to water resources due to its fixed share of the Nile water, and scarcity of rainfall, groundwater and desalination capacities. Climate change causes an additional challenge for water availability and accessibility in Egypt. The Nile Basin upstream developments especially (the Grand Ethiopian Renaissance Dam) will lead to more water shortage that would threaten the Country's water security.

The Egyptian water resources system, including conventional and non-conventional water sources, is complex and multifaceted. The main challenge is to close the gap between the water resources available and the rising demand for freshwater, much more in the light of the current and future challenges that include unilateral upstream development, population increase, climate change and its implications within and outside the borders (variability in Nile flood regime, rainfall pattern, sea-level raise and seawater intrusion in the coastal groundwater aquifers).

The current regional perspective represents a major challenge to Egypt's water security since the Nile Basin countries have their individual plans to develop their water resources. The amount of water required to implement these developing projects (hydropower dams and irrigated agriculture) exceeds the available water in the Nile River. This necessitates a basin-wide water agreement to coordinate such activities over the Nile.

Keywords Egypt · Climate change · Nile River · Water security · Grand Ethiopian Renaissance Dam

---

M. A. Aty (✉)
Government of Egypt, Cairo, Egypt
e-mail: abdelaty@mwri.gov.eg

© The Author(s), under exclusive license to Springer Nature Singapore Pte Ltd. 2022
A. K. Biswas and C. Tortajada (eds.), *Water Security Under Climate Change*,
Water Resources Development and Management,
https://doi.org/10.1007/978-981-16-5493-0_15

15.1 Introduction

The water situation in Egypt is critical and reached a state where the quantity of water available is imposing limits on its national economic development. It will be evident with population predictions for 2025 that per capita water share might go down to less than 500 m^3 with indicators of rapid deterioration of surface and groundwater quality.

Being the most downstream country in the Nile Basin and yet depending almost totally on the Nile River originating outside its borders, Egypt is the world's driest country. It has a total renewable water resources dependency ratio of 97% according to (FAO 2016).

The gap between the needed and available water resources is about 21 billion cubic metres (BCM) per year. This gap is fulfilled with drainage and treated wastewater reuse helping the Nile system in Egypt to reach an overall efficiency exceeding 88%. Moreover, Egypt imports annually around 34 BCM of virtual water to balance its food gap. Egypt has developed a long-term strategy for the development and management of water resources up to the year 2050. It relies on four main pillars which are: water quality protection, improvement of water use efficiency, development of water resources and creation of the enabling environment. Furthermore, Egypt has prepared its National Water Resources Plan for the period 2017–2037. This plan comprises several ambitious measures including implementation of various efficient irrigation water improvement programs, multiple water recycling mechanisms and higher agriculture productivity techniques that contribute to Egypt's food security.

It should be noted the importance of the transboundary dimension in addressing climate change adaptation (Block and Strzepek 2010), given the need for dialogue and cooperation among the Nile Basin states. Cooperation can help to reconcile the water use and development priorities of all riparian countries, including their capacity to adapt to any reductions or other changes in Nile flows due to climate change (Jeuland and Whittington 2014).

Accordingly, Egypt has consistently called upon its upstream co-riparians to follow the principles of international water law with regards to prior notification and planned measures, to be notified and consulted on any upstream water project in order to minimise their adverse effects on Egypt and to devise coordination mechanisms to ensure that these projects are operated in a manner that takes the interests of downstream states into consideration. One principal purpose of these coordination mechanisms would be to assist the various riparian states to manage their water resources sustainably and to enable them to adapt to drought conditions that might occur in the future.

15.2 Hydrological Regime of the Nile River Basin

As shown in Fig. 15.1, the Nile River is a north-flowing international watercourse that is shared by eleven riparian states. The Nile obtains its flows from three main watersheds: (a) the basin of the Equatorial Lakes plateau, (b) the Ethiopian highland plateau and (c) the Bahr el Ghazal Basin. Almost 85% of the annual natural flow that reaches Aswan originates from the Ethiopian Highlands, from which the three

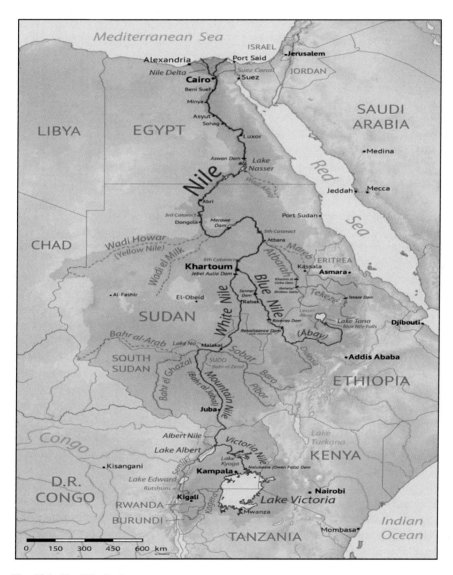

Fig. 15.1 The Nile Basin

main tributaries of the Nile River, the Sobat River, the Blue Nile and the Atbara River, emanate. The average annual natural flow of the Blue Nile as measured at the Grand Ethiopian Renaissance Dam (GERD) site is around 49 BCM. The remaining 15% of the Nile River flow originate from the Equatorial Lakes through the White Nile. The contribution of Bahr el Ghazal basin is almost negligible. From the confluence of Atbara River north of Khartoum to the Mediterranean Sea, the Nile receives no effective inflow. Although the Nile is the world's longest river with a total length of 6853 km and the total amount of rainfall on the basin varies between 1600 and 2000 BCM/year, the yield of the Nile river represents 4–5% of this total rainfall with an average flow of only 84 BCM measured at Aswan (Hurst et al. 1965; Sutcliffe and Parks 1999).

Ethiopia, on the other hand, is an immensely water-rich country. In terms of surface water, Ethiopia has 12 river basins, almost all of which originate in the central Ethiopian plateau. Ethiopia also has 11 freshwater lakes and 9 saline lakes, 4 crater lakes as well as over 12 major wetland areas. Moreover, Ethiopia receives an estimated average of 936 BCM/annually of rainwater. Hydrological studies have revealed that Ethiopia has the potential to dramatically increase its water resources by harvesting the abundance of available rainwater, but that "this potential is not fully utilised and translated into development because of many factors including limited financial resources, technical challenges and lack of good governance in the water sector".

These hydrological realities are the principal determinant of the positions and policies of Egypt in matters relating to the use and management of the Nile River. As a matter of principle, Egypt has never objected to the construction of waterworks or development projects based on the resources of the Nile River that could contribute to economic growth in the upstream states. On the contrary, Egypt has supported the construction of dams in different Nile Basin countries for either hydropower or rainfall harvesting purposes. For example, Egypt has supported South Sudan, Uganda and Tanzania to build new dams. Starting in 1949, Egypt has supported its sister Uganda for construction of Owen Falls dam. The dam helped to provide hydropower for development and regulating the flow towards downstream in the southern regions of Uganda. Recently, the Egyptian company Arab Contractors announced it will start building the 2100 MW Stiegler's Gorge hydroelectric dam in Tanzania. This project is endorsed and supported by the Egyptian government. However, given its dependency and vulnerability, Egypt is particularly sensitive to projects that would cause tangible disruptions to the Nile River system or that would alter the quantity or quality of waters flowing through its tributaries.

The hydrology of the Nile is characterised by high inter-annual variability, stark differences in geography and climate and flows modified by natural features and water infrastructure (Abu-Zeid and Biswas 1991). As future flows in the Nile cannot be predicted with certainty, it has become common practice in reservoir simulation models to stress test systems using synthetic flows, sometimes incorporating future changes that are projected from climate models. However, results from these simulation studies are often not considered sufficiently credible by decision makers (Loucks 2020). The probability, severity, and timing of specific sequences of low flows are unknowable, especially as climate change unfolds (Wheeler et al. 2020). It should be recognised that the future conditions will not replicate those of the past and that more severe conditions could materialise, especially with a changing climate.

15.3 Water Resources in Egypt

Egypt is a unique country with respect to water resources as Egypt's entire gamut of economic and service activities is extremely reliant on the River Nile which is originated outside its territory, with a dependency ratio estimated at 97%.

A fixed 55.5 BCM/year passes through the High Aswan Dam, Egypt's quota, according to the 1959 treaty between Egypt and Sudan, constituting 97% of the country's total renewable water resources; the remaining 3% being small quantities of rainfall. The total renewable resources currently available for use in Egypt are 56.8 BCM/year, while water usage is 80.25 BCM/year. This gap between the needs and availability of water is managed by reuse from drainage water, shallow groundwater, desalinated water and treated wastewater. The country's total water demand is 114 BCM/year to meet a steady population that has exceeded 100 million capita for the year 2020. Therefore, 34 BCM/year of virtual water is being utilised.

The overall efficiency of the irrigation system in Egypt is one of the highest in the world and the highest in Africa (88%), despite the reliance on traditional irrigation methods in some areas this is due to the fact of recycling of drainage and treated wastewater after treatment more than once at the amount of 21 BCM in the year.

Downstream from the High Aswan Dam, the River Nile is 100% regulated using a number of large barrages and irrigation structures with the result that the water resource system in Egypt is partially closed, and thus has a tendency to retain pollutants (Strzepek et al. 2008).

Since 1990, Egypt has reached the so-called water poverty line with respect to the per capita share of water of almost 1000 m^3/year. In 2017, the per capita share decreased to almost 600 m^3/year, and it is expected to fall to less than 500 m^3/capita before the year 2030, when the population is expected to become 130 million. Meanwhile, an increase in cultivated land in Egypt is vital to produce the necessary quantities of food to feed the growing population and secure social and political stability in the country. This will add more challenges to improve Egypt's utilisation of its limited water resources and to develop new resources to cover the needs of the cultivated areas as well as other uses.

In light of the above, the water sector is considered one of the most important pillars of Egypt's national security. All comprehensive sustainable development plans in all fields depend on the state's ability to provide the necessary water resources required to implement these plans, and the state strives to preserve its water resources and maximise the efficiency of their utilisation.

The country has also adopted an ambitious programme to double the quantities of desalinated water that will be utilised in the drinking water sector with investments amounting to 135 Billion Egyptian Pounds (EGP) until 2030.

The selection of the water treatment plant of Al-Mahsma drain in Ismailia being the best construction project in the world in 2020 is the culmination of the country's effort in adopting many water reuse projects that contribute to bridging the water deficit the country suffers from.

15.4   Risks and Impacts of Climate Changes on Egypt and Adaptation Measures

Egypt's vulnerability to climate change touches several sectors and the common cause of this vulnerability is water. The results of studies on climate change impact show that Egypt will face numerous threats to its economy, social and environmental sustainability, agriculture and food security, water resources, energy, human health, coastal zones and physical infrastructure.

Climate change in Egypt is witnessed through different phenomena that include the increase in the intensity and frequency of extreme weather events, such as:

- Hot or cold heat waves (the rise and fall in temperature from their normal levels) and their consequences including the increase of water needs for agriculture.
- Increase the drought and desertification rates due to dust and sandstorms.
- Flash floods in highlands such as Sinai and the Red Sea mountain range.
- Heavy rains on the northern coast.

Climate change studies anticipate that the productivity of the main crops in Egypt will decrease by 11–51% due to temperature increase by 1.5–3.5°C. Moreover, 12–15% of the most fertile arable land in the Nile Delta is negatively affected by a rise in sea level and salt water intrusion, deteriorating ground water quality. Also, the increase in both seawater temperature and salinity in the coastal lakes is negatively affecting fish species, with a serious impact on low-level lands in the delta and the adjacent highly populated cities such as Alexandria and Port Said. The impact includes the destruction of weak sections of the sand belt, inundation of valuable agricultural land, damage to the ecosystems and communities of the Northern Lakes and the endangering of recreational tourism beach facilities. It is believed that up to six million people and 4500 km^2 of land may be affected, resulting in a more significant challenge—the migration of people from the affected areas.

Egypt is the furthest downstream of the countries in the Nile Basin, the most dependent on the basin's resources and the most arid. Studies using global and regional climate models show a high degree of uncertainty regarding the impact of climate change on the Nile Basin's precipitation and runoff.

Accordingly, the main objective of Egypt's National Strategy for Adaptation to Climate Change and Disaster Risk Reduction is to increase the flexibility of Egyptian communities when dealing with the risks and disasters that might be caused by climate change and its impact on various sectors and activities. The Ministry of Water Resources and Irrigation (MWRI), in coordination with all ministries concerned with the management and use of water, has therefore prepared a National Water Resources Plan (NWRP) which looks as far ahead as 2037, and is based on the principle of integrated water resources management (IWRM). NWRP 2037 is not the first Egyptian national water resources plan. MWRI had already developed a plan covering the period 2005–2017, following the same methodology, and in full cooperation and coordination with all stakeholders. With the approach of 2017, it was necessary to develop a new strategy, with the lessons learned from the implementation of the first plan used as one of the main inputs. NWRP 2037 will pursue four objectives:

- Improving water quality.
- Enhancing the management of water use.
- Increasing the availability of freshwater resources.
- Improving the enabling environment for IWRM, planning and implementation.

In addition, the new plan takes into consideration the objectives of Egypt's Sustainable Development Strategy 2030 and the latest circumstances surrounding the water sector. As part of the government's overall orientation towards decentralisation, the NWRP supports this trend in the water sector, as well as supporting the required interaction between public policies and determinants at national level and the actual needs and priorities at local level. Governorate water resources plans have been developed in full cooperation and coordination between the stakeholders at governorate level in order to support the required interaction.

In the NWRP, several measures are currently being considered to adapt to climate change impact on water resources. These efforts include, but are not limited to: improving irrigation and drainage systems; changing cropping patterns and farm irrigation systems; reducing surface water losses by redesigning and lining canals; rain water harvesting; research and development for low-cost desalination techniques; treated wastewater recycling; developing new water resources through upper Nile projects; and developing new species of crops to cultivate in high temperatures.

Adaptation options for coastal zones are highly site dependent. However, changes in land use, integrated coastal zone management and proactive planning for protecting coastal zones are necessary adaptation policies.

In another context, MWRI has developed a comprehensive master plan for rehabilitation and replacement of the major hydraulic structures on the Nile and the main canals and rayahs, which distribute water to large- and medium-sized canals. New Esna, Naga Hammadi and new Assiut barrages have been constructed, while the executive procedures have started for the new Dairout group of regulators. Meanwhile, irrigation improvement and IWRM projects have been developed in the Nile Delta, with construction of flood protection works. Coastal flood protection works have been completed in Rosetta, Baltim, Ras El Bar and Alexandria in addition to the implementation of an integrated management plan for the protection of Northern Lakes from sea-level rise and salt water intrusion.

Moreover, public awareness is being raised on the need for rationalising water use, enhancing precipitation measurement networks and encouraging data exchange among Nile Basin countries as well as developing regional circulation models to predict the impact of climate change on national and regional water resources.

15.5   The Transboundary Dimension in Egypt's Policy

Egypt believes in the inevitability of cooperation between the Nile Basin states in the utilisation of the basin's water resources. Egypt has participated in the establishment of the existing institutional framework that governs the relations between the riparian states. It also played a leading role in establishing several cooperation initiatives, including the Nile Basin Initiative (NBI) in 1999. Egypt then suspended its participation in NBI activities in 2010 in response to the non-consensual decision taken by some upstream states to open for signature the unfinished draft of the Cooperative Framework Agreement (CFA), in breach of the NBI rules of procedures and those of the negotiating committee. Since then, the NBI has continued to function as a non-inclusive and non-consensual framework (Wheeler et al. 2018).

Nevertheless, driven by its belief in the inclusiveness of the initiative, Egypt has engaged in a consultative process to further address its concerns and exchange views with other NBI member states in order to seek ways for Egypt to resume its participation in NBI activities on a permanent basis.

Despite the challenges and difficulties facing this process, Egypt is determined to continue its efforts to restore inclusiveness to the NBI in order to manage the transboundary waters of the Nile Basin in accordance with the principles of international law, paving the way to enhancing cooperation on the basin level. To this end, Egypt believes that genuine cooperation among the Nile Basin states should be based on the following pillars:

- All states should respect and uphold existing obligations under international law, including the existing bilateral, plurilateral and multilateral agreements.
- The consensual decision-making process should form the basis of the management of transboundary waters.
- Riparian states should abstain from unilateral actions that could harm other riparian states. The principle of no-harm and timely prior notification should be respected where the construction of projects has cross-border effects.
- Riparian states should exert their utmost efforts to reach an agreed definition of the equitable and reasonable utilisation that avoids causing harm to any of those states.

Finally, development partners and private sector actors should promote the establishment of consensual institutional arrangements to facilitate the management of transboundary water resources. The United Nations should play a more active role in facilitating and enhancing cooperation among riparian states to enable them to achieve the agreed SDGs.

15.6 Impact of GERD

15.6.1 Main Characteristics of the GERD

The GERD, which is projected to become the largest dam in Africa, is located on the Blue Nile River approximately 20 km upstream from the Ethiopian–Sudanese border. It is a Roller Compacted Concrete dam with a projected height of 145 m above its foundation and a crest length of 1780 m. The main dam is complemented by a 50m high and 5 km long rock-fill saddle-dam that confines the reservoir of the main dam. The full supply level of the GERD is 640 a.m.s.l with a total storage volume of 74 BCM. The GERD reservoir is expected to cover 1874 km^2 and will extend for 264 km.

The purpose of the GERD is the generation of hydropower. Indeed, Article 2 of the 2015 Agreement on Declaration of Principles (DoP) states the following: "The purpose of GERD is for power generation, to contribute to economic development, promotion of transboundary cooperation and regional integration through generation of sustainable and reliable clean energy". The total electric power production capacity of the GERD is over 5150 MW with an average annual energy generation capacity of 15,692 GWh/yr. The power stations of the dam are positioned on the right and left banks of the river and include 13 hydropower turbines.

Several studies were previously undertaken on the feasibility of constructing a major dam in the area where the GERD is located. One of these was undertaken by the U.S. Bureau of Reclamation during a Blue Nile survey conducted between 1956 and 1964. That study proposed constructing a dam with a storage capacity of 11 BCM, which is significantly lower than the planned storage capacity of the GERD. Furthermore, a study undertaken in 2007 by the Eastern Nile Technical Regional Office (ENTRO) titled "Prefeasibility Study of Border Hydropower Project, Ethiopia" concluded that the optimum storage capacity for a hydropower dam at the location of the GERD is 14.47 BCM.

Despite the fact that sufficient energy would have been efficiently generated at the levels proposed by the U.S. Bureau of Reclamation and the Eastern Nile Technical Regional Office (ENTRO), the technical specifications of the GERD were altered repeatedly and its storage capacity was progressively increased from the originally proposed 11 BCM–62 BCM and then to 67 BCM, till it was increased to 70 BCM and finally to 74 BCM. This dramatic increase in the volume of the storage reservoir of the GERD is unjustified and raises questions about the actual purpose of the dam and its projected uses. Indeed, technical studies and hydro-logical modelling have shown that retaining 19 BCM in the GERD reservoir would have been a sufficient volume to generate electric power.

Several experts, including an Ethiopian expert, demonstrated that the GERD is a highly inefficient and oversized project for the purposes of power generation (Beyene 2011). According to this study, the hydropower generated from the GERD will be equivalent to that produced by a power plant with the much lower capacity of 2872 MW that operates at 60% efficiency. Therefore, the total cost of the GERD could have been reduced by at least 40–45% by building a smaller dam with a higher efficiency to generate the same amount of hydropower. Due to such solid scientific analyses, the Ethiopian Ministry of Water, Irrigation and Energy has recently announced that three turbines have already been omitted to increase the power plant factor.

15.6.2   Trilateral Negotiations on GERD

Egypt has engaged in intensive negotiations on the Grand Ethiopian Renaissance Dam (GERD) for almost a decade. Since Ethiopia unilaterally commenced the construction of the GERD in 2011, Egypt has negotiated in good faith and with a genuine political commitment to reach a fair and balanced agreement on the GERD. These negotiations went through several phases and were undertaken in numerous forums. Regrettably, in each and every round of talks, Ethiopia adopted a policy of time-wasting that undermined these negotiations.

Despite the fact that an International Panel of Experts issued a deeply troubling report on the GERD (May 2013) and recommended undertaking studies on its transboundary and environmental effects, Ethiopia has effectively thwarted every attempt to conduct these studies. It undermined the work of a Tripartite National Committee (TNC) that was overseeing the completion of these studies. It violated an agreement reached by the Nine-Party Meeting reached during a meeting of the ministers of foreign affairs and water affairs and the heads of the intelligence agencies of the three countries on the necessary steps to enable an international consultancy firm that was hired to conduct these studies. Ethiopia's policies and positions also prevented the National Independent Scientific Research Group (NISRG), which was an independent group of scientists who were tasked with agreeing on the technical modalities of the filling and operation of the GERD, from fulfilling its mandate.

In an attempt to facilitate the reaching of an agreement on the GERD, Egypt concluded an international treaty with Ethiopia and Sudan titled the Agreement on Declaration of Principles on the GERD (DoP) on 23 March 2015. This agreement obliges Ethiopia to reach an agreement on the rules governing the processes of the filling and operation of the GERD. Pursuant to this treaty, Ethiopia is under an obligation not to commence the impoundment of waters for the purposes of filling the GERD reservoir without an agreement with Egypt.

Since the conclusion of the DoP, negotiations have been held with Ethiopia in various settings and formats. Throughout all of these negotiations, Egypt showed immense flexibility and sought to address Ethiopia's concerns and presented numerous technical proposals that were designed to enable Ethiopia to achieve the objective of the GERD, which is the generation of hydropower, while preventing the infliction of significant harm on downstream states.

Unfortunately, more than five years of talks proved futile. Every effort to complete the studies on the GERD failed, and trilateral discussions aimed at agreeing on the rules on the filling and operation of the dam did not lead to fruition. Moreover, attempts by African states to exercise good offices to assist in bridging the gap between the three countries were unsuccessful. Therefore, in accordance with article ten of the DoP, Egypt called for international mediation to facilitate discussions between the three countries. This led to the launch, in November 2019, of a new process of negotiations in which the USA and the World Bank Group participated.

After 12 rounds of meetings, including at the ministerial and expert levels, that were attended by our American partners and by representatives of the World Bank Group, the U.S. administration, in coordination with the World Bank, formulated a final agreement on the filling and operation of the GERD. This agreement is fair, balanced and mutually beneficial and was prepared on the basis of the positions adopted by the three countries during the discussions. This agreement satisfies Ethiopia's priority, which is the expeditious and sustainable generation of hydropower, while protecting downstream states against the adverse effects of the GERD. Accordingly, on 28 February 2020, Egypt accepted and initialled this agreement, which further demonstrates our goodwill and good faith commitment to reach an agreement on the GERD.

Regrettably, Ethiopia decided not to attend the ministerial meeting that the U.S. administration called for on 27–28 February 2020 to conclude an agreement on the GERD, and refused to sign the final agreement prepared by the U.S. and the World Bank. This position is entirely consistent with Ethiopia's longstanding posture of obstructionism and its overall desire to establish a fait accompli that enables it to exercise unfettered and unrestrained control over the Blue Nile.

This was followed by the session held on 29 June 2020 by the U.N. Security Council on the question of the GERD, which demonstrates the recognition by the international community of the seriousness of this issue and the necessity of reaching an amicable solution that prevents the further destabilisation of an already troubled region.

Despite of the outcomes of the Extraordinary meetings of the Bureau of the Assembly of the African Union held on 26 June and 21 July 2020, where all parties agreed to refrain from making statements or taking any action that may undermine the AU-led process, and taking into considerations that the negotiation between the three countries about the Agreement on Rules for the Filling and Operation the GERD is not concluding yet since no consensus reach on some important technical and legal issues. Although the three countries agreed not to take any unilateral actions that may contradict with the current negotiations spirit particularly the first filling of the GERD, Ethiopia announced the accomplishment of the first stage filling of 4.9 BCM on July 2020.

Since late June 2020, we have been negotiating under the auspices of the African Union. Unfortunately, this process has not led to fruition. Instead of engaging in intensive substantive negotiations, our talks have been stalled for most of the past six months over procedural issues and minor organisational matters.

At no point in history has Egypt sought to obstruct the implementation of water projects by its co-riparians. This reflects Egypt's firm commitment to supporting its fellow African states, especially the Nile Basin states, in their endeavours to achieve development, peace and prosperity. However, in pursuing these developmental objectives and in utilising the resources of the Nile, Egypt believes that, in keeping with the established rules of international law, riparian states are required to consult their co-riparians on planned projects and to ensure that these projects are undertaken in a manner that is both reasonable and equitable and that minimises the harm that may be inflicted on other states.

We call upon the international community to encourage Ethiopia to reconsider its position and to impress upon Ethiopia the importance of signing the agreement on the filling and operation of the GERD that was prepared by the U.S. and World Bank. As a shared resource that is co-owned by all the riparian states, Ethiopia must not undertake any unilateral measures, including the impoundment of waters for the purposes of filling the GERD, without an agreement with its co-riparians.

15.6.3   Potential Impacts of the GERD

The effects of the GERD on the Blue Nile system, including its impacts on existing water uses and waterworks, the largest of which is the High Aswan Dam (HAD), are hard to predict with precision. The exact impact of the GERD will depend on a wide range of variables, including the annual yield of the Blue Nile, the effects of climate change, and the filling plan and operational rules that will be applied by the GERD.

Nonetheless, numerous studies have been undertaken by academics and specialised institutions on the possible hydrological economic, social and environmental impacts of the GERD (van der Krogt and Ogink 2013). Several stochastic models have been designed to assess the range of impacts of the GERD on the Blue Nile system. Stochastic models are a globally recognised tool that helps to identify the possible impacts of major waterworks by estimating the probability of potential outcomes in the light of variables such as the river yield, the storage volume of new dams and the nature of current water uses.

The impacts of water shortages in Egypt caused by projects undertaken by Ethiopia could be catastrophic. Millions of jobs would be lost, thousands of hectares of arable land would disappear, cultivated land would experience increased salinisation, the cost of food imports would increase dramatically, and urbanisation would sky rocket due to rural depopulation, which will lead to an increase in unemployment, crime rates and transnational migration. Indeed, a decrease of only 1 billion cubic metres of water would lead, in the agricultural sector alone, to 290,000 people losing their incomes, a loss of 130,000 hectares of cultivated land, an increase of $150 million USD in food imports and a loss of $430 million USD of agricultural production. As water shortages increase and continue over an extended period, the ripple effects on every sector of Egypt's economy and its socio-political stability are inestimable (DELTARES and MWRI 2018).

However, regardless of the filling plan or operational rules that will be applied, the GERD will have significant effects on the Blue Nile system. Some of the most serious effects of the GERD are outlined here:

1. **Hydrological Effects**—The GERD will significantly alter the Blue Nile system due to the abstraction of part of the flow for the first impoundment of the dam, additional evaporation losses, and flow regulation. This will cause significant changes in the quantity and distribution of the flow of the river to the HAD.

   The Blue Nile river will be transformed downstream of the GERD from a free-flowing river ecosystem to an artificial water canal. This is a major change that will occur regardless of the specific rules for the filling and operation of the GERD that will be applied. Hydrological models have revealed that the water velocity of the Blue Nile will be reduced by 5, 11, 20, or 42%, depending on the exact operational rules of the GERD. The water temperature in the river is expected to drop by 0.5–1.5 °C. In addition, the chemical composition, dissolved oxygen levels and the physical properties of the waters of the Blue Nile are expected to be altered, thereby changing the natural habitat of aquatic plants and animals.

2. **Water Shortages in Egypt**—A water shortage is a situation where HAD is unable to release sufficient water to satisfy existing water needs and uses of 104 million Egyptians. The possibility and severity of water shortages that might occur in Egypt due to the GERD is dependent on the filling and operational rules that will be implemented and on the hydrological conditions of the Blue Nile. Hence, if the filling coincides with a drought, water shortages will occur in Egypt that will have significant, and even disastrous, socio-economic effects. These effects will undermine Egyptian food security and will hinder Egypt's ability to fulfil the United Nations Sustainable Development Goals. Figure 15.2 illustrates the negative impacts of the Ethiopian proposal for filling of the GERD. This will impact the HAD dramatically, since GERD will be filled to the full reservoir level while the HAD may reach the shut down level and then may continue to drop until the dead storage level.

   Fig. 15.3 illustrates how the Egyptian proposal is more cooperative since it decreases the negative impacts from filling of the GERD on the HAD, and secures more than 80% of the planned hydropower generation of the GERD as a win–win solution.

   Studies that have been undertaken on the possible impacts of the filling of the GERD on the water availability in Egypt have shown that, depending on the exact filling plan of the GERD, the risk that the entire live storage of the HAD

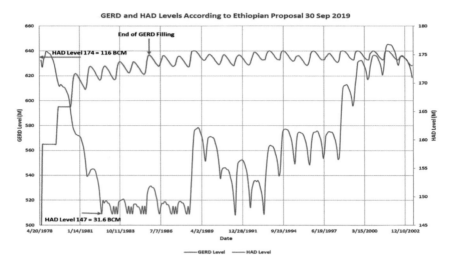

Fig. 15.2 The Ethiopian proposal filling impacts of GERD on the HAD

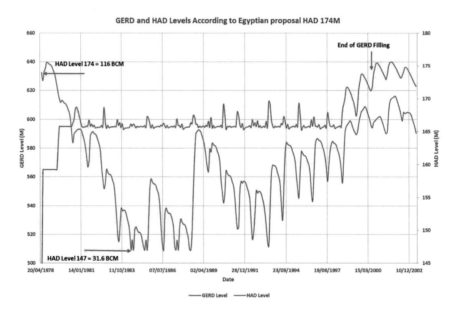

Fig. 15.3 Egyptian proposal filling impacts of GERD on the HAD

will be depleted and the shortage will increase by as much as 25 times compared with the normal conditions before the construction of the GERD. Moreover, simulations using actual data from the drought period that occurred in the 1980s showed that Egypt could experience water shortages in four or more years of a prolonged drought period due to the introduction of the GERD into the Blue Nile system and depending on the filling strategy that will be implemented. Figure 15.4 shows the impacts of GERD filling on water resources for dry periods.

As to the medium and long-term impacts of the GERD, virtually every simulation and model have demonstrated that HAD will become significantly more vulnerable to water shortages and power shutdowns during periods of low yield of the Blue Nile due to the operation of the GERD. Indeed, the possibility that water deficits might occur in Egypt due to the operation of the GERD may increase to 41–61% depending on the operational plans to be executed by the GERD and the hydrological conditions of the Blue Nile.

3. **Power Generation**—Studies have shown that when the water level of Lake Nasser is decreased due to the filling and operation of the GERD, there will be a reduction in power generated by the HAD. While the exact amount of reduction will depend on numerous variables, it is expected to be between 34 and 50% during the filling and operation of the GERD, especially during drought periods.

4. **Socio-Economic Impacts**—The exact socio-economic adverse impacts of the GERD on Egypt will depend on various variables, including hydrological conditions, future droughts and the filling and operation plans of the GERD and are difficult to predict with precision. However, simulations have shown that the cost of adapting to water shortages in Egypt due to the filling and operation of the GERD could range from $11.3–$46.6 billion/annually, depending on the exact operational plans and the hydrological conditions of the Blue Nile. These costs would grow exponentially during periods of drought that would significantly reduce the flow of the Blue Nile. Figure 15.5 shows detailed socio-economic impacts of GERD filling in dry periods for both Ethiopian and Egyptian proposals.

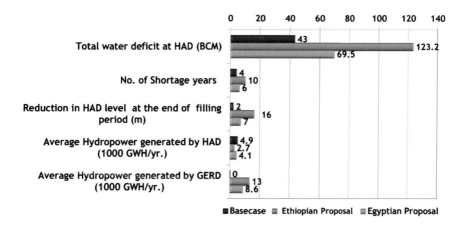

Fig. 15.4 Impacts of GERD filling on water resources for dry period

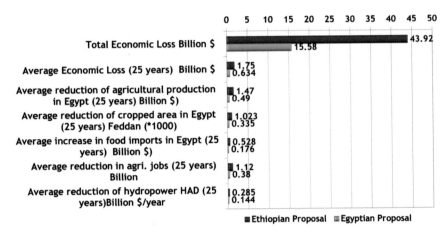

Fig. 15.5 Socio-economic impacts of GERD filling for dry period

5. **Effects on River Sediments**—During the annual flood, the Nile carries between 80 and 130 million tonnes of sediment to Egypt and Sudan from the Ethiopian Highlands. These sediments provide indispensable natural replenishment to the river ecosystem. Before the construction of the GERD, the accumulated sediment at the HAD was increasing progressively. This will be invariably altered after the completion of the construction of the GERD. A vast amount of sediment will be retained upstream of the GERD, and the released amount of sediment to downstream of the dam side will be significantly smaller. Indeed, one study has shown that the amount of sediment from 2020 to 2060 will decrease by more than 90%, thereby dramatically reducing the soil fertility, especially in Sudan. Accordingly, the use of chemical fertilisers will significantly increase in Sudan to compensate for the reduction in soil fertility, which will adversely impact the water quality in Egypt.

6. **Soil Salinisation and Sea Water Intrusion**—It is highly likely that the reduction in the freshwater release from the HAD due to water shortages caused by the GERD will advance seawater intrusion along the Egyptian Nile Delta. Due to the shortages in freshwater, farmers in the Nile Delta will be forced to use groundwater that will destabilise the freshwater–seawater interface. This will lead to an increase in groundwater table depletion as well as soil salinisation and other associated environmental impacts. Overall, soil salinisation will increase the probability of earth subsidence, geotechnical engineering problems, and agricultural degradation.

These potential adverse effects of the GERD demonstrate the need for reaching an agreement that covers both the filling and operation of the GERD. The negative impacts of the GERD, in the absence of such an agreement that preserves the rights and interests of the downstream states, could be catastrophic. It must include measures to minimise the effects of the GERD on the downstream states, especially

Egypt, that is totally reliant on the Nile to satisfy its water needs. Moreover, the filling and operation rules of the GERD must include mechanisms to adapt to the changing hydrological conditions of the Blue Nile and must account for the possibility of future droughts that might cause severe reductions in the annual yield of the Blue Nile. These measures must be based on burden sharing and cooperation to enable both the GERD and the downstream states, especially Egypt, to withstand and limit the effects of droughts.

15.7   Conclusions and Recommendations

- Climate change is a main determinant of the development and management of the water resources systems. Therefore, there is a need to have enough information, knowledge to plan and negotiate necessary adaptations to climate change. Furthermore, due to considerable regional climate variability, a high spatial resolution of future climate scenarios seems advisable to support decision-making. Also, there is an urge to increase the institutional capacities in this field.
- On the other hand, regional climate modelling reveals the expectation that the Nile Basin is a sensitive region. Accordingly, given its dependency and vulnerability on the Nile, Egypt is particularly sensitive to transboundary projects in upstream countries that would cause palpable disruptions to the Nile River system such as the GERD as its impacts related to water shortages in Egypt could be catastrophic cause a decrease of only 1 billion cubic metres of water would lead, in the agricultural sector alone, to 290,000 people losing their incomes, a loss of 130,000 hectares of cultivated land, an increase of USD \$150 million in food imports and a loss of USD \$430 million of agricultural production. As water shortages increase and continue over an extended period, the ripple effects on every sector of Egypt's economy and its socio-political stability are inestimable.
- Accordingly, a legal and binding agreement for the filling and operation rules of GERD should be reached by the three countries (Egypt, Ethiopia and Sudan) in order to harness and manage their shared water resources and to mitigate the negative impacts resulting from the filling and operation of GERD.

References

Abu-Zeid MA, Biswas AK (1991) Some major implications of climatic fluctuations on water management. Int J Water Resour Dev 7:74–81. https://doi.org/10.1080/07900629108722497

Beyene M (2011) How efficient is the grand Ethiopian renaissance dam? International Rivers. https://archive.internationalrivers.org/resources/how-efficient-is-the-grand-ethiopian-renaissance-dam-2452. Accessed 21 Feb 2021

Block PJ, Strzepek K (2010) Economic analysis of large-scale upstream river basin development on the Blue Nile in Ethiopia considering transient conditions, climate variability, and climate change. J Water Resour Plan Manag 136:156–166

DELTARES and MWRI (Ministry of Water Resources and Irrigation) (2018) Impacts of GERD on Egypt. Giza, Egypt

FAO (Food and Agriculture Organization) (2016) FAO Aquastat. http://www.fao.org/aquastat/en/countries-and-basins/country-profiles/country/EGY. Accessed 21 Feb 2021

Hurst HE, Black RP, Simaika YM (1965) Long-term storage: an experimental study. Constable, London

Jeuland M, Whittington D (2014) Water resources planning under climate change: assessing the robustness of real options for the Blue Nile. Water Resour Res 50:2086–2107. https://doi.org/10.1002/2013WR013705

Loucks DP (2020) From analyses to implementation and innovation. Water 12:974. https://doi.org/10.3390/w12040974

Strzepek KM, Yohe GW, Tol RSJ, Rosegrant MW (2008) The value of the High Aswan Dam to the Egyptian economy. Ecol Econ 66:117–126. https://doi.org/10.1016/j.ecolecon.2007.08.019

Sutcliffe J, Parks Y (1999) The hydrology of the Nile. IAHS Press, Oxford

van der Krogt W, Ogink H (2013) Development of the Eastern Nile water simulation model (Report No. 1206020-000-VEB-0010). Author report to Nile Basin Initiative, Delft

Wheeler KG, Hall WJ, Abdo GM, Dadson SJ, Kasprzyk JR, Smith R, Zagona EA (2018) Exploring cooperative transboundary river management strategies for the Eastern Nile Basin. Water Resour Res 54:9224–9254. https://doi.org/10.1029/2017WR022149

Wheeler KG, Jeuland M, Hall JW, Zagona E, Whittington D (2020) Understanding and managing new risks on the Nile with the Grand Ethiopian Renaissance Dam. Nat Commun 11:5222

Chapter 16
Water Security Under Conditions of Increased Unpredictability: A Case Study

Marius Claassen

Abstract The supply of sufficient water of good quality as well as the safe management of wastewater manifests as contemporary challenges for sustained growth in large cities around the world. South Africa experienced a severe drought in 2015–2016, with the City of Cape Town case demonstrating the value of joint efforts from different tiers of government, residents, the private sector and agriculture in averting the crisis. Although the combined efforts of role players prevented a "Day Zero" situation from occurring, the system could be made more resilient by implementing broad-based scenario planning. Long-term predictions at broader geographic scales provide useful perspectives and scenarios to inform planning, but downscaling to local levels at shorter time scales greatly reduces predictability. The management of such a complex system will benefit from an analysis of the context for decision-making according to the Cynefin Framework to increase water security under conditions of increased unpredictability.

Keywords Water security · Day Zero · Unpredictability · Scenarios · Decision-making

16.1 Introduction

In the 4th millennium B.C., the so-called trenched villages were encircled within drainage ditches which also made water available both for drinking and for irrigation (De Feo et al. 2010). The supply of sufficient water of good quality as well as the safe management of wastewater has since manifested as contemporary challenges for sustained growth in large cities around the world (Grey and Sadoff 2007; van Ginkel et al. 2018; Kinouchi et al. 2019).

In South Africa, increased climate variability and climatic extremes affect water quality and availability through changes in rainfall patterns, with more intense

M. Claassen (✉)
Centre for Environmental Studies, Department of Geography Geoinformatics
and Meteorology, University of Pretoria, Pretoria, South Africa
e-mail: mclaasse86@gmail.com

© The Author(s), under exclusive license to Springer Nature Singapore Pte Ltd. 2022   331
A. K. Biswas and C. Tortajada (eds.), *Water Security Under Climate Change*,
Water Resources Development and Management,
https://doi.org/10.1007/978-981-16-5493-0_16

storms, floods and droughts leading to changes in soil moisture, runoff and evap-oration and also altering water temperature in aquatic systems (DEA 2017). The country experienced a severe drought in 2015–2016, which coincided with the El Niño Southern oscillation conditions. The drought was accompanied by extremely high near-surface temperatures with the mean temperature anomalies for 2015 being 0.86°C above the 1981–2010 reference period, making it the warmest year on record since 1951 (DEA 2017). Over this period, the average level of dams in South Africa decreased from 93% of full storage capacity in March 2014 to 53% by mid-2016 (DEA 2017).

A systems approach can be helpful to understand the complexity of the urban system and its linkages with its global environment (Hoekstra et al. 2018). An integrated approach to urban dynamics and urban design includes water-sensitive design, rainwater harvesting, recycling, reuse, pollution prevention and other innovative urban water approaches (Hoekstra et al. 2018). In an analysis of the relationship between resolution and predictability, Costanza and Maxwell (1994) found that while higher resolution provides more descriptive information about the patterns in data, it also increases the difficulty of accurately modelling those patterns. Water security is not only an issue of scarcity to be addressed by technical advances in water supply, but also a process of contestation and coordination embedded in hydro-political and hydro-social interactions (Wang and Dai 2021).

The integration of climatic patterns, natural systems, engineered systems, governance processes, economic drivers and societal aspirations to advance water security necessitates a robust approach to decision-making. Snowden and Boone (2007) proposed the Cynefin Framework, which takes account of complexity and uncertainty in decision-making contexts. The framework distinguishes simple, complicated, complex and chaotic contexts and proposed differentiated approaches to determine the context and making decisions in each of the contexts. A key characteristic for determining the decision-making context is the clarity of the cause–effect relationships. In a complicated context, this relation is discoverable by experts, whereas "emergent instructive patterns" define a complex context. In a complicated context, the decision-making approach is to "sense" (collect information), "analyse" (determine the best option) and "respond" (implement decision). In a complex context, the decision-making approach is to "probe" (administer an impulse to the system), "sense" (determine the response of the system to the impulse) and "respond" (implement the best option).

16.2   City of Cape Town, South Africa

The City of Cape Town, being situated at the south-western tip of the African continent, is part of the Western Cape Province in the Republic of South Africa. It had a population of 4,174,510, a GDP of 7692 USD per capita and a life expectancy of 64 years in 2017 (City of Cape Town 2018a). In terms of access to basic services, 95.6% of people had access to water, 92.3% to sanitation, 91.8% to an energy source and 96.4% to refuse removal in 2016 (Western Cape Government 2018). The City of Cape Town contributes 72.5% to the Provincial GDP (2016), which in turn contributes 14% to the national economy of the Republic of South Africa (Stats SA 2019).

Water scarcity is nothing new in Cape Town, with the need for water security having prompted the construction of the Waegenaere's Dam in 1661 (Kotzé 2010). As the City's demand for water increased, more infrastructure was put in place, including Reservoir No. 1 with a capacity of 2.5 million gallons that was built in 1849 and Reservoir No. 2, with a capacity of 12 million gallons, having been constructed in 1856 (Kotzé 2010). Rainfall variability has been a hallmark of Cape Town. The Molteno Reservoir, which was designed to hold 40 million gallons, was completed in 1880 but stood empty until 1882 due to exceptionally low winter rainfall (Kotzé 2011). Water quality has also been of concern since the early settlement by Europeans, with the Fresh River having been reported to be unfit for drinking purposes in 1714, due to waste being dumped into it (Kotzé 2010).

The average annual temperature deviation for Cape Town illustrates the inter-annual climatic variability and the directional change (Fig. 16.1). The cumulative annual rainfall from 1981 to 2020 at the Steenbras monitoring point highlights the dry years between 2015 and 2018 (Fig. 16.2), while also illustrating rainfall variability within the Western Cape Water Supply System (WCWSS). The combined storage of the biggest six dams in the WCWSS as well as the urban water

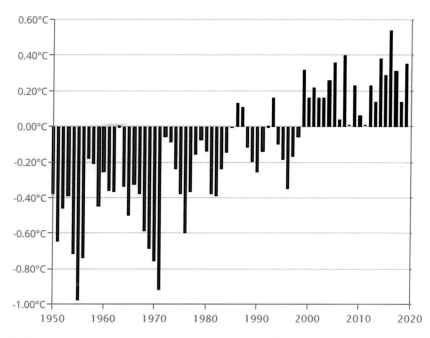

Fig. 16.1 Average annual temperature deviation from the 1981–2010 reference period for Cape Town (redrawn from NOAA 2020)

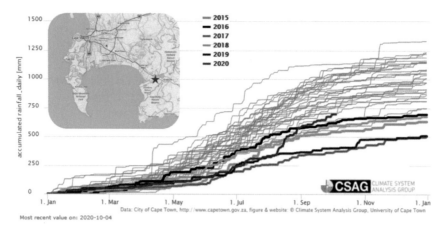

Fig. 16.2 Historical accumulated daily rainfall at Steenbras (adapted from Climate Systems Analysis Group 2020)

demand for a 10-year period from 2008 shows the impact of the 2015–2016 drought on water storage and the impact of interventions on water demand (Fig. 16.3).

The WCWSS Reconciliation Strategy (DWAF 2007) set out to ensure the reconciliation of future water requirements. The strategy makes provision for a regular review of future water requirement scenarios and reconciliation interventions to meet the water requirement up to 2030 (DWAF 2007). According to the strategy, the water demand from the WCWSS in 2006 was 310 million m^3/a for urban use and 154 million m^3/a for agricultural use. The strategy predicted a high scenario for total water demand of around 930 million m^3/a and a low scenario for total water

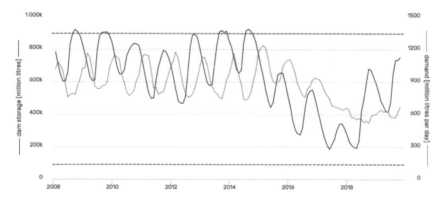

Fig. 16.3 Historical levels of the big six dams in the WCWSS with urban water demand over the same period (adapted from Climate Systems Analysis Group 2020)

demand of 670 million m^3/a by 2030. However, in 2007, the 1:50 year yield of the WCWSS was specified to have been 556 million m^3/a (DWAF 2007).

The decision-making context for engineered water supply in Cape Town meets the requirements of a complicated system according to the Cynefin Framework (Snowden and Boone 2007). The climatic variability and directional change bring about additional uncertainty and thus diminishes the predictability of supply, which defines the decision-making context as a complex system. The reconciliation strategy makes provision for monitoring system changes to impulses and adaptive management, which is in line with the recommendations of the Cynefin Framework in a complex context (Snowden and Boone 2007).

16.3   Alternative Supplies

Groundwater extraction is the least costly and fastest to implement option for the production of new drinking water, but it is not sustainable in the long-term unless groundwater recharge, water reuse and desalination projects are designed to recharge the exploited aquifers at the rate of water extraction (Olivier 2017). In a study, based on the 2004–2005 drought, Wright and Jacobs (2016) found that although a single groundwater access point (GAP) could supply the water demand for an entire residential property, due to household plumbing complexities and water quality concerns, GAPs are commonly used exclusively to meet outdoor needs, mainly garden irrigation. Wright and Jacobs confirmed that residential properties with GAPs use markedly less water from the municipal water distribution system. Pengelly (2018) reflected that many users went off grid, mostly through accessing groundwater, which impacts both the viability of municipal finances and the sustainability of the groundwater resource.

Vairavamoorthy et al. (2019) reported that the City of Cape Town introduced measures to reduce the impact of future droughts and new infrastructure, such as rainwater harvesting and desalination plants. However, according to Voutchkov (2017), new containerised desalination plants will provide minimal to no relief in terms of public water supply challenges and will produce water at unreasonably high costs. Vouchkov recommended the building of two or three permanent desalination plants and placing these at the Harbour Site (with a capacity of 125 ML/d), False Bay and Atlantic Ocean coast (with a capacity of 100–150 ML/d each) and to not build desalination plants of capacity larger than 200 ML/d. Pankratz (2018) also recommended that the City develop a permanent seawater desalination plant with a production capacity of no less than 50 ML/day, with a preferred capacity range of 100–150 ML/d. Andréassian (2018) concluded that desalination remains a very costly solution, at around €0.50 per m^3 for very large plants and consumes a great deal of energy (from 3.5 to 18kWh per m^3). Pankratz (2018) reported the costs of a basic 100 ML/d seawater desalination plant at 1.52 million ZAR, which provides a baseline comparison with a plant located at the

Cape Town Harbour site (2.139 million ZAR) and a plant located at a site requiring an offshore, non-tunnelled, intake and outfall (2.589 million ZAR).

Towing icebergs from Antarctica has been touted as an option, but there is some disagreement about the feasibility hereof. Malan (2018) stated that it will be prohibitively expensive, impractical to harvest fresh water from the ice, it will lead to significant negative social and environmental impacts and that it will probably be illegal under international treaties. In response to the paper, Orheim (2018) provides arguments that counter the claims of Malan (2018), stating that the environmental and social impacts would be minor, that the legal implications are diminished if the iceberg is collected outside of the boundary covered by the Antarctic treaty, and that it *"would be less than the cost per cubic metre of desalinated water schemes presently approved by the Western Cape Government."*

Whereas the City of Cape Town promotes the implementation of alternative supplies (City of Cape Town 2011a, b, c, d, 2018b, 2019a), there are no reliable data on the volume of water drawn from alternative supplies and/or the savings from municipal supplies.

16.4   System Efficiency

Cape Town loses only 15% of its water to "non-revenue water", which is much lower than in other cities in South Africa and close to being one of the best systems in the world (Oliver 2017; City of Cape Town 2019b). The low percentage of non-revenue water is achieved through proactive leak detection, pipe and metre replacement informed by a sophisticated asset management strategy, whereas the time from leak alert to completion of a repair was also substantially improved through a first responder system (City of Cape Town 2019b; Flower 2019). The (then) Director of the New Water Programme in the City of Cape Town stated that pressure control was also used as an effective means to reduce losses (Flower 2019; Pengelly 2018). The water supply system was organised in 212 zones with independent pressure control for each, such that a maximum pressure of 1.5 Bar (equivalent to a 15 m head) was maintained (Flower 2019). Whereas the engineering complexity of the configuration might be complicated, the overarching decision to focus on pressure control falls in the simple decision-making context of the Cynefin Framework (Snowden and Boone 2007), since the cause-and-effect relationship is immediately clear.

16.5   Demand Management

City planners had been bracing for Day Zero (when water supplies would run dry) to arrive by April 2018 after annual rainfall dropped from an average of 1100 mm in 2013 to just 500 mm in 2017, devastating a provincial supply system reliant almost entirely on surface water (Cotterill 2018). In response to the drought, Cape Town residents have more than halved their water use from up to 1.2 billion L a day in 2015, to just over 500 million L at the start of 2018 (Fig. 16.4), through enforcing suburban restrictions of up to 50 L a day per person (versus a global average use of 185 L) with key sectors such as tourism and agriculture bearing the brunt of scarcity (Cotterill 2018). Companies that implemented aggressive demand management strategies include Virgin Active, PPC, The Beverage Company, JG Afrika, One&Only Resort, the Vineyard Hotel and Old Mutual (Virgin Active 2019; Green Cape 2018a; Green Cape 2019; The Beverage Company 2019; Explore South Africa 2018; Green Cape 2018b; Old Mutual 2018). An article in *The Conversation* supported this approach in stating that cutting water use is the most important, fastest and most cost-effective way to avoid Day Zero (Olivier 2017). The three main lines of response to maintain the supply of water to Cape Town has been to develop an emergency programme of investments to augment supplies, to introduce aggressive demand management and to establish relations with other water users in the region to increase the availability of bulk water (Voutchkov 2017). The approach to improving water security went beyond technical solutions to water supply, but embraced a systems perspective (Hoekstra et al. 2018) across hydro-political and hydro-social interactions (Wang and Dai 2021) to mitigate against increased unpredictability.

The sustainability of water demand management outcomes is important for Cape Town's water resilience. The water use increased from around 500 ML/day between March and September 2018, to around 600 ML/day between January and September 2019 (Fig. 16.5). During the same periods, the restrictions were eased from a target of 450 ML/day to a target of 650 ML/day (City of Cape Town 2019c). Water use has subsequently increased to 731 ML/day in March 2020, compared to around 600ML/day during the same period in 2019 (City of Cape Town 2019c). The 2020 summer consumption has increased by around 22% from 2019 summer consumption levels and exceeds the target of 650ML/day, but the consumption

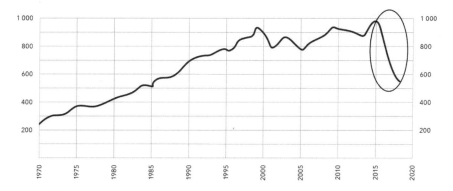

Fig. 16.4 Average annual water use for Cape Town between 1970 and 2018, as measured by the treated water supplied (including losses) in million litres per day (City of Cape Town 2019b)

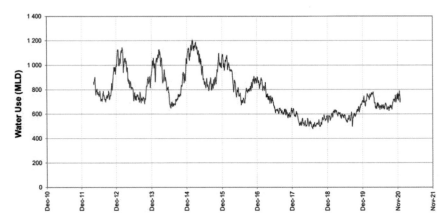

Fig. 16.5 Daily water use for Cape Town between 2011 and 2020 (City of Cape Town 2020a)

levels are still significantly lower than pre-drought levels, which touched on 1200 ML/day in 2015. The City of Cape Town publishes a Dam Levels Report on a weekly basis to keep residents and other stakeholders up to date on the state of the water supply system (City of Cape Town 2019d, 2020a).

The City of Cape Town implemented water restrictions and a differentiated pricing structure to complement other approaches to demand management. The levels and associated key restriction are summarised in Table 16.1.

Although there are no large wet industries in Cape Town, the City has been working closely with industries (Solomon 2019). The City developed a system for star grading in 2015 to attract commercial and industrial sectors to do water audits and monitor effluent quality, with awards for good performance (Flower 2019). Under the star rating system, participants who are compliant with the minimum requirements of relevant water legislation and by-laws receive one star, whereas participants can be awarded between two and five stars for implementing different degrees of additional measures to further reduce water consumption and limit water pollution (City of Cape Town 2019f). Those who achieve five stars are certified as champion innovators that have found unique/extraordinary ways in which to conserve water and limit water pollution, thereby protecting the environment. The City engaged with the tourism and hospitality industry on an ongoing basis, with around 800 people participating in the Water Wise Expo on 23 March 2018, where the City paid for stands for small business and stakeholders to display their efforts in water saving. These exhibitions were also hosted at all major shopping malls (Flower 2019). The City also developed a range of pamphlets, brochures and guidelines to promote water-saving approaches and technologies (City of Cape Town 2019g), that amongst others, promote the use of water-saving devices (such as installing low flow shower heads) and practices (such as taking short stop-start showers). The voluntary implementation of shower timers has also been reported by several groups in the hospitality industry, as reported in the following sections.

Table 16.1 Water restriction levels, dates and restriction details (Green Cape 2018c; Cape Talk 2018; City of Cape Town 2010–2018)

Level	Date	Key restrictions (cumulative)	Target
1	2005	No irrigation 10:00–16:00 Spray nozzles for hosepipes No hosing down hard surfaces No dampening of building sand	10% saving
2	1 January 2016	Irrigation for 1 h on Tue, Wed, Thurs No irrigation 9:00–16:00	20% saving
3	1 November 2016	Car washing with buckets only Pool covers must be installed	30% saving
3B	1 February 2017	No private car washing with municipal water	30% saving
4	1 June 2017	No irrigation with municipal water No topping up of private pools with municipal water	100 L/person/day
4B	1 July 2017	No topping up of public pools	87 L/person/day
5	3 September 2017	Residential: Fines if use > 20 kl/month Commercial: 20% less than same month previous year	87 L/person/day
6	1 January 2018	Residential: Fines if use > 10.5 kl/month Non-residential properties: 45% less than in 2015	87 L/person/day
6B	1 February 2018	Residential: Fines if use > 6 kl/month Non-residential properties: 45% less than in 2015	50 L/person/day
5	1 October 2018	Residential: Fines if use > 20 kl/month Commercial: 20% less than same month previous year	87 L/person/day
3	1 December 2018	Car washing with buckets only Pool covers must be installed	105 L/person/day
1	1 July 2019 (on revised 3-level system)	Car washing with buckets or high-pressure low volume cleaner Pool covers must be installed	120 L/person/day

Due to the marked income inequality in the City of Cape Town, the water crisis posed a possible threat to social order, with the City of Cape Town having been rated Baa3 (the lowest level of investment grade) and having been on review for a downgrade in January 2018 (Africa Research Bulletin 2018). The economic impact of the drought includes the loss of 30 000 jobs in the agricultural sector in the Western Cape region, which was caused by a 20% decrease in agricultural production (Knight 2019).

The Group Head of Sustainability for Woolworths Holdings reflected in an interview under the Drought Response Learning Initiative that the impact on Woolworths Holdings were multifaceted and varied (Koor 2018). The impacts were felt across the value chain, including logistics, distribution, operations, suppliers as well as the customer interface at store level. Koor said the initial planning by the government had not fully taken account of the extent of societal disruption that would have occurred under a Day Zero scenario (Koor 2018).

The coordinated engagement with water users in the various sectors supports the view of Wang and Dai (2021) that water security is also related to contestation and coordination embedded in hydro-political and hydro-social interactions.

16.6   Lessons Learned

Having worked with the City of Cape Town for more than a decade on water supply modelling (Rossouw 2019), Cullis and Fisher-Jeffes from Aurecon (Cullis and Fisher-Jeffes 2019) reflected that the National Department of Water and Sanitation (DWS) made reasonable decisions about delaying projects to increase the storage capacity of the Western Cape Water Supply System. This should be seen against the background of the dams overflowing in 2015 and that reasonable models of below average rainfall indicated a high level of water security. However, the drought that followed was statistically a 1 in 600-year event. Eberhard contended that government can only be exonerated to some extent on the basis that the system was not designed to cope with such a rare and severe drought, but that it should take some responsibility because if the system had been operated in terms of its rules, the impact of the drought would not have been nearly as severe (Eberhard 2018).

The Director of the New Water Programme for the City of Cape Town, Mr. Peter Flower, argued that the dam levels should not have gone down to 38% in November 2017, but could have been kept at 56% if decision-making had been better (Flower 2019). This was confirmed in an interview with Rolfe Eberhard in an interview under the Drought Response Learning Initiative, saying that if the system had been managed according to agreed rules, the levels of dams at the end of winter 2017 would have been substantially higher than they were (Eberhard 2018). Flower (2019) suggested that the water restriction levels of 20% could have been implemented in 2015 already and that water could have been transferred from the Berg River to Theewaterskloof Dam. This view was confirmed by Cullis and Fisher-Jeffes (2019) from Aurecon that does water supply modelling for the City of Cape Town. At the same time, there should have been better controls over illegal abstractions. With the threat of a drought looming, the agriculture sector withdrew access water to "stock up" their storage (Flower 2019).

A study of the water governance processes required to transition towards water-sensitive urban design showed that information transparency and access to information by the public, through social media, posters in public spaces and on the City's Website, plays an important role in educating the public about water challenges, and it can be used as a tool to encourage behavioural change in relation to water scarcity (Cameron and Katzschner 2017). Applying and implementing Water Sensitive Urban Design (WSUD) principles in South African cities is challenging due to fragmented institutional structures within municipalities (e.g. local government departments working in silos), social constraints, as well as financial and human resource limitations (Cameron and Katzschner 2017). Madonsela et al. (2019) used 27 indicators, in social, environmental and financial categories, to assess water governance processes in the context of WSUD in the City of Cape Town. The study concluded, among others, that consideration of the uncertainty and complexity in the governance of water scarcity has been lacking over the years. An example of this was the City's attempts to implement augmentation schemes in a short time span of six to eighteen months during 2017–2018 to address the water crisis (Madonsela et al. 2019).

In an interview under the Drought Response Learning Initiative, George Gabriel stated that the process of changing behaviour should start by engaging with the people whose behaviour you are trying to change (Gabriel 2018). The delivery of good service is a worldwide challenge, which requires resources, but in a country with limited resources you need to be very creative in how you go about solving these problems, for instance through effective engagement with business and civil society. *"We are not doing this in Cape Town. We need a rethink of how citizens and municipalities engage with each other, a rethink of how the different sectors of society work together"* (Gabriel 2018).

The Cape Town Water Map was launched in 2018 to make a dent in the 200,000 homes going over the limits on water usage imposed by the city (City of Cape Town 2019h). The Water Map shows the street address and the plot number as well as the water usage level of individual properties to highlight those who are sticking to the water restrictions. According to the City of Cape Town, the Water Map was a behavioural change tool introduced, in January 2018, at the height of the Cape Town drought crisis in order to encourage water-saving (City of Cape Town 2019i). The main purpose of the map was to publicly acknowledge or "reward" households that saved water, thereby normalising and incentivising water conservation behaviour. The number of green dots "awarded" to households for achieving water-saving targets increased dramatically from January 2018, peaking at over 400,000 green dots in June 2018 (Table 16.2). The Water Map was discontinued in January 2019 following the recovery of the Cape Town dams and the introduction of more relaxed level 3 water restrictions from 1 December 2018 (City of Cape Town 2019i).

The Water Map applies a theory developed by the University of California in 2015 that households will drastically reduce their water use when they realise they are the odd one out in a water-saving neighbourhood (Olivier 2018). Cape Town seems to be one of the first municipalities to apply the theory. Arguments against the Water Map include that it is a violation of privacy, that it is divisive and that it could promote harassment (City of Cape Town 2019h). The City of Cape Town's media campaign has been lauded for keeping citizens informed and driving the ethos of water conservation (City of Cape Town 2019j).

The City of Cape Town publishes the Dam Levels Report on a weekly basis to keep residents and other stakeholders up to date on the state of the water supply system (City of Cape Town 2019f). An extract from the 18 March 2019 report is provided in Fig. 16.6. The Dam Levels Report also reflected the historical water

Table 16.2 Number of green dots "awarded" to households for achieving water-saving targets (City of Cape Town 2019h)

	<6000 L per month	<10,500 L per month	Total dots awarded
January 2018	153,819	159,743	313,562
February 2018	203,144	166,184	369,328
March 2018	218,705	167,008	385,713
April 2018	211,497	171,640	383,137
May 2018	217,271	182,404	399,675
June 2018	217,254	183,284	400,538
July 2018	211,487	185,697	397,184
August 2018	212,720	186,631	399,351
September 2018	203,620	189,663	393,283
October 2018	190,165	191,974	382,139

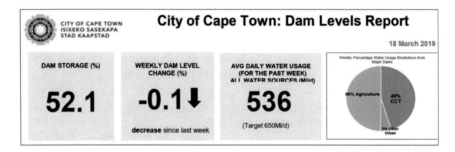

Fig. 16.6 Extract from the City of Cape Town Dam Levels Report (18 March 2019)

Table 16.3 Historical water storage and supply from the Dam Levels Report (18 March 2019)

Major dams	Storage						
	Capacity MI	% 18 March 2019	% Previous week	% 2018	% 2017	% 2016	% 2015
Berg River	130,010	73.3	73.4	47.8	37.7	28.0	64.3
Steenbras lower	33,517	50.2	50.8	40.6	33.4	46.1	47.6
Steenbras upper	31,767	70.0	70.0	85.1	56.1	58.4	80.8
Theewaterskloof	480,188	40.8	40.4	10.7	24.3	36.9	63.5
Voëlvlei	164,095	62.9	64.1	14.8	27.3	20.7	57.4
Wemmershoek	58,644	58.2	59.7	43.8	35.9	50.8	61.5
Total stored	898,221	467,603	468,782	204,295	260,650	311,476	560,502
% Storage		52.1	52.2	22.7	29.0	34.7	62.4

storage and supply, indicated in Table 16.3 and Fig. 16.5 (City of Cape Town 2019e, 2020a).

Ziervogel (2019a) stated that *"The severe drought experienced by Cape Town in 2017 and 2018 and the near miss of "Day Zero" should serve as a warning for other cities as to what climate impacts might look like in future."* Ziervogel (2019b) identified three phases in a chronological reflection of the drought, being the "new normal" phase early in 2017, followed by the "Day Zero" and disaster management phase from late 2017 and drought recovery in 2018. Whereas the above represents response phases, the drought started from a meteorological perspective in 2015 (Ziervogel 2019b). The lessons learned through the Cape Town case study relate to technical aspects as well as governance and societal aspects.

The above analysis indicates that demand management was an effective component of the drought experience. Consumption habits was also identified by Andréassian (2018) as key areas for long-term solutions. The analysis, however, shows that effective demand management as an outcome depends on a number of other issues. Eberhard (2018) highlighted, among other factors, institutional strengthening with a need to maintain focus on underlying systemic issues rather than crisis response. Cullis and Fisher-Jeffes (2019) emphasised the need to understand the difference between long-range planning and drought response. The good cooperation between the different spheres of government saved the system, whereas the relationships between individuals played an important part. There is, however, a need to invest in partnerships beyond the city (Ziervogel 2019a). There was a clear picture emerging that civil servants tried to depoliticise the issue. Whereas commentators emphasised the failure of the national government to ensure bulk supply, referring to political motives (Turton 2018), the civil servants focussed on a more constructive collaborative approach.

The data indicated that there was not an immediate need to augment the system, but rather to manage the existing capacity well (Flower 2019). Eberhard (2018) suggested that supply should be increased and diversified by adding more expensive groundwater, reuse and desalinated water, combined with surface augmentation and catchment management. Different views did create tension between the national and local tiers of government. The response to the crisis demonstrated the power of collaboration across industry, between competitors, with government, with civil society. It highlighted the need to pull stakeholders together when facing this kind of large issues to achieve shared focus and cooperation across society, which makes it possible to move forward (Koor 2018). Lessons related to governance include the need to build systems and relationships of mutual accountability for effective water management between spheres of government and strengthen horizontal/transversal management between municipal departments and entities (Ziervogel 2019a).

Local governments are in a better position to take decisive action and act at a local scale, whereas national governments are slow to intervene, and when they do, their actions are often not at the right scale or timely enough (Winter 2018). To achieve this, cities need more autonomy to act decisively, although proactive, inter-governmental support and cooperation are both helpful and necessary. This perspective is in line with the Cynefin Framework's approach to managing complex systems (Snowden and Boone 2007), where the design should explicitly allow for managed impulses to the complex system and the effective assessment of the response of the system to the impulses. Management options and action can then be tailored according to the emergent patterns of the biophysical social and political subsystems.

With the vastly reduced consumption levels, the City of Cape Town has lost a lot of revenue. At the same time, interventions that were implemented came at a high cost. When the City announced its intention to implement a drought levy, there was significant public outcry, resulting in the City having to change its strategy. The issue is still unresolved. Ziervogel (2019a) identified the need to recognise the limitations of the current financial model for water as a key lesson from the crisis.

The City of Cape Town's public awareness Website has been recognised worldwide as one of the best, but more effort is required to contain the level of misinformation shared in the public domain and media (Winter 2018). Public trust has been identified as being key to encouraging water saving and helping to establish confidence in managing the crisis (Winter 2018). Trust is strengthened by honest, credible messaging, whereas trust gains momentum when citizen voices are heard and when politicians and officials respond accordingly. This view is supported by Pengelly (2018). Communication from the City was however not always perfect. For example, there was a need to communicate the fact that "*we can't build our way out of it*" and "*we are in this together.*". The early messages were not strong enough, with the initial 10% saving not being not serious enough. The Levels 6 and 6B restrictions that were implemented later did not achieve results across the board. There was also a need to communicate risk more effectively, for instance a one in 50-year event does not mean it will happen once every 50 years (Cullis and Fisher-Jeffes 2019).

There was limited capacity within the city government to handle stories on social media (Ziervogel 2018). The main problems in the workings of the democratic process in relation to the Cape Town water crisis were silo-based thinking with ego-based decision-making, a lack of meaningful engagement in modern democracy, a lack of acceptance of a change in conditions, a lack of the necessary willingness and ability to confront the crisis with a sense of urgency and a lack of trust. "W*e need an entrepreneurial-driven, decentralised, self-organising, tech-enabled dynamic reinvention of how government, business, academia, civil society, and the tech and creative industries engage and implement*" (Gabriel 2018).

Water use management in agriculture was also identified as an important contribution to the response (Andréassian 2018). From Western Cape Water Supply Scheme perspective, licences were issued by DWS for agriculture and urban development that should not have been issued. These licences could also have an impact on future schemes. In this regard, there was a disconnect between the regional and national offices of the Department (Cullis and Fisher-Jeffes 2019).

Climate signals and associated changing hydrology could have an impact, but more data is needed to confirm this. Counter to this requirement, there is less monitoring than in the past, with the long-term trends not clearly showing a decline (Cullis and Fisher-Jeffes 2019). Adaptation to climate change requires better preparation to deal with a prolonged drought (Winter 2018). A water- resilient city should reduce their risk by diversifying water sources to include supplies from groundwater, storm water, reused water, treated effluent and desalination and integrate the whole urban water cycle into its water resource management system (Winter 2018). A city without reliable data will struggle to implement strategic plans and priorities, but not all data are useful, and more data adds little value in the absence of robust analytical and reporting systems (Winter 2018). The WCWSS was preconditioned to have a higher risk of failure than people thought, because it assumed the climate was not changing, whereas if the impact of climate change was taken into account more robustly, we would have had a much better and more reliable estimate of what needs to happen in terms of planning for the evolution of the water supply system (New 2018). The way we will feel climate change is through extreme events, so we need some redundancy built into our system (and expect and pay for that redundancy) and that we have to be ready and respond to events much quicker (Pengelly 2018).

The City of Cape Town incorporated scenarios in the 2020 water strategy, with the key variables being future water demand (without restrictions) and future changes to water availability (climate change affecting rainfall, temperature and wind) (City of Cape Town 2020b). This approach is however limited since it only considers known drivers of change. The scenarios for the South African water sector for instance identified the commitment to sustainable development and institutional capacity to deal with complexity (Funke et al. 2013), whereas the scenarios for the maritime sector included sector unification and technology uptake as key drivers (Claassen et al. 2014).

An adaptive water-sensitive urban system needs to be geared towards recognising and working with change. This includes the need to engage with the various parts of the water system, such as social, ecological and physical, while remaining agile and adaptive (Ziervogel 2019a). No single actor can tackle this scale of complex challenge alone, which is why partnerships and leadership are essential (Ziervogel 2019a). The first step of adaptation planning is understanding the sensitivity or vulnerability of the system and incorporating this by means of scenario planning, whereas the scenario planning that was happening at the end of the crisis should really become standard practice (New 2018). We should focus on adaptation pathways rather than a master plan that drives the building of a massive system that might become redundant. An adaptive pathway should lead to flexibility in the decisions as we learn more and more about the impacts of climate change (Ziervogel 2019a).

16.7   Conclusions

The City of Cape Town case demonstrated the value of joint efforts from different tiers of government, residents, the private sector and agriculture in averting the crisis. The response was pivoted on a good understanding of the water supply system, with a pressure control system in place. The City employed an extensive communication strategy coupled with water restrictions and a tiered pricing structure, with residents reducing their water use substantially in response. Organised agriculture reached an agreement with the national government to release stored water on condition that future allocations would not be curtailed, so that their reserves could be replenished after the drought. The business and tourism sectors implemented extensive water demand management strategies to reduce their water use. Although the combined efforts of role players prevented a "Day Zero" situation from occurring, the system could be made more resilient by implementing broad-based scenario planning. Whereas drivers such as population growth, climate change and pollution are known factors to respond to, the COVID-19 pandemic (WHO/UNICEF 2020) demonstrated that we cannot rely on our predictive capacity to achieve desired outcomes, but need a learning system that considers a wider array of drivers of change, designed for efficient feedback and learning and ultimately allows for effective adaptive management.

Key lessons that can be drawn from the above perspectives relate to predictability and an appropriate decision-making approach. It is clear that long-term predictions at broader geographic scales provide useful perspectives and scenarios to inform planning. When such predictions are however downscaled to the local level at annual and seasonal time scales, the reliability of the predictions are greatly reduced. The management of such a complex system requires a clear analysis of the context for each decision according to the Cynefin Framework (Snowden and Boone 2007). Such analyses will direct resources to information and expertise that is appropriate for the uncertainty and risk associated with each decision and thus contribute to water security under conditions of increased unpredictability.

References

Africa Research Bulletin (2018) Cape Town water supply crisis: a ratings agency raises red flags over the city's desperate shortage. Afr Res Bull 22001 (Wiley)

Andréassian V (2018) 'Day Zero': from Cape Town to São Paulo, large cities are facing water shortages. The conversation, June 18

Cameron R, Katzschner T (2017) Every last drop: the role of spatial planning in enhancing integrated urban water management in the City of Cape Town. S Afr Geogr J 99(2):196–216

CapeTalk (2018) Waterwatch: city of Cape Town relaxes water restrictions to Level 5, 10 Sept 2018

City of Cape Town (2011a) Alternative water resources: introduction to alternative water resources (pamphlet no 1 of 4)

City of Cape Town (2011b) Alternative water resources: Boreholes/wellpoints (pamphlet no 2 of 4)

City of Cape Town (2011c) Alternative water resources: Greywater re-use (pamphlet no 3 of 4)
City of Cape Town (2011d) Alternative water resources: rainwater harvesting (pamphlet no 4 of 4)
City of Cape Town (2018a) State of Cape Town. Research branch: organisational policy & planning
City of Cape Town (2018b) Guidelines for the collection and use of sea water for household purposes during the Cape Town water crisis
City of Cape Town (2019a) Guidelines for installation of alternative water systems
City of Cape Town (2019b) Our shared water future: Cape Town's water strategy
City of Cape Town (2019c) Water dashboard, 28 Oct 2019
City of Cape Town (2019d) Dam levels report, 30 Mar 2020
City of Cape Town (2019e) Dam levels report, 18 Mar 2019
City of Cape Town (2019f) Water star rating certification. www.capetown.gov.za/City-Connect/Apply/Municipal-services/Water-and-sanitation/apply-for-water-star-rating-certification. Accessed 25 Mar 2019
City of Cape Town (2019g) Water saving resources. http://www.capetown.gov.za/Familyandhome/Education-and-research-materials/Graphics-and-educational-material/water-saving-resources. Accessed 25 Mar 2019
City of Cape Town (2019h) The city water map. https://citymaps.capetown.gov.za/waterviewer/. Accessed 25 Mar 2019
City of Cape Town (2019i) Water map. http://www.capetown.gov.za/Familyandhome/Residential-utility-services/Residential-water-and-sanitation-services/cape-town-water-map. Accessed 25 Mar 2019
City of Cape Town (2019j) Water saving toolkits. http://www.capetown.gov.za/Familyandhome/Education-and-research-materials/Graphics-and-educational-material/water-saving-resources. Accessed 25 Mar 2019.
City of Cape Town (2020a) Dam levels report, 28 Dec 2020
City of Cape Town (2020b) Our shared water future: Cape Town's water strategy. https://resource.capetown.gov.za/documentcentre/Documents/City%20strategies,%20plans%20and%20frameworks/Cape%20Town%20Water%20Strategy.pdf. Accessed 30 July 2020
City of Cape Town (2010–2018) Water and sanitation annual reports
Claassen M, Funke N, Lysko MD, Ntombela C (2014) Scenarios for the South African maritime sector. In: Funke N, Claassen M, Meissner R, Nortje K (eds) Reflections on the state of research and development in the marine and maritime sectors in South Africa, pp 53–64. Council for Scientific and Industrial Research, Pretoria
Climate Systems Analysis Group (2020) Online graphs. http://www.csag.uct.ac.za/current-seasons-rainfall-in-cape-town/. Accessed 3 Apr 2020.
Costanza R, Maxwell T (1994) Resolution and predictability: an approach to the scaling problem. Landscape Ecol 9(1):47–57
Cotterill J (2018) South Africa: how Cape Town beat the drought. Financial Times, May 2
Cullis J, Fisher-Jeffes L (2019) Personal communication. 18 Apr
De Feo G, Laureano P, Drusiani R, Angelakis A (2010) Water and wastewater management technologies through the centuries. Water Sci Technol Water Supply 10:337–349
DEA (2017) South Africa's 2nd annual climate change report. Department of Environmental Affairs, Pretoria
DWAF (2007) Western Cape water supply system: Reconciliation strategy study. Volume 1 of 7: Reconciliation Strategy
Eberhard R (2018) Interview. Cape Town drought response learning initiative. www.drought-response-learning-initiative.org/film-library/. Accessed 3 Apr 2020
Explore South Africa (2018) #WaterCrisis: SewTreat produces unbelievable water savings for Cape Town's One&Only resort. 20 June 2018
Flower P (2019) Personal communication

Funke N, Claassen M, Nienaber S (2013) Development and uptake of scenarios to support water resources planning, development and management: examples from South Africa. In: Wurbs R (ed) Water resources planning, development and management. Intech publications, Rijeka, pp 1–27

Gabriel G (2018) Interview. Cape Town drought response learning initiative. www.drought-response-learning-initiative.org/film-library/. Accessed 3 Apr 2020

Green Cape (2018a) Case study: reducing water use in offices. JG Afrika, Cape Town

Green Cape (2018b) Case study: reducing water wastage in the hospitality industry. Vineyard Hotel, Cape Town

Green Cape (2018c) Water: market intelligence report 2018. Green Cape, Cape Town

Green Cape (2019) Case study: reducing water use in cement manufacturing. PPC Cement, De Hoek, Cape Town

Grey D, Sadoff CW (2007) Sink or Swim? Water security for growth and development. Water Policy 9(6):545–571

Hoekstra AY, Buurman J, Ginkel KCH (2018) Urban water security: a review. Environ Res Lett 13(5):053002

Kinouchi T, Nakajima T, Mendoza J, Fuchs P, Asaoka Y (2019) Water security in high mountain cities of the Andes under a growing population and climate change: a case study of La Paz and El Alto, Bolivia. Water Secur 6:100025

Knight J (2019) Cape Town has a plan to manage its water. But there are big gaps. The Conversation, 27 Feb

Koor F (2018) Interview. Cape Town drought response learning initiative. www.drought-response-learning-initiative.org/film-library/. Accessed 3 Apr 2020

Kotzé P (2010) Cape Town—water for a thirsty city (part 1). Water Wheel November/December 2010. Published by the Water Research Commission

Kotzé P (2011) Cape Town—water for a thirsty city (part 2). Water Wheel January/February 2011. Published by the Water Research Commission

Madonsela B, Koop S, van Leeuwen K, Carden K (2019) Evaluation of water governance processes required to transition towards water sensitive urban design—an indicator assessment approach for the City of Cape Town. Water 11(292):1–14

Malan N (2018) Are icebergs a realistic option for augmenting Cape Town's water supply? Water Wheel March/April 2018

Old Mutual (2018) Old mutual Cape Town will go "off the water grid" with launch of water filtration plant. www.oldmutual.co.za/media-centre. Accessed 3 Apr 2020

New M (2018) Interview under the drought response learning initiative. www.drought-response-learning-initiative.org/film-library/. Accessed 3 Apr 2020

NOAA (2020) Climate at a Glance. https://www.ncdc.noaa.gov/cag/global/time-series/. Accessed 3 Apr 2020

Olivier DW (2017) Cape Town water crisis: 7 myths that must be bust. The Conversation, 7 Nov

Olivier DW (2018) Cape Town's map of water usage has residents seeing red. The Conversation, 17 Jan

Orheim O (2018) Response to article—Iceberg harvesting IS a possibility. Water Wheel March/April 2018

Pankratz T (2018) Seawater desalination in Cape Town, South Africa. Technical memorandum from Water Consultants International

Pengelly C (2018) Interview under the drought response learning initiative. www.drought-response-learning-initiative.org/film-library/. Accessed 3 Apr 2020

Rossouw N (2019) Personal communication

Snowden DJ, Boone ME (2007) A leader's framework for decision making. Harvard Bus Rev (Nov 2007)

Solomon N (2019) Personal communication

Stats SA (2019) http://www.statssa.gov.za. Accessed 20 May 2020

The Beverage Company (2019) World Water Day on 22 March: "Leaving no one behind". www.thebeveragecompany.co.za/sustainability/initiatives/. Accessed 20 May 2020

Turton AR (2018) Interview in DW documentary entitled "South Africa: Cities without water" (17:26–18:20)

Vairavamoorthy K, Matthews N, Brown P (2019) Building resilient urban water systems for an uncertain future—the source magazine. IWA Publishing, 15 Mar 2019

van Ginkel KCH, Hoekstra AY, Buurman J, Hogeboom RJ (2018) Urban water security dashboard: Systems approach to characterizing the water security of cities. J Water Resour Plan Manag 144 (12):04018075

Virgin Active (2019) Water stewardship at virgin active in South Africa. www.virginactive.co.za/blog/water-wise. Accessed 15 June 2020

Voutchkov N (2017) Critical review of the desalination component of the WRP. Technical memorandum from Water Globe Consultants

Wang RY, Dai L (2021) Hong Kong's water security: a governance perspective. Int J Water Resour Dev 37(1):48–66

Western Cape Government (2018) Provincial economic review and outlook 2018. https://www.westerncape.gov.za/assets/departments/treasury/Documents/Research-and-Report/2018/2018_pero_revised.pdf. Accessed 20 June 2020

WHO/UNICEF (2020) Water, sanitation, hygiene and waste management for the COVID-19 virus: technical brief. WHO/2019-nCoV/IPC_WASH/2020.1

Winter K (2018) Five key lessons other cities can learn from Cape Town's water crisis. The Conversation, 3 Apr

Wright T, Jacobs HE (2016) Potable water use of residential consumers in the Cape Town metropolitan area with access to groundwater as a supplementary household water source. Water SA 42(1):144–151

Ziervogel G (2018) Interview. Cape Town drought response learning initiative. www.drought-response-learning-initiative.org/film-library/. Accessed 20 June 2020

Ziervogel G (2019a) What the Cape Town drought taught us: 4 Focus areas for Local governments. Cities Support Programme/National Treasury

Ziervogel G (2019b) Unpacking the Cape Town drought: lessons learned. Cities Support Programme/National Treasury

Chapter 17
Temperature–Rainfall Anomalies and Climate Change: Possible Effects on Australian Agriculture in 2030 and 2050

R. Quentin Grafton and Glyn Wittwer

Abstract Australia is a food-surplus country with much of its landmass located in arid or semi-arid areas subject to extreme variation in both precipitation and summer temperatures. The possible economic effects of climate change on water in relation to Australian agriculture are analysed by region and sector, including its possible impacts in Australia's 'food basket', the Murray–Darling Basin, using precipitation and temperature data from 2011 to 2020. Three scenarios are evaluated that include: (1) the decade 2011–2020; (2) a '2030' scenario in which farm productivity falls by 10% relative to the 2011–2020 scenario in five of the 10 growing seasons; and (3) a '2050' scenario in which farm productivity falls by 20% relative to the 2011–2020 scenario in five of the 10 growing seasons. The welfare impacts, on a national basis, of the first scenario relative to a baseline without year-on-year seasonal variations is minus $35 billion in net present value terms. The welfare impact in the second scenario is minus $46 billion and minus $59 billion in the third scenario. The findings support the selection and implementation of particular adaptation pathways in response to climate change for Australian agriculture.

Keywords Drought · Productivity · Climate variability · CGE models · Murray–Darling Basin

R. Q. Grafton (✉)
Crawford School of Public Policy, The Australian National University,
Crawford Building (132), Lennox Crossing, Acton, ACT 2601, Australia
e-mail: quentin.grafton@anu.edu.au

G. Wittwer
Victoria University (Melbourne), Footscray, Australia
e-mail: glyn.wittwer@vu.edu.au

17.1 Introduction

The Australian Bureau of Meteorology reported that the mean temperature in Australia in the decade ending 2020 was 0.33°C above the previous decade, with the average surface air temperature increasing by more than 1°C since 1910. This warming trend is accelerating with seven of the ten warmest years on record in Australia having occurred since 1998 (Bureau of Meteorology and CSIRO 2018) with the 2018–2019 summer being the hottest on record for Australia by a margin of 0.86°C (Bureau of Meteorology 2019)..

Higher Australian temperatures have multiple impacts and, thus, require multiple responses (Australian Academy of Science 2021). Climate change impacts in Australia include, but are not limited to, public health (Hughes and McMichael 2011), labour productivity (Zander et al. 2015), tourism (Climate Council 2018), agriculture (Howden and Stokes 2010), biodiversity (Howden et al. 2003), the Great Barrier Reef (Great Barrier Reef Marine Park Authority 2019) and the frequency and severity of wildfires (Abram et al. 2021). High temperatures and dry conditions in 2019–2020 contributed to extensive Australian wildfires that burnt some 30 million hectares based on satellite data (Bowman et al. 2020), destroyed more than 2,400 buildings and resulted in the immediate loss of 34 deaths (Filkov et al. 2020), with hundreds suffering early mortality as the result of smoke exposure (Arriagada et al. 2020). Thermal expansion of the oceans due to rising sea-surface temperatures has resulted in sea-level rise of about 3 cm per decade (Australian Academy of Science 2015). In turn, this has contributed to increased coastal flooding in Australia during storm surges and may also contribute to saltwater intrusion (Costall et al. 2020) in freshwater coastal aquifers.

Here, we focus on the effects of climate change on Australian agriculture only and in relation to three scenarios (1) Scenario One, the decade 2011 to 2020 seasons; (2) Scenario Two, a '2030' scenario in which farm productivity falls by 10% relative to Scenario One in five of the 10 growing seasons; and (3) Scenario Three, a '2050' scenario in which farm productivity falls by 20% relative to Scenario One in five of 10 growing seasons.

Our analysis quantifies climate change impacts by region and by agricultural sectors in terms of real Gross Domestic Product (GDP), employment and real wages. This is done by using temperature and rainfall anomalies and including their possible effects on agricultural productivity noting that three scenarios we examine do not account for seasonal variation within each year. Our approach is to undertake Computable General Equilibrium (CGE) modelling which is a method that has been applied previously in Australia, and other countries, to assess the economic impacts of water scarcity effects (Liu et al. 2016). Our results are obtained from a dynamic, multi-regional CGE model, VU-TERM (Wittwer et al. 2005; Wittwer and Griffith 2011, 2012), that is an Australian economy-wide model that includes small-region representation.

17.2   Modelling the Effects of Climate Change on Australian Agriculture

In this section, we present a summary of Australian rainfall and temperature trends. Based on anomalies over the period 2011–2020, we model three scenarios in relation to climate change that account for annual weather variability, regional impacts and sectoral variations for Australian agriculture. A summary of impacts in Australia's 'food bowl', the Murray–Darling Basin (MDB), located in south-eastern Australia is also highlighted.

17.2.1   Anomalies in Rainfall and Temperatures

Climate is highly variable in Australia relative to many other countries which makes it difficult to discern time trends. Nevertheless, an observable and statistically significant trend is evident in both maximum and mean surface temperatures with an average increase in mean temperature for Australia of about 0.1°C per decade (Ukkola et al. 2019). This increasing trend exceeds the observed variability during all seasons, is consistent with anthropogenic climate change (Karoly and Braganza 2005a) and is observable for all regions, including south-eastern Australia (Karoly and Braganza 2005b). By contrast, national average annual rainfall and atmospheric water demand (as measured by pan evaporation) trends remain within historical bounds and appear to be non-stationary. Nevertheless, at a regional level, in southwestern Australia, winter rainfall is less than the historical mean and the last decade has been the driest on record (Abram et al. 2021) while there also appears to be a possible shift towards summer rainfall in the MDB (Ukkola et al. 2019) that, all else equal, would reduce stream and river flows.

While historical and recent trends are not necessarily an indication of the future climate, they do provide a basis for evaluating the possible effects of climate change on Australian agriculture. These effects include rising mean temperature and an increased frequency of high temperature (above 35°C) days (Hennessy et al. 2010). More extreme heat days, all else equal, increases the severity and length of agricultural droughts (McDonell et al. 2020). At a global scale, Lesk et al. (2016) show that extreme heat events reduce national cereal yields, in the year that they occur, by about 9% (8.4–9.5%, 95% confidence interval). The effects of climate change on grain yields are particularly important because, should there be a global average crop yield increase for grains of *less* than 1% per year to 2050, there would be large global food deficits before 2050 (Grafton et al. 2017).

17.2.2   Climate Change Scenarios for Australian Agriculture

We investigate three possible climate scenarios with full details provided in Wittwer (2021). Scenario One (2020) is a stylised version of 2020 conditions and is based on actual mean rainfall and temperature recorded from 2016 to 2020. In the model, Year 1 agricultural productivity represents 2016 conditions, Year 2 2017 conditions and so on until year 5 noting that much of south-eastern Australia was in drought from 2016 to 2020. Thereafter, Year 6 represents 2011 conditions, up to year 10 that represents 2015 conditions. Scenario Two (2030) is a stylised depiction of 2030 with agricultural productivity 10% lower in years 1, 5, 7, 9 and 10 than in Scenario One, and productivity levels are unchanged in years 2, 3, 4, 6 and 8. Scenario Three (2050) is a stylised depiction of 2050 with agricultural productivity 20% lower in years 1, 5, 7, 9 and 10 than in Scenario One, and productivity levels are unchanged in years 2, 3, 4, 6 and 8. Our modelling captures the climate variability over the past decade but includes a productivity shock that we attribute to more extreme heat events in 2030 and 2050.

Given the large area and climate variations in Australia, we capture regional, sectoral and temporal (annual) variations when evaluating the possible effects of climate change. Spatial variation is needed because inland farming regions may be more vulnerable than coastal regions to rising temperatures and are subject to greater changes in rainfall anomalies.

17.2.3   National Impacts

Scenario One models two phases in the 2011–2020 decade. First, years 1–5 (i.e. representing 2016–2020) use observed regional annual rainfall and temperature differences from average to infer farm productivity levels for each region. Year 6–10 are based on historical annual rainfall and temperature differences for 2011–2015. Details and maps of these annual rainfall and temperature anomalies are provided in Wittwer (2021).

In years of drought, productivity falls sharply relative to the base in agricultural sectors in farm-affected regions. A key measure of the effects of drought is via income, as defined by GDP, which is a function of the primary factors of production (capital K and employment L) and the underlying technology ($1/A$), defined by Eq. (17.1):

$$\text{GDP} = f\left(K, L, \frac{1}{A}\right) \tag{17.1}$$

While agriculture accounts for little more than 2% of Australian GDP, wide variations in seasonal conditions can make substantial contributions to national income fluctuations. This is shown in Fig. 17.1, real GDP falls to 1% below base in year 4 (equivalent to the drought year of 2019) that represents the hottest year on record for Australia. Some of the decline in capital used in agriculture, which falls to as much as 0.2% below base, arises from the culling of livestock in drought years, and reductions in investment in drought years (Fig. 17.2).

The assumption underlying Scenario Two (2030) is that the 'normal' climate years of 2011–2020 are hotter with a consequent reduction in agricultural productivity of 10% relative to Scenario One during drought years. Scenario Three

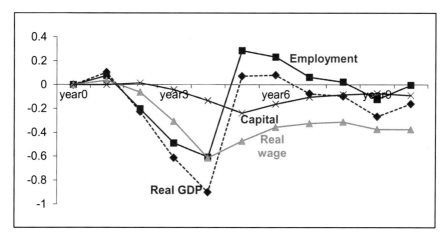

Fig. 17.1 Australian Real GDP, wages, employment and capital variations for Scenario One (2020) decade (*% deviation from base*)

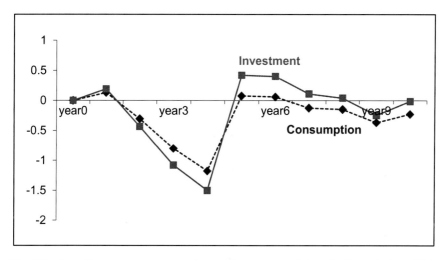

Fig. 17.2 Australian aggregate consumption and investment variations for Scenario One (2020) decade (*% deviation from base*)

(2050) assumes an even larger productivity decline of 20% relative to Scenario One during drought years, an assumed consequence of climate change. Thus, both Scenario Two and Scenario Three result in weaker recovery years and larger income losses than Scenario One. Employment is slightly lower in both Scenarios Two (2030) and Three (2050) than in Scenario One (2020).

Drought-induced productivity losses also impose costs in relation to the labour market. In particular, Fig. 17.1 shows productivity losses depress both real wages and employment nationally relative to the base. In year 5 (equivalent to 2020), which is a recovery year relative to year 4 and associated with a higher average rainfall and lower temperature, aggregate investment and employment rise (Figs. 17.1 and 17.2).

In years 3 and 4, the percentage fall in real GDP is larger than the percentage fall in labour or utilised capital at the national level and in many of the model's regions. Australia-wide, employment falls about 0.7% below the base in the drought-affected year 4 (equivalent of 2019). By assumption in the model, real wages adjust slowly at the regional level. Thus, in severe regional downturns, real wage adjustment is limited: any workers who are relatively mobile leave a drought-affected region. In addition, labour force participation rates may fall temporarily in drought. Consequently, to avoid unrealistic downward wages adjustment, exogenous temporary downward labour supply shifts are imposed in several regions in year 4 such that no regional real wage falls more than 2% below the base in any year.

17.2.4 Regional Impacts

The regions for which VU-TERM impacts are provided are shown in Fig. 17.3. Table 17.1 shows that the impacts in relatively agriculture-intensive regions are proportionally much greater than those at the national level. In particular, the regions of Western New South Wales (NSW); Far West-Orana, New England-North West-Grafton, Darling Downs-Maranoa-Granite Belt, and also Outback Queensland, there are prolonged periods in which real GDP falls more than 8% below base in the Scenario One (2020). At a regional scale, in Outback Queensland, in which years 9 and 10 turn from moderate drought in Scenario One (2020) to severe drought in Scenario Three (2050), a labour supply shift reduces the real wage drop but worsens the employment outcome, which is 1.3% below base in year 10, because labour leaves the region. While the VU-TERM model allows for an endogenous decrease in labour supply over time, it falls to 0.6% below base in year 10 and this is insufficient to prevent real wages from dropping more than 2% below base in Outback Queensland in that year. In Far West-Orana and New England-North West–Grafton, employment is around 4% below base in year 4, reflecting an exogenous downward labour supply shift.

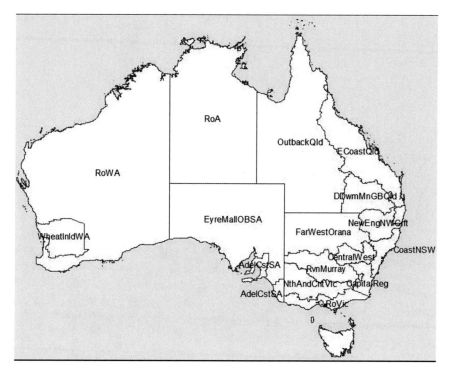

Fig. 17.3 Regions of the VU-TERM model. *Source* Glyn Wittwer

Figure 17.4 shows agriculture's percentage share of income in each VU-TERM region. These relative agricultural income shares allow us to estimate how regional GDPs differ in any given year for Scenario Two (2030) and Scenario Three (2050), relative to Scenario One (2020). For example, in Far West–Orana, agriculture's share of GDP is 11.4% in year 0. In year 1, in which agriculture's productivity in the Scenario Three (2050) is 20% lower than in the Scenario One (2020), our initial estimate is that real GDP will fall by 2.3% (=0.2 × 11.4%, based on Eq. 17.1) relative to the Scenario One (2020). The year 1 column (a year of above average rainfall) of Table 17.1 shows that real GDP in Far West–Orana is 0.8% above base in the Scenario One (2020), 0.7% below based in Scenario Two (2030), and 1.2% below base in Scenario Three (2050). In subsequent years in which agricultural productivity is lower, such as in Scenario Three (2050), the modelled difference is smaller than implied by this calculation because of the gradual decline in agriculture's share of regional GDP over time and changing database weights due to endogenous output price changes.

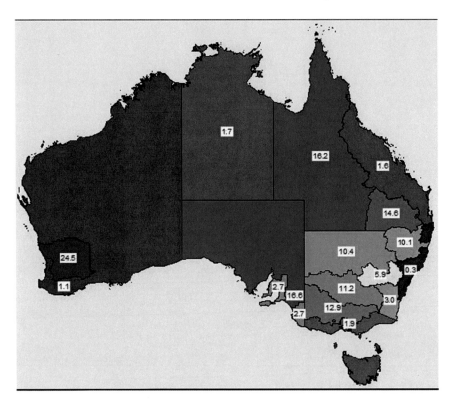

Fig. 17.4 Agricultural Share of Income by Region, Year 0. *Source* VU-TERM database

17.2.5 Sectoral Changes in Agriculture

VU-TERM allows for substitutability between land and fodder (*HayCerealFod*) in livestock production and assumes that fodder produced in one year may be used in another, thereby providing a means of managing seasonal risk. Intermediate inputs are assumed to be used in the same period as they are produced but *HayCerealFod* productivity shocks are not imposed in Scenario One (2020) because fodder produced in good weather years may be later used in drought years. In rangeland production in Outback Queensland, we assume that such substitution is not possible and that output (i.e. herd numbers) must diminish during prolonged dry spells.

Table 17.2 provides VU-TERM model results for the drought years 2 to 4. The model finds that beef cattle output in New England—North West—Grafton rises relative to base because fodder (including the use of feedlots) inputs substitute for land inputs throughout the region. Beef cattle output, however, falls dramatically in Outback Queensland due to an inability to switch to fodder in the region. The modelled outcome is that even though land productivity in beef cattle production declines in New England—North West—Grafton, there is an increase in output because of switch from Outback Queensland to the relatively favoured northern NSW region. In Scenario Two (2030) and Scenario Three (2050), productivity losses further decline in Outback Queensland and substitution of production to the NSW region is increased.

17.2.6 Murray–Darling Basin (MDB)

Located in south-eastern Australia, the MDB is commonly described as Australia's 'food basket' because it accounts for almost 40% (2005–2006) of the country's gross value of agricultural production (Australian Bureau of Statistics 2008, Table 4.21). The MDB encompasses multiple jurisdictions including the Australian Capital Territory, parts of southern Queensland, most of New South Wales, northern Victoria and southern South Australia. The northern MDB includes the Barwon–Darling River, and its catchments, and it is where much of the large stream flows are generated by cyclonic activity from the Pacific. The southern MDB includes the Murrumbidgee and the Murray Rivers, and it is where much of the stream flow is generated from winter rains. In the northern MDB, water is principally kept in private storages and extracted directly by individual irrigators from the streams and rivers. In the southern MDB, inflows are principally captured in large publicly owned storages and water is extracted as part of well-developed large water infrastructure that would, typically, service a group of irrigators.

Water in the MDB is jointly managed by state governments and the federal government. The federal government instituted a Basin Plan in 2012, due to expire in 2026, that had as one of its goals to ensure sustainable level of extractions of both surface and groundwater (Grafton 2019). The sustainable diversion limits (SDLs) in the current Basin Plan were developed based on the historical climate record (1895–2006) and, thus, did not account for climate change projections developed in the preparation of the Plan by the federal government's own science body, CSIRO (2008).

As in the rest of Australia, average surface temperatures in the MDB have increased and there has also been a substantial increase in extreme daily heat events since 1990. There is, however, no statistically significant trend in average rainfall beyond the historical climate variability (Bureau of Meteorology 2020). Using the climate change projections from CSIRO (2008) for both a median, and the 'best' estimate climate for 2030, Jiang and Grafton (2011) simulated the effects on irrigated agriculture using an 18-catchment hydro-economic model of the MDB. Under a median 2030 scenario, they found a 1% reduction in irrigation profits from a projected decline in water extraction of 4%. By comparison, their simulation of the effects of an extensive decade-long drought, known as the Millennium Drought that ended in 2010 was that a decline in water extractions for irrigation of 13% resulted in a much lower decline of 5% of irrigated agricultural profits.

Over the period from 2006 to 2008, the headwater regions of the southern Murray–Darling Basin suffered record rainfall deficits. These rainfall deficiencies resulted in a dramatic reduction in inflows to MDB rivers and, hence, unprecedented cuts in irrigation water allocations in the southern MDB. Wittwer and Griffith (2011) modelled impacts of this drought using TERM-H2O, a model including water accounts and factor mobility between irrigated and dry-land farm activities. Their model generated a decline in real GDP of more than 5% relative to a normal year base in the southern MDB for three consecutive years, plus a decline in employment in the region exceeding 6000 jobs. Their model also included sales of water from annual to perennial crop producers in the southern MDB in response to diminished water allocations and a virtual cessation of rice production.

A lesser proportional decline in the gross value of irrigated agricultural production (GVIAP) from an equivalent reduction in water extractions was also found using observed data during the Millennium Drought. Kirby et al. (2014) found that, despite a two-third decline in water extractions for irrigated agriculture from 2000–2001 to 2008–2009, there was only a 14% fall in GVIAP. This ability of farmers to reduce the impacts of reduced water availability was attributed by the authors to; crop switching, water trading, substitution of purchased feed for pasture and increased crop yields.

Using the VU-TERM model, we find little difference between Scenarios One, Two and Three in either the northern or the southern MDB except that fluctuations in GDP are more pronounced for Scenario Three (2050) followed by Scenario Two (2030), with the least variation for Scenario One (2020). As shown in Fig. 17.5, the northern MDB suffers the greatest impact in drought year (year 4) with a reduction in GDP of about 12% but the region is still able to recover quickly in a non-drought year (year 6). Real wages and employment decline by around 2% and 4%, respectively, by year 4. In the southern MDB (Fig. 17.6) declines in GDP are around 4.5% by year 4 but the impact on both employment and real wages is minimal.

A partial explanation for the bigger negative effect on the northern MDB is that the impacts of rainfall deficiencies are more severe than in the southern Basin. This is because they are based on the decade ending in December 2019: the final three years of this decade resulted in the record rainfall deficiencies over much of New South Wales and southern Queensland, spanning most of the northern MDB. Rainfall in the southern MDB was also below average over this three-year period but drought was less severe than in the north.

17.2.7 Welfare Impacts

Multiple factors should be included when assessing welfare impacts of climate change. An example of a CGE model that has been used to quantify the effects on GDP from climate damages associated with multiple factors, including sea-level rise, losses in agricultural productivity, labour productivity and human health, energy demands and tourism, was developed by Kompas et al. (2018). In their CGE modelling of 139 countries, they find that for 3 °C warming the long-term decline in

Scenario One (2020)

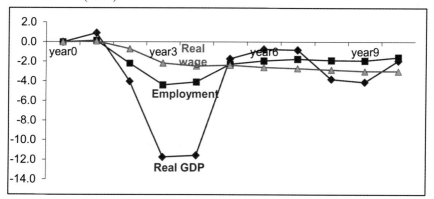

Scenario Two (2030)

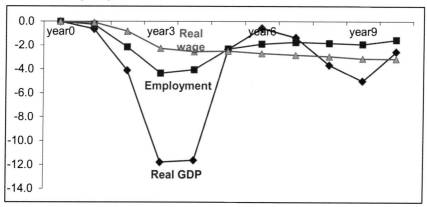

Scenario Three (2050)

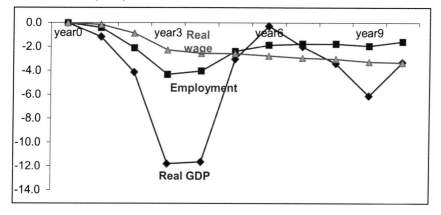

Fig. 17.5 Climate change and agriculture impacts in the Northern Murray–Darling Basin *(% deviation from base)*

Scenario One (2020)

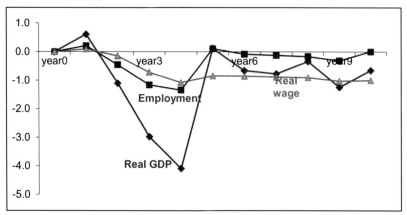

Scenario Two (2030)

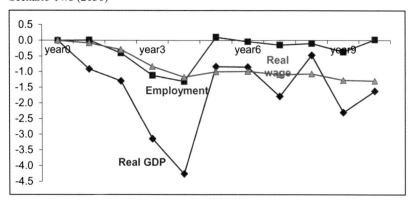

Scenario Three (2050)

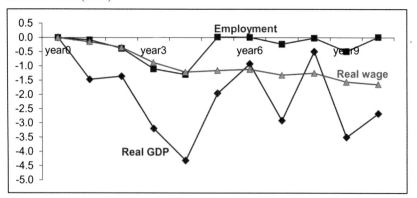

Fig. 17.6 Climate change and agriculture impacts in the Southern Murray–Darling Basin (*% deviation from base*)

Australian GDP is about 1% but the GDP losses for some African countries (Cote d'Ivoire and Togo) approaches 20% of GDP. Notably, GDP losses increase proportionally more than the proportional increase in the average surface temperature such that GDP losses are much higher with 4°C average warming.

Here, using the VU-TERM model, and only for Australia, we calculate the deviation in welfare (*dWELF*), as per Eq. (17.2), a measure of the overall welfare of climate change via Australian agriculture.

$$dWELF = \sum_d \sum_t \frac{dCON(d,t) + dGOV(d,t)}{(1+r)^t} - \frac{dNFL(z)}{(1+r)^z} + \frac{dKstock(z)}{(1+r)^z} \quad (17.2)$$

where *dCON* and *dGOV* are the deviations in real household and government spending in region *d* and year *t*; *dNFL* is the deviation in real net foreign liabilities in the final year (*z*) of the simulation; and *dKstock* is the deviation in value of capital stock in the final year (*z*) of the simulation; and *r* is the discount rate.

In Scenario One (2020), based on rainfall and temperature anomalies of 2011–2020, the net present value of the national welfare loss is A$35 billion. In Scenario Two (2030), this declines to A$46 billion, while in Scenario Three (2050), the national losses are calculated to be A$59 billion. If these losses are converted to annuities with a 2.5% discount rate (*r*), a net present value of minus $59 billion equates to an annuity of minus $1.5 billion or around $60 per Australian resident. While these welfare losses from Australian agriculture are small at a national level, they are large at a regional and sectoral level. For instance, in Outback Queensland, livestock production dramatically falls with droughts and this has a substantial impact on regional GDP because livestock production accounts for a relatively large share of its economy.

17.3 Adaptation Pathways

17.3.1 Climate Adaptation in Australian Agriculture

Two key issues stand out in relation to climate adaptation pathways in Australian agriculture. First, observations supported by modelling show that farmers are resilient to climate variability and, thus, can implement a range of adaptations in response to higher temperatures associated with climate change that include but are not limited to: earlier planting, change in crop type, change in varietals, plant breeding, better weather forecasting, smart metering for soil moisture, use of irrigation and moisture conservation (Howden et al. 2010). Second, in terms of rainfall variability, governments and farmers have adapted by building water storages (private and public) to reduce inter-temporal variations, utilised government subsidies to increase irrigation efficiency (Grafton 2019), unbundled water from land rights, and supported water markets to allow water to be reallocated to higher value uses (Grafton and Horne 2014).

While climate change is already imposing costs on Australian agriculture, other factors that are more amenable to control within Australia have an even larger immediate and on-going impact. For example, in the MDB, the projected effects of climate change on stream flows are much less than the impact of current irrigation water extractions on stream flows (Grafton et al. 2013). Importantly, failures in water governance, rather than water availability per se, are more important in the creation of water crises globally than is water availability (OECD 2011). In the MDB, documented failures or weaknesses in water governance (Murray–Darling Basin Royal Commission 2019; Productivity Commission 2018; Grafton and Williams 2020) are also the single most important factor in relation to reduced stream flows and environmental poor outcomes (Wentworth Group of Concerned Scientists 2020) and '…point to serious deficiencies in governance and management, which collectively have eroded the intent of the Water Act 2007 and implementation of the Murray-Darling Basin Plan (2012) framework' (Australian Academy of Science 2019, p. 2).

17.3.2 Adaptation Pathways

Adaptation pathways in response to climate change risk may be characterised by two distinct approaches. First, by 'top-down' approaches that, typically, involve the use of down-scaled climate model projections for public planning and investment purposes. Second, 'bottom-up' approaches that use the historical record and extreme events (such as droughts and floods) for planning and to promote resilience (Grafton et al. 2019) to climate shocks, especially to vulnerable populations.

Azhoni et al. (2018) illustrate (Fig. 17.7) the key components of adapting to climate changes in the water context that include: (1) impacts; (2) adaptation enablers; (3) institutional networks; and (4) implementation. This figure highlights that climate change generates multiple impacts (reduced water quality, higher water demand, etc.) that, in turn, need multiple and different enablers (technology, infrastructure, etc.) which operate in diverse institutional networks and that, ultimately, deliver adaptation actions or pathways. What these pathways should be, and when they should be implemented and sequenced, depends on the local or regional circumstances.

An approach for selecting among possible adaptation pathways to climate change, in the context of water, was developed Wilby and Dessai (2010) and is illustrated in Fig. 17.8. This figure highlights the need to identify as many adaptation options as possible and then undertake both a social acceptability and economic and risk appraisal to determine a sub-set of options for further investigation. They propose that the evaluation of possible adaptation be evaluated using methods that ensure options deliver desired outcomes in multiple possible states of the world. In this framework, the selected options are then regularly reassessed based

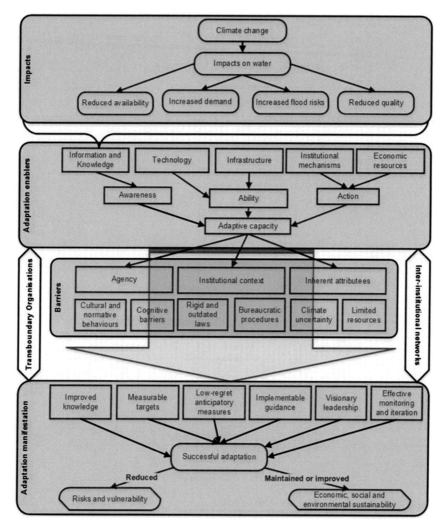

Fig. 17.7 Key components of adaptation to climate change in relation to water. *Source* Azhoni et al. (2018, p. 742)

on updated data and evidence. Ideally, the selected adaptation actions or pathways should be robust rather than optimal, perform well in a multiplicity of plausible futures (Groves et al. 2019) and, where possible, be 'no regret' or 'low regret' such that they be worth undertaking regardless of the state of the world. Further, such methods should include more than experts and seek to encompass all relevant stakeholders within participatory climate risk management processes (Döll and Romero-Lankao 2017).

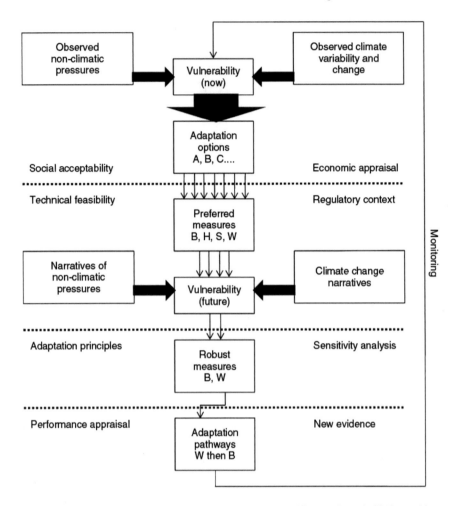

Fig. 17.8 Selection of climate adaptation pathways. *Source* Wilby and Dessai (2010, p. 183)

An overarching strategy for adapting to climate change that complements the approaches in Figs. 17.7 and 17.8 has been proposed by Matthews and Le Quesne (2008) in relation to water. This climate adaptation strategy comprises seven overarching tactics: (1) improve governance and institutional capacity; (2) support flexible water reallocations among sectors and between use and non-uses of water; (3) mitigate non-climate change pressures such as current over-extractions from water sources; (4) facilitate the movement of communities from more to less vulnerable locations; (5) assess water infrastructure (such as dams) to ensure they do result in maladaptation; (6) develop sustainable and risk-based flood management; and (7) promote climate change awareness and adaptation planning to improve risk management. Across all sectors, and in general, the Australian federal government highlights four key priorities for climate adaptation: (1) Understand and Communicate; (2) Plan and Act; (3) Check and Reassess; and (4) Collaborate and Learn (Commonwealth of Australia 2015, p. 18). Special attention must also be given to vulnerable communities, especially those in remote locations and First Peoples' communities which are the most exposed to heat events, reduced water availability and lower water quality.

Combined the approaches in Figs. 17.7 and 17.8 and the adaptation strategy of Matthews and le Quesne (2008) provide a basis for responding to climate change, in the context of water, in relation Australian agriculture. A first step in this process is to establish an ensemble of the multiple possible outcomes at regional and sectoral level, and their linkages to the rest of the economy, and that would include results from the VU-TERM model. The second step is to evaluate a range of adaptation actions from the farm scale to state and federal government initiatives, defined over time, by location and by actors, under alternative, plausible futures and then select among the possible alternatives those actions that are robust adaptation pathways. The third step is to evaluate the selected robust strategies within participatory processes to prioritise and to define the socially acceptable set of robust strategies. The fourth step is to leverage private and public sector funding and expertise to deliver, with well-defined objectives, performance measures and evaluation, the prioritised actions.

17.4 Conclusions

Climate change is already having a profound impact on the global economy and imposing costs on key sectors, such as agriculture. Australia, as the world's driest inhabited continent, is subject to large climate variability in terms of rainfall and summer temperatures. Thus, climate change has the potential to impose large costs on Australia and, in particular Australian agriculture.

Rainfall and temperature anomalies for the period 2011–2020 were used to model the impacts of climate change on Australian agricultural regions and sectoral changes. On a national scale, climate change impacts are small and reduce GDP by about 1% of GDP and investment by about 1.5% following a series of drought years based on data from an extreme drought period from 2016 to 2020. On a regional scale, however, our model results show that regional income losses and fluctuations are much larger than the national impact, especially in locations where agriculture accounts for a relatively large share of a region's income base.

In our analyses, we only considered the drought dimension of climate change. More extreme weather events arising from climate change are likely to result in worse flood events, more damaging bushfires, stronger sea surges and more severe cyclones than otherwise. Some regions, such as the northern coast of New South Wales, have experienced the extremes of fires and floods within a period spanning little more than a year (Nicholas and Evershed 2021). Each of these extreme weather events has the potential to disrupt heavily populated regions, particularly

Table 17.1 Deviations in real GDP: by region, by year and by scenario (% deviation from base)

	Year 1	Year 2	Year 3	Year 4	Year 5	Year 6	Year 7	Year 8	Year 9	Year 10
Scenario One (2020)										
CapitalReg	0.2	-0.2	-1.0	-1.4	-0.2	-0.2	-0.2	-0.2	-0.4	-0.2
CoastNSW	0.0	-0.1	-0.4	-0.6	0.1	0.1	0.0	0.0	-0.1	0.0
CentralWest	0.5	-0.3	-2.2	-2.3	0.2	0.2	-0.2	-0.3	-0.3	0.1
FarWestOrana	0.8	-3.1	-9.3	-9.5	-0.4	-0.5	-1.3	-1.6	-2.1	-0.9
RvnMurray	0.9	-1.1	-5.5	-5.7	0.2	0.3	0.1	-0.9	-1.3	-0.1
NewEngNWGrft	0.8	-3.9	-10.8	-11.0	-0.7	-0.2	0.0	-3.0	-3.6	-0.9
NthAndCntVic	0.3	-0.9	-1.1	-2.7	0.0	-1.1	-1.1	-0.2	-1.2	-0.9
RoVic	0.1	-0.1	-0.3	-0.4	0.1	0.1	0.0	0.1	-0.1	-0.1
ECoastQld	0.0	-0.1	-0.4	-1.2	0.0	0.1	0.0	-0.1	-0.3	-0.2
DDwmMnGBQld	0.8	-3.4	-9.9	-9.6	-2.3	-0.4	-0.2	-4.0	-3.9	-2.1
OutbackQld	2.0	-2.1	-11.4	-13.2	-4.0	-0.8	-0.8	-12.5	-10.2	-10.2
AdelCstSA	0.2	-0.2	-0.5	-0.8	0.1	0.1	-0.1	0.2	-0.2	-0.2
EyreMallOBSA	0.9	-0.8	-0.9	-1.1	0.1	0.3	-0.4	1.1	-0.2	-0.3
WheatInldWA	0.0	-2.4	-3.4	-7.2	-1.6	1.2	-5.7	-1.7	-2.3	-2.2
RoWA	0.0	-0.1	-0.2	-0.3	0.1	0.3	-0.1	0.1	0.1	0.1
RoA	0.1	-0.3	-0.4	-0.7	0.2	0.2	0.1	0.1	-0.3	-0.2
Scenario Two (2030)										
CapitalReg	-0.4	-0.2	-1.0	-1.3	-0.5	-0.2	-0.4	-0.1	-0.6	-0.4
CoastNSW	-0.2	-0.1	-0.3	-0.6	0.0	0.1	-0.1	0.0	-0.2	-0.1
CentralWest	-0.2	-0.2	-2.1	-2.3	-0.1	0.3	-0.5	-0.2	-0.6	-0.1
FarWestOrana	-0.7	-3.2	-9.3	-9.5	-0.8	-0.4	-1.8	-1.4	-2.8	-1.3
RvnMurray	-0.5	-1.1	-5.4	-5.6	-0.5	0.3	-0.7	-0.8	-2.2	-0.9
NewEngNWGrft	-0.6	-4.0	-10.9	-11.1	-1.2	0.0	-0.4	-2.8	-4.4	-1.4

(continued)

Table 17.1 (continued)

	Year 1	Year 2	Year 3	Year 4	Year 5	Year 6	Year 7	Year 8	Year 9	Year 10
NthAndCntVic	-0.9	-1.0	-1.2	-2.7	-0.7	-1.2	-1.9	-0.2	-1.9	-1.6
RoVic	-0.3	-0.1	-0.3	-0.4	-0.1	0.1	-0.2	0.1	-0.4	-0.2
ECoastQld	-0.4	-0.1	-0.4	-1.1	-0.2	0.1	-0.2	-0.1	-0.4	-0.3
DDwmMnGBQld	-0.7	-3.5	-9.9	-9.6	-2.9	-0.3	-0.7	-4.0	-4.7	-2.8
OutbackQld	0.8	-2.3	-11.4	-13.2	-5.6	-0.7	-2.4	-12.4	-11.5	-11.5
AdelCstSA	-0.3	-0.3	-0.7	-1.0	-0.4	0.0	-0.5	0.1	-0.7	-0.6
EyreMallOBSA	-0.6	-1.5	-1.6	-1.8	-0.9	-0.3	-1.4	0.5	-1.1	-1.2
WheatInldWA	-2.5	-2.5	-3.5	-7.3	-3.2	1.1	-7.1	-1.7	-3.7	-3.6
RoWA	-0.2	0.0	-0.2	-0.2	0.0	0.3	-0.2	0.2	0.0	0.0
RoA	-0.4	-0.2	-0.4	-0.7	-0.1	0.2	-0.2	0.2	-0.5	-0.4
Scenario Three (2050)										
CapitalReg	-0.6	-0.2	-0.9	-1.3	-0.8	-0.1	-0.7	-0.1	-0.9	-0.6
CoastNSW	-0.3	-0.1	-0.3	-0.6	-0.2	0.2	-0.2	0.1	-0.4	-0.2
CentralWest	-0.4	-0.2	-2.1	-2.3	-0.5	0.4	-0.9	-0.1	-1.1	-0.5
FarWestOrana	-1.2	-3.2	-9.3	-9.5	-1.3	-0.1	-2.2	-1.1	-3.6	-1.9
RvnMurray	-1.0	-1.1	-5.4	-5.6	-1.6	0.4	-1.7	-0.7	-3.3	-1.9
NewEngNWGrft	-1.1	-4.0	-10.9	-11.1	-1.7	0.3	-1.0	-2.5	-5.4	-1.9
NthAndCntVic	-1.4	-1.0	-1.2	-2.7	-1.6	-1.2	-2.8	-0.1	-2.8	-2.5
RoVic	-0.4	-0.1	-0.3	-0.4	-0.4	0.1	-0.5	0.1	-0.6	-0.5
ECoastQld	-0.5	-0.1	-0.4	-1.1	-0.4	0.2	-0.4	0.0	-0.7	-0.5
DDwmMnGBQld	-1.2	-3.5	-9.9	-9.6	-3.6	-0.1	-1.3	-3.9	-5.7	-3.5
OutbackQld	0.3	-2.3	-11.4	-13.2	-7.4	-0.6	-4.2	-12.4	-13.0	-12.9
AdelCstSA	-0.5	-0.4	-0.7	-1.0	-0.8	-0.1	-1.0	0.1	-1.1	-1.0
EyreMallOBSA	-1.2	-1.7	-1.8	-2.0	-1.4	-0.5	-1.9	0.3	-1.7	-1.7

(continued)

Table 17.1 (continued)

	Year 1	Year 2	Year 3	Year 4	Year 5	Year 6	Year 7	Year 8	Year 9	Year 10
WheatInldWA	−3.3	−2.6	−3.6	−7.3	−4.8	1.2	−8.6	−1.7	−5.3	−5.2
RoWA	−0.3	0.0	−0.2	−0.2	−0.2	0.3	−0.3	0.2	−0.2	−0.1
RoA	−0.5	−0.2	−0.4	−0.7	−0.4	0.2	−0.5	0.2	−0.8	−0.6

on the east coast of Australia, thereby magnifying economic damage from the national perspective.

We highlight that climate change will likely affect every sector and all Australians and these multiple impacts, beyond agriculture, must be comprehensively analysed. Such impacts in relation to labour productivity, biodiversity loss, public health, sea-level rise, among others, would be over and above the costs of climate change we calculate in relation to Australian agriculture.

To respond to the possible impacts of climate change on agriculture, and in other sectors, Australia needs a well-developed and accepted strategy for climate adaptation. Such a strategy must be complemented by approaches that respond to the needs of agriculture, and water governance priorities, including ensuring sustainable levels of water extractions. These approaches should seek to find robust adaptation actions that perform well, relative to alternatives such as 'business as usual', over a wide range of plausible futures, and prioritise vulnerable communities. Such adaptation actions must be delivered if Australia is to effectively respond to climate change.

Acknowledgements We gratefully acknowledge the research assistance of Mai Nguyen in the preparation of this chapter.

References

Abram NJ, Henley BJ, Sen GA et al (2021) Connections of climate change and variability to large and extreme forest fires in southeast Australia. Commun Earth Environ 2(1):1–17. https://doi.org/10.1038/s43247-020-00065-8

Academy of Science (2019) Investigation of the causes of mass fish kills in the Menindee Region NSW over the summer of 2018–2019. https://www.science.org.au/supporting-science/science-policy-and-sector-analysis/reports-and-publications/fish-kills-report. Accessed 31 Mar 2021

Arriagada NB, Palmer AJ, Bowman DM, Morgan GG, Jalaludin BB, Johnston FH (2020) Unprecedented smoke-related health burden associated with the 2019–20 bushfires in eastern Australia. Med J Aust 213(6):282–283. https://doi.org/10.5694/mja2.50545

Australian Academy of Science (2015) How are sea levels changing? https://www.science.org.au/learning/general-audience/science-climate-change/6-how-are-sea-levels-changing. Accessed 31 Mar 2021

Australian Academy of Science (2021) The risks to Australia of a 3C Warmer World. https://www.science.org.au/files/userfiles/support/reports-and-plans/2021/risks-australia-three-deg-warmer-world-report.pdf. Accessed 31 Mar 2021

Australian Bureau of Statistics (2008) Water and the Murray Darling Basin—a Statistical Profile. https://www.ausstats.abs.gov.au/ausstats/subscriber.nsf/0/451801352604FCA9CA2574A50015764F/$File/4610055007_2000-01%20to%202005-06_ch4.pdf. Accessed 31 Mar 2021

Azhoni A, Jude S, Holman I (2018) Adapting to climate change by water management organisations: enablers and barriers. J Hydrol 559:736–748

Bowman D, Williamson G, Yebra M, Lizundia-Loiola J, Pettinari ML, Shah S, Bradstock R, Chuvieco E (2020) Wildfires: Australia needs national monitoring agency. Nature 584:188–191. https://www.nature.com/articles/d41586-020-02306-4

Bureau of Meteorology (BoM) and CSIRO (2018) State of the climate 2018. http://www.bom.gov.au/state-of-the-climate/. Accessed 31 Mar 2021

Bureau of Meteorology (BoM) (2019) Special climate statement 68—widespread heatwaves during December 2018 and January 2019. http://www.bom.gov.au/climate/current/statements/scs68.pdf. Accessed 31 Mar 2021

Bureau of Meteorology (BoM) (2020) Trends and historical conditions in the Murray-Darling Basin. A report prepared for the Murray-Darling Basin Authority. https://www.mdba.gov.au/sites/default/files/pubs/bp-eval-2020-BOM-trends-and-historical-conditions-report.pdf. Accessed 31 Mar 2021

Commonwealth of Australia (2015) National climate resilience and adaptation strategy 2015. https://www.environment.gov.au/system/files/resources/3b44e21e-2a78-4809-87c7-a1386e350c29/files/national-climate-resilience-and-adaptation-strategy.pdf. Accessed 31 Mar 2021

Costall AR, Harris BD, Teo B, Schaa R, Wagner FM, Pigois JP (2020) Groundwater throughflow and seawater intrusion in high quality coastal aquifers. Sci Rep 10:9866

Climate Council (2018) Icons at risk: climate change threatening Australian tourism. https://www.climatecouncil.org.au/uploads/964cb874391d33dfd85ec959aa4141ff.pdf. Accessed 31 Mar 2021

CSIRO (2008) Water Availability in the Murray-Darling Basin. A report to the Australian Government from the CSIRO Murray-Darling Basin Sustainable Yields Project. CSIRO, Canberra

Döll P, Romero-Lankao P (2017) How to embrace uncertainty in participatory climate change risk management—a roadmap. Earth's Future 5:18–36. https://doi.org/10.1002/2016EF000411

Filkov AI, Ngo T, Matthews S, Telfer S, Penman TD (2020) Impact of Australia's catastrophic 2019/20 bushfire season on communities and environment. Retrospective analysis and current trends. J Saf Sci Resilience 1:44–56

Grafton RQ (2019) Policy review of water reform in the Murray-Darling Basin, Australia: the "do's" and "do nots." Aust J Agric Resour Econ 63(1):116–141

Grafton RQ, Horne J (2014) Water markets in the Murray-Darling Basin. Agric Water Manag 145:61–71

Grafton RQ, Williams J (2020) Rent-seeking and regulatory capture in the Murray-Darling Basin, Australia. Int J Water Resour Dev 36:484–504

Grafton RQ, Pittock J, Davis R, Williams J, Fu G, Warburton M, Udall B, McKenzie R, Yu X, Che N, Connell D, Jiang Q, Kompas T, Lynch A, Norris R, Possingham H, Quiggin J (2013) Global insights into water resources, climate change and governance. Nat Clim Chang 3:315–321

Grafton RQ, Williams J, Jiang Q (2017) Possible pathways and tensions in the food and water nexus. Earth's Future 5:449–462

Grafton RQ, Doyen L, Béné C, Borgomeo E, Brooks K, Chu L, Cumming GS, Dixon J, Garrick DE, Helfgott A, Jiang Q, Katic P, Kompas T, Little LR, Matthews N, Ringler C, Squires D, Steinshamn SI, Villasante S, Wheeler S, Williams J, Wyrwoll P (2019) Realizing Resilience for Decision-Making. Nat Sustain 2:907–913

Great Barrier Reef Marine Park Authority (2019) Position statement climate change. https://elibrary.gbrmpa.gov.au/jspui/retrieve/9ad23e79-5915-4866-941a-ff6e566cbe71/v1-Climate-Change-Position-Statement-for-eLibrary.pdf. Accessed 31 Mar 2021

Groves DG, Molina-Perez E, Bloom E, Fischbach JR (2019) Robust decision making (RDM): Application to water planning and climate policy In: Marchau AVAWJ, Walker WE, Bloemen PJTM, Popper SW (eds) Decision making under deep uncertainty from theory to practice, pp 23–51. Springer, Cham

Hennessy KJ, Whetton PH, Preston B (2010) Climate projections. In: Stokes CJ, Howden M (eds) Adapting agriculture to climate change. CSIRO Publishing, Collinwood, Vic, Australia, pp 13–20

Howden SM, Stokes CJ (2010) Introduction. In: Stokes CJ, Howden M (eds) Adapting agriculture to climate change. CSIRO Publishing, Collinwood, Vic, Australia, pp 1–12

Howden SM, Gifford RG, Meinke H (2010) Grains. In: Stokes CJ, Howden M (eds) Adapting agriculture to climate change. CSIRO Publishing, Collinwood, Vic, Australia, pp 21–48

Howden M, Hughes L, Dunlop M, Zethoven I, Hilbert D, Chilcott C (eds) (2003) Climate change impacts on biodiversity in Australia. Outcomes of a workshop sponsored by the Biological Diversity Advisory Committee, 1–2 October 2002, Commonwealth of Australia, Canberra. https://www.environment.gov.au/system/files/resources/3f374cfd-3eaa-4c56-a2d3-92fd4bee 286e/files/greenhouse.pdf. Accessed 31 Mar 2021

Hughes L, McMichael T (2011) The critical decade: climate change and health. Climate Commission. https://www.climatecouncil.org.au/uploads/1bb6887d6f8cacd5d844fc30b0857931. pdf. Accessed 31 Mar 2021

Jiang Q, Grafton RQ (2011) Economic effects of climate change in the Murray-Darling Basin, Australia. Agric Syst 110:10–16

Karoly DJ, Braganza K (2005) A new approach to detection of anthropogenic temperature changes in the Australian region. Meteorol Atmos Phys 89:57–67

Karoly DJ, Braganza K (2005) Attribution of recent temperature changes in the Australian region. J Clim 18:457–464

Kirby M, Bark R, Connor J, Qureshi ME, Keyworth S (2014) Sustainable irrigation: how did irrigated agriculture in Australia's Murray-Darling Basin adapt in the Millennium Drought? Agric Water Manag 145:154–162

Kompas T, Ha PV, Che TN (2018) The effects of climate change on GDP by country and the global economic gains from complying with the Paris Climate Accord. Earth's Future 6(8): 1153–1173

Lesk C, Rowhani P, Ramankutty N (2016) Influence of extreme weather disasters on global crop production. Nature 529(7584):84–87. https://doi.org/10.1038/nature16467

Liu J, Hertel T, Taheripour F (2016) Analyzing future water scarcity in computable general equilibrium models. Water Econ Policy 2(04):1650006. https://doi.org/10.1142/S2382624 X16500065

Matthews J, Le Quesne T (2008) Adapting water to a changing climate. WWF International: Gland, Switzerland

McDonnell R, Fragarzy S, Sternberg T, Veeravalli S (2020) Drought policy and management. In: Dadson SJ, Garrick DE, Penning-Rowsell EC, Hall JW, Hope R, Hughes J (eds) Water science, policy, and management: a global challenge. Wiley, Oxford, pp 233–254

Murray-Darling Basin Royal Commission (2019) Murray-Darling Basin Royal Commission Report. https://www.environment.sa.gov.au/topics/river-murray-new/basin-plan/murray-darling-basin-commission. Accessed 31 Mar 2021

Nicholas J, Evershed N (2021) For some areas hit by NSW flood crisis, it's the fourth disaster in a year. The Guardian. https://www.theguardian.com/news/datablog/2021/mar/24/for-some-areas-hit-by-nsw-flood-crisis-its-the-fourth-disaster-in-a-year. Accessed 31 Mar 2021

OECD (2011) Water governance in OECD countries: a Multi-level approach, OECD studies on water. OECD Publishing, Paris. https://doi.org/10.1787/9789264119284-en

Productivity Commission (2018) Murray-Darling basin plan: five-year assessment. Final report no. 90, Productivity Commission, Canberra

Ukkola AM, Roderick ML, Barker A, Pitman AJ (2019) Exploring the stationarity of Australian temperature, precipitation and pan evaporation records over the last century. Environ Res Lett 14:124035

Wentworth Group of Concerned Scientists (2020) Assessment of river flows in the Murray-Darling Basin: Observed versus expected flows under the Basin Plan 2018/19. Wentworth Group of Concerned Scientists, Sydney

Wilby RL, Dessai S (2010) Robust adaptation to climate change. Weather 65(7):180–185

Wittwer G, Griffith M (2011) Modelling drought and recovery in the southern Murray-Darling basin. Aust J Agric Resour Econ 55(3):342–359

Wittwer G, Griffith M (2012) The economic consequences of prolonged drought in the southern Murray-Darling Basin, Chapter 7. In: Wittwer G (ed) Economic modeling of water, the Australian CGE experience. Springer, Dordrecht, Netherlands, pp 119–141

Wittwer G, Vere D, Jones R, Griffith G (2005) Dynamic general equilibrium analysis of improved weed management in Australia's winter cropping systems. Aust J Agric Resour Econ 49(4): 363–377

Wittwer G (2021) Modelling the economy-wide marginal impacts due to climate change in Australian agriculture. The Centre of Policy Studies, Victoria University Working Paper No. G-312. https://ideas.repec.org/p/cop/wpaper/g-312.html. Accessed 31 Mar 2021

Zander KK, Botzen WJW, Oppermann E, Kjelstrom T, Garnett ST (2015) Heat stress causes substantial labour productivity loss in Australia. Nat Clim Chang 5:647–651

Index

© The Editor(s) (if applicable) and The Author(s), under exclusive license to
Springer Nature Singapore Pte Ltd. 2022
A. K. Biswas and C. Tortajada (eds.), *Water Security Under Climate Change*,
Water Resources Development and Management,
https://doi.org/10.1007/978-981-16-5493-0